HELP YOUR KIDS WITH

S√iEnce

A UNIQUE STEP-BY-STEP VISUAL GUIDE

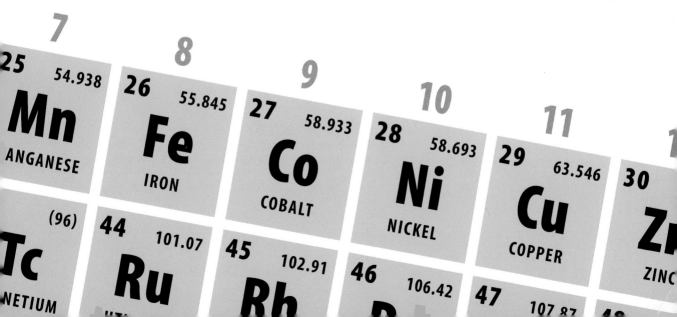

DK

LONDON, NEW YORK, MELBOURNE,
MUNICH, AND DELHI

DORLING KINDERSLEY
Senior Editor Carron Brown
Project Editors Steven Carton,
Matilda Gollon, Ashwin Khurana
US Editors Jill Hamilton, Rebecca Warren

Senior Designer Jim Green
Project Art Editor Katie Knutton
Art Editor Mary Sandberg
Designer Mik Gates
Packagers Angela Ball, David Ball

Managing Editor Linda Esposito
Managing Art Editor Diane Peyton Jones

Category Publisher Laura Buller

Senior Production Controller Erika Pepe
Production Editor Adam Stoneham

Jacket Editor Manisha Majithia
Jacket Designer Laura Brim

Publishing Director Jonathan Metcalf
Associate Publishing Director Liz Wheeler
Art Director Phil Ormerod

DORLING KINDERSLEY INDIA
Illustrations:
Managing Art Editor Arunesh Talapatra
Deputy Managing Art Editor Priyabrata Roy
Chowdhury
Senior Art Editor Chhaya Sajwan
Art Editors Shruti Soharia Singh, Anjana Nair,
Priyanka Singh, Shipra Jain
Assistant Art Editors Payal Rosalind Malik,
Nidhi Mehra, Niyati Gosain, Neha Sharma,
Jomin Johny, Vidit Vashisht

Editorial Assistance:
Deputy Managing Editor Pakshalika Jayaprakash
Senior Editor Monica Saigal
Project Editor Roma Malik

First published in 2012 by
Dorling Kindersley Limited,
80 Strand, London WC2R 0RL
13 14 10 9 8 7 6 5 4 3 2
006—181318—June/2012

Published in Great Britain by Dorling Kindersley Limited

A catalog record for this book is available from the Library
of Congress.

ISBN 978-0-7566-9268-1

DK books are available at special discounts when
purchased in bulk for sales promotions, premiums,
fund-raising, or educational use. For details, contact:
DK Publishing Special Markets, 375 Hudson Street,
New York, New York, 10014 or SpecialSales@dk.com

Printed and bound by South China Printing Co. Ltd, China

Discover more at
www.dk.com

TOM JACKSON has written nearly 100 books and contributed to many more about science, technology, and natural history. Before becoming a writer, Tom spent time as a zookeeper, worked in safari parks in Zimbabwe, and was a member of the first British research expedition to the rain forests of Vietnam since the 1960s. Tom's work as a travel writer has taken him to the Sahara Desert, the Amazon jungle, the African savanna, and the Galápagos Islands—following in the footsteps of Charles Darwin.

DR. MIKE GOLDSMITH has a Ph.D. in astrophysics from Keele University, awarded for research into variable supergiant stars and cosmic dust formation. From 1987 until 2007 he worked in the Acoustics Group at the UK's National Physical Laboratory and was Head of the group for many years. His work there included research into automatic speech recognition, human speech patterns, environmental noise and novel microphones. He still works with NPL on a freelance basis and has recently completed a project to develop a new type of environmental noise mapping system. He has published more than forty scientific papers and technical reports, primarily on astrophysics and acoustics. Since 1999, Mike has written more than thirty science books for readers from babies to adults. Two of his books have been short-listed for the Aventis prize (now the Royal Society prize) for children's science books.

DR. STEWART SAVARD is the Science Head Teacher and district eLibrarian/eResource teacher in British Columbia's Comox Valley, Canada. Stewart has published papers on the role of Science Fiction and Science collections in libraries and helped edit 18 Elementary Science books. He is actively developing a range of school robotics programs.

ALLISON ELIA graduated from Brunel University in 1989, with a BSc (Hons) in Applied Physics. After graduating, she worked in Public Sector finance for several years, before realizing that her true vocation lay in education. In 1992 she undertook a PGCE in Secondary Science at Canterbury Christ Church College. For the past 18 years, Allison has taught Science in a number of schools across Essex and Kent and is currently the Head of Science at Fort Pitt Grammar School in Kent, UK.

Introduction

Science is vital to understanding everything in the Universe, from what makes the world go around to the workings of the human body. It explains why rainbows appear, how rockets work, and what happens when we flick a light switch. These may seem difficult subjects to get to grips with, but science needn't be complex or baffling. In fact, much of science depends on simple laws and principles. Learn these, and how they can be applied, and even the most complicated concepts become more straightforward and understandable.

This book sets out to explain the essentials of three key sciences—biology, chemistry, and physics. In particular, it focuses on the curricula for these subjects taught in schools worldwide for students between the ages of 9 and 16. This is often a crucial time for developing an understanding of science. Many children become confused by the terminology, equations, and sheer scale of some of the topics. Inevitably, parents—who themselves often have a limited understanding of science—are asked to help with homework. That is where this book can really come to the rescue.

Help Your Kids with Science is designed to make all aspects of science easy and interesting. Beginning with a clear overview of what science is, each of the three sections is broken down into single-spread topics covering a key area of that science. The text is presented in short, easy-to-read chunks and is accompanied by clear, fully annotated diagrams and helpful equations. Explanations have been kept as simple as possible so that anyone—parent or child—can understand them.

Another problem children often have with science is relating scientific concepts to real life. To help them make a connection, "Real World" panels have been introduced throughout the book. These give the reader a look at the practical applications of the science they've been reading about, and the exciting ways it can be used. Cross-references are used to link related topics and help reinforce the idea that many branches of science share the same basic principles. A useful reference section at the back provides quick and easy facts and explanations of terms used in the text.

As a former research scientist, I am only too aware of how science can seem bewildering. Even scientists can get stuck if they stray into an unfamiliar discipline or are the first to investigate a new line of study. The trick is to get a firm grasp on the basics, and that is exactly what this book sets out to provide. From there you can go on to investigate how the world around you works and explore the endless possibilities that science has to offer mankind.

DR. MIKE GOLDSMITH

Contents

1 BIOLOGY

2 CHEMISTRY

3 PHYSICS

What is science?

A SYSTEM INVOLVING OBSERVATIONS AND TESTS USED TO FIGURE OUT THE MYSTERIES OF THE UNIVERSE AND EXPLAIN HOW NATURE WORKS

The word "science" means "knowledge" in Latin, and a scientist is someone who finds out new things. Scientific knowledge is the best way of describing the Universe—how it works and where it came from.

Science is...

...a collection of knowledge that is used to explain natural phenomena. The knowledge is arranged so that any fact can be confirmed by referring to other previously known facts.

...a way of uncovering new pieces of knowledge. This is achieved using a process of observation and testing that is designed to confirm whether a proposed explanation of something is true or false.

Answering questions

Science is an effective method of explaining natural phenomena. The way of doing this is known as the scientific method, which involves forming a theory about an unexplained phenomenon and doing an experiment to test it. Strictly speaking, the scientific method can only show whether a theory is false or not false. Once tested, a false theory is obviously no good and is discarded. However, a "not false" theory is the best explanation of a phenomenon we have—until, that is, another theory shows it to be false and replaces it.

ice cream changes states from a solid to a liquid with heat

◁ **Solving problems**
Much of science is driven by practical problems that need answers, such as "Why does ice cream melt?" However, scientific breakthroughs also come about from pure curiosity about the Universe.

Measurements

Scientists need to make measurements as they gather evidence of how things behave. Saying a snake "was as long as an arm" is less useful than giving a precise length. Scientists use a system of measurements called the SI (Système International) units (see p.200), which include meters for length, kilograms for mass, seconds for time, and moles for measuring the quantity of a substance. All other units of measurement (eg, for force, pressure, or speed) are derived from the SI units. For this reason, metric units are given first throughout the book, with imperial equivalents in parentheses.

the mercury gauge on a thermometer rises in degrees with the heat of the Sun

◁ **Setting a scale**
The degrees marked on a thermometer show the temperature rising and falling. However, like all units, the difference between one degree and the next is not something that is set by nature. The sizes of the units are generally set because they are practical to use in scientific calculations.

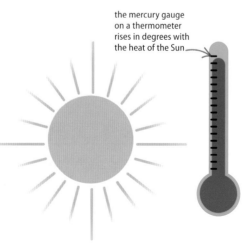

Backing up knowledge

The reason science is such a reliable way of describing nature is because every new piece of knowledge added is only accepted as true if it is based on older pieces of knowledge that everyone already agrees upon. Few scientific breakthroughs are the work of a single mind. When outlining a discovery, scientists always refer to the work of others that they have based their ideas on. In so doing, the development of knowledge can be traced back hundreds, if not thousands, of years.

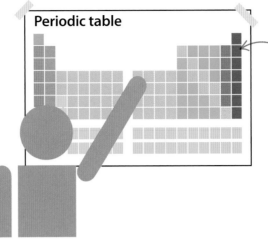

Periodic table

the periodic table lists the world's elements, which are arranged according to their atomic structure

◁ **Laying out the table**
The Russian Dmitri Mendeleev is credited with formulating the periodic table in 1869, but in reality it was the culmination of many centuries of investigation into the nature of elements.

Specialists

Modern science has been practiced for around 250 years, and in that time great minds have revealed a staggering amount about the nature of life, our planet, and the Universe. Early scientists investigated a wide range of subjects. However, no one alive today can have an expert understanding of all areas of scientific knowledge. There is just too much to know. Instead, scientists specialize in a certain field that interests them, devoting their working lives to unlocking the secrets of that subject.

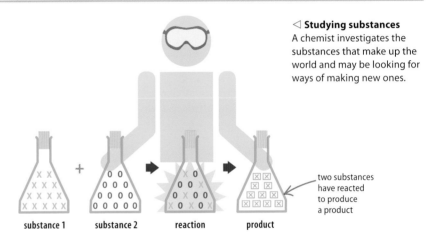

◁ **Studying substances**
A chemist investigates the substances that make up the world and may be looking for ways of making new ones.

two substances have reacted to produce a product

substance 1   substance 2   reaction   product

Applying science

Some scientists find explanations for natural phenomena because they are curious—they just like knowing. However, other scientists figure out how the latest understanding of nature might be put to practical use. Applied science and engineering is perhaps the best example of why science is such a powerful tool. If the knowledge discovered by scientists was not correct, none of our high-tech machines would work properly.

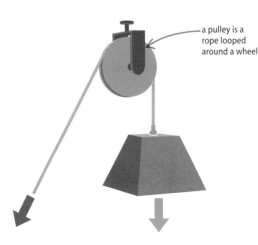

a pulley is a rope looped around a wheel

◁ **Using force**
Understanding forces and energy explains how it is easier to lift weights with a set of pulleys. For example, lifting a weight with two pulleys requires only half the force needed when using just one.

The scientific method

THE PROCESS BY WHICH IDEAS ABOUT NATURAL PHENOMENA
ARE PROVEN TO BE LIKELY OR INCORRECT

All scientific investigations follow a process called the scientific method.
They all begin with a flash of inspiration, where a scientist has a new
idea about how the Universe might work.

Ask a question
All science begins with a person wondering why a natural phenomenon
occurs in the way that it does. This may be in response to a previous
discovery that gives rise to new areas of investigation.

Do background research
The next step is to observe the phenomenon, recording its
characteristics. Learning more about it will help the scientist
form a possible explanation that fits the acquired evidence.

Try again
No experiment is ever
a failure. When results
disprove a hypothesis,
the scientist can use that
knowledge to reconsider
the question, and provide
a new hypothesis that
supports the evidence.

Construct hypothesis
At this stage, the scientist sets out a theory for the phenomenon.
This is known as a hypothesis. As yet, there is no proof for
the hypothesis.

Test the hypothesis
The scientist now designs an experiment to test the hypothesis, and
uses the hypothesis to predict the result. The experiment is repeated
several times to ensure that the results are generally the same.

Hypothesis is proven
The results of the
experiment show that the
hypothesis is a good way
of describing what is
happening during the
natural phenomenon.
It can therefore be used
as an answer to the
original question.

Draw a conclusion
If the results of the experiment are not what is predicted by the
hypothesis, then the theory about it is disproven. If the results match
the prediction, then the hypothesis has been proven (for now).

PROVEN   **DISPROVEN**

Hypothesis is disproven
The experiment shows that
the natural phenomenon
being investigated behaves
in a different way from
the one predicted by the
hypothesis. Therefore this
explanation cannot be
not correct and the original
question remains
unanswered.

Report results
It is important for positive results to be announced publicly so other
scientists can repeat the experiment and check that it was performed
correctly. The results are reviewed by experts before the findings
are accepted. This new knowledge then becomes a foundation
on which to investigate even more ideas.

Question

Does adding salt to water have any effect on how fast it evaporates (turns from liquid into vapor)?

Background research

Saltwater's freezing point is lower than 0°C (the normal freezing point of pure water) because the dissolved salt gets in the way of the water molecules, making it harder for them to form into solid ice crystals.

Hypothesis

Salt makes it harder for water to form ice, lowering the freezing point. Therefore, does salt also lower the boiling point of water, making it easier to form water vapor? If so, saltwater will evaporate faster than freshwater.

Test the hypothesis

Divide some freshwater into two cups. Add some salt to one cup to make a salt solution. Weigh out 5 ml (0.17 fl oz) of each liquid and pour each amount into two identical shallow dishes. The water should be about 1 mm (0.04 in) deep. Leave the dishes in direct sunlight. Monitor them over a few hours to see which dish dries out first. The hypothesis predicts that the saltwater will evaporate first.

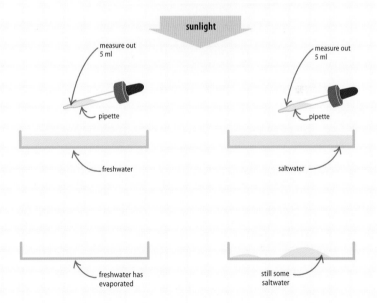

Results

The freshwater dish dries out first. What is the conclusion? Is the hypothesis false or not false?

Conclusion

The hypothesis is false.
Salt in the water does not make it evaporate faster.

Fields of science

SCIENCE IS DIVIDED INTO A NUMBER OF DISCIPLINES THAT EACH
FOCUS ON INVESTIGATING SPECIFIC AREAS OF THE SUBJECT.

Modern scientists are all specialists who belong to one of dozens of
disciplines. Some fields fall under the main subjects of biology, chemistry,
and physics, while others combine knowledge of all three to uncover facts.

Biochemistry
Studying the chemical
reactions that take place
inside cells and which
keep organisms alive.

Genetics
Understanding the way
chemicals can carry coded
instructions for making new
cells and whole bodies.

Forensic science
Using scientific evidence
to link criminals with
crime scenes to help
prove their guilt.

BIOLOGY

Any science that is concerned
with living things is described as
biology. Biologists investigate every
aspect of life, from the working
of a cell to how animals behave
in large groups.

CHEMISTRY

This science investigates the
properties of atoms and the many
different substances atoms produce
when combined in different ways.
Chemistry forms a link between
physics and biology.

Zoology
The area of biology
that investigates
everything there is
to know about animals.

Botany
The area of biology
that is concerned
wholly with the
study of plants.

Organic chemistry
Investigating
carbon-based
compounds, mostly
derived from organic
(once-living) sources.

Microbiology
The field of biology
concerned with cell
anatomy, using
microscopes to see
the structure of cells.

Ecology
Looking at
communities of
organisms and how
they survive together
in a habitat.

Electrochemistry
A field of chemistry
that uses the
energy in chemical
reactions to produce
electric currents.

Medicine
Applying knowledge
of biochemistry,
microbiology, and anatomy
to diagnose and treat illnesses.

Paleontology
Studying fossilized
remains of extinct
animals and relating
them to modern species.

Inorganic chemistry
Investigating the
properties of all
nonorganic (nonliving)
substances.

Until the 17th century, scientists were known as **"natural philosophers."** Today's philosophers contend with subjects such as ethics, which cannot be tested by the scientific method.

Geology
Investigating the processes that form rocks and shape our planet's landscape.

Nuclear chemistry
Studying the behavior of unstable atoms that break apart and release powerful radiation.

PHYSICS
With its name meaning "nature" in Greek, physics is the basis of all other sciences. It provides explanations of energy, mass, force, and light without which other sciences would not make sense.

Particle physics
Studying the particles that make up atoms and carry energy and mass throughout the Universe.

Thermodynamics
Studying the way energy flows through the Universe according to a series of unbreakable laws.

Mechanics
Understanding the motion of objects in terms of mass and how energy is transferred between them by forces.

Optics
Studying the behavior of beams of light as they reflect off or shine through different substances.

Wave theory
Explaining sound and other natural phenomena using an understanding of the behavior of waves.

Electromagnetism
Investigating electric currents and magnetic fields, and their uses in electronic devices.

Astronomy
Studying objects, such as planets, stars, and galaxies, in space.

Meteorology
Understanding the conditions that produce weather.

Social sciences
These sciences are not linked directly with the "natural sciences" (eg, biology, chemistry, or physics). Instead, they apply scientific methods to investigate humanity. Examples include:

Anthropology
Studying the human species, especially how societies and cultures from around the world differ from one another.

Archaeology
Studying ancient civilizations from the remains of their homes and cities.

Economics
Developing theories as to how people and companies spend their money.

Geography
Researching the natural landscape and how humans use the land, such as where they build cities.

Psychology
Investigating the way the human mind works using scientific methods.

Applied science
This area of work takes pure scientific knowledge and uses it for practical purposes. Some applied sciences can be described as types of engineering. Examples include:

Biotechnology
Using the knowledge of genetics and biochemistry to make artificial organisms and biological machines.

Computer science
Building microchip processors and writing software instructions to build faster and smarter computers.

Materials science
Developing new materials with properties suited to a particular application.

Telecommunications
Making use of electromagnetism, radiation, and optics to send signals and information over long distances.

Biology

What is biology?

THE SCIENCE THAT INVESTIGATES EVERY FORM OF
LIFE—HOW IT SURVIVES AND WHERE IT ORIGINATED.

**Biology, or life science, is a vast subject that studies
life at all scales, from the inner workings of a microscopic
cell to the way whole forests behave.**

What is life?

All life shares seven basic characteristics.
These are not exclusive to life, but
only living things have all seven. For
example, a car can move, it "feeds" on
fuel, excretes exhaust, and may even
sense its surroundings, but these four
characteristics do not make the car alive.

▷ **The seven characteristics**
Living things, or organisms, are incredibly
varied. Even so, they all share the same
seven characteristics that set them
apart from nonliving things.

THE SEVEN REQUIREMENTS FOR LIFE	
Requirement	**Description**
movement	altering parts of its body in response to the environment
reproduction	being able to make copies of itself
sensitivity	able to sense changes in the surroundings
growth	increasing in size for at least a period of its life
respiration	converting fuels (eg, food) into useful energy
excretion	removing waste materials from its body
nutrition	acquiring fuel to power and grow its body

Taxonomy

The field of biology that organizes, or
classifies, organisms is called taxonomy.
Modern taxonomy groups organisms
according to how they are related to each
other (rather than just how they look).
It involves placing all organisms in groups,
or taxons, arranged in this hierarchy:
domain, kingdom, phylum (or division
in the plant kingdom), class, order, family,
genus, and species. Animals and plants
are part of the largest domain, Eukaryota.

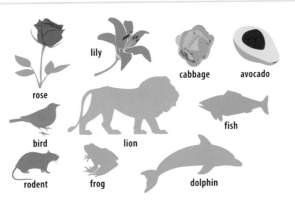

lily

cabbage   avocado

rose

fish

bird

lion

rodent   frog   dolphin

◁ **Classification**
Taxonomy (see pages
20–21) shows us that
some of these organisms
are more closely related
than others. For example,
animals belong to the
animal kingdom, whereas
plants belong to the
plant kingdom.

Microbiology

A cell is the smallest unit of life and that is what
microbiologists study. They use microscopes
to see inside cells and investigate how their
minute inner machinery, often called organelles,
functions to keep the cells alive. Microbiology
has shown that not all cells are the same, which
helps explain how bodies work and gives clues
to how life started and has since evolved.

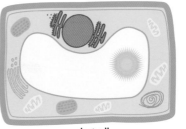

plant cell

◁ **Seeing in detail**
This cutaway artwork shows the
inner structures of a plant cell.
Microbiologists (see page 23)
view the finest details using
powerful electron microscopes,
which use a beam of electrons
instead of light to magnify cells.

Physiology

Biologists are interested in the anatomy of living things—how bodies are made from tissues and organs. Physiology is the study of how an organism's anatomical features relate to a particular function. Physiologists may even study the fossils of extinct animals, such as dinosaurs, to make discoveries about their lives and deaths.

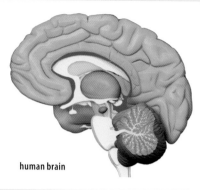

human brain

◁ **Nerve center**
The brain is a complex organ (a body part that has a specific function and is made of two or more kinds of tissue). The mass of nerve tissue is the main control center for the body (see page 68).

Ecology

The field of biology that investigates how communities of organisms live together is called ecology. Ecologists group wildlife into ecosystems, which occupy a specific living space or habitat. Scientists try to figure out the complex relationships between the members of an ecosystem. They may use their findings to help protect the habitat and its inhabitants from harmful human activities.

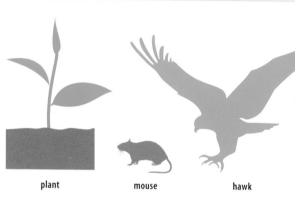

plant   mouse   hawk

◁ **Food chains**
One way that ecologists define an ecosystem is by a food chain, which tracks how plants are eaten by herbivorous animals, which in turn are preyed upon by predators (see pages 76–77).

Evolution

Biologists have discovered that living things can change, or evolve, to adapt to new habitats. The process is very slow, but it explains why the fossils of extinct organisms share features with today's wildlife. Evolution also explains how similar animals such as these finches have become slightly different from each other in order to suit how they live.

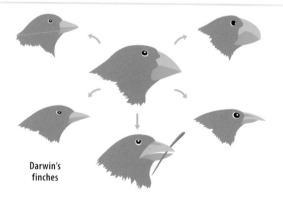

Darwin's finches

◁ **Bill shapes**
These species of Darwin's finch each target specific types of food, such as seeds or insects. As a result, their bills have all evolved into different shapes (see page 82).

Conservation

The more biologists reveal about the natural world, the more they find that many species are under threat of extinction. While extinction is a normal part of evolution, it appears that human activities, such as farming and industry, are making species die out much faster than normal. Conservationists use their knowledge of biology to protect endangered species and prevent unique habitats from being destroyed.

giant panda

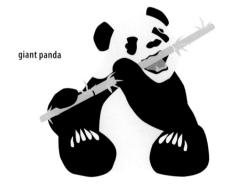

◁ **Saving species**
Without conservation, the giant panda, a bamboo-eating bear from China, may have become extinct. It was threatened by hunting and loss of its mountain habitat.

Variety of life

LIFE ON EARTH IS ORGANIZED INTO RELATED GROUPS.

Scientists have attempted to make sense of Earth's biodiversity—its enormous variety of life—by classifying living things into different groups, according to how they look and how they are related.

Three domains of life

Biologists estimate that there are about eight million species of living things on Earth today. The field of biology that organizes all these species into an understandable system is called taxonomy. Taxonomy arranges organisms in a hierarchy of groups. The largest groups are called domains. Most biologists divide life into three domains: Bacteria, Eukaryota, and the Archaea.

The word **"dolphin"** means "womb fish"—early biologists thought dolphins were related to fish, and not land mammals.

Bacteria

These simple-celled organisms live in all parts of Earth, from deep inside rocks to the guts of most eukaryotes. A few bacteria infect eukaryotes, causing diseases.

Eukaryota

This domain includes plants, animals, fungi, and some single-celled organisms. The Eukaryota is the only domain to contain multicellular organisms, where body cells work together to do different jobs.

Archaea

These are the oldest living things on Earth. They evolved more than 3.8 billion years ago out of the extreme conditions on Earth back then, and can still be found today in conditions too harsh for other life.

Archaea come in many different shapes, from strands like these, to cubes, and even spherical varieties

Classification

Taxonomists group organisms according to how they are related to each other. Group members have all evolved from a common ancestor at some point in the past. The further you go down the groups, the closer the similarities are between species.

Animal kingdom

Every animal belongs to this group. They all have multicellular bodies, must feed on other organisms to survive, and are usually able to respond rapidly to threats and problems.

Fungi kingdom

Until the middle of the 20th century, these organisms were considered a branch of the plant kingdom. Fungi are molds and mushrooms that live in damp habitats, and grow on their food, which they break down and absorb outside themselves.

Protist kingdom

The protists are a diverse group of eukaryotes that do not develop into specialized multicellular bodies. Instead they survive as single, solitary cells. However, a few species develop into clusters or colonies of individual cells.

Plant kingdom

Plants are multicellular organisms that make their own food by photosynthesis. Most plants are terrestrial or live in freshwater, and live in one place during their lifetime, although they can move in response to their environment.

△ KINGDOM
Eukaryota is the largest domain and it is the only one that is subdivided into kingdoms.

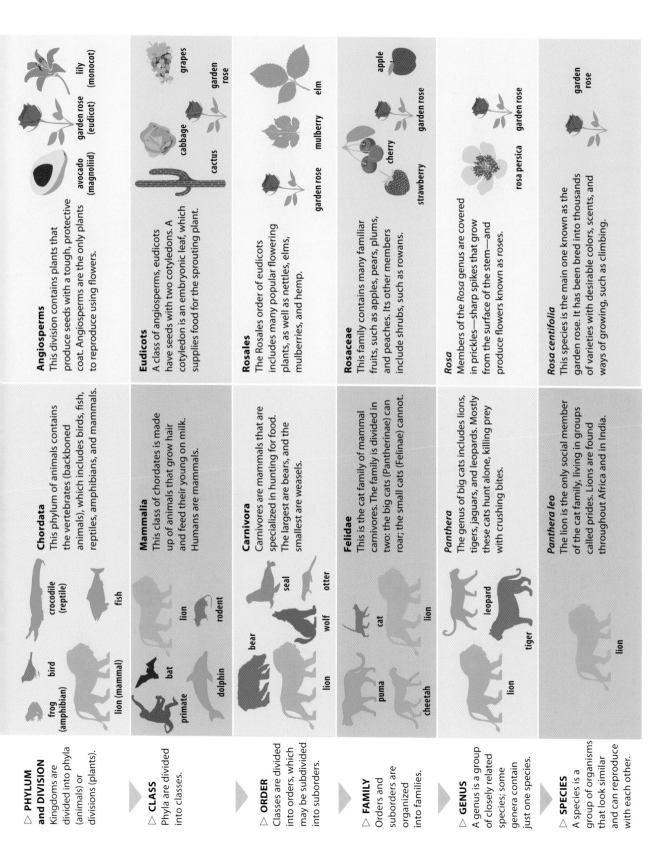

PHYLUM and DIVISION
Kingdoms are divided into phyla (animals) or divisions (plants).

Angiosperms
This division contains plants that produce seeds with a tough, protective coat. Angiosperms are the only plants to reproduce using flowers.

avocado (magnoliid)   garden rose (eudicot)   lily (monocot)

Chordata
This phylum of animals contains the vertebrates (backboned animals), which includes birds, fish, reptiles, amphibians, and mammals.

frog (amphibian)   bird   crocodile (reptile)   fish   lion (mammal)

CLASS
Phyla are divided into classes.

Eudicots
A class of angiosperms, eudicots have seeds with two cotyledons. A cotyledon is an embryonic leaf, which supplies food for the sprouting plant.

cactus   cabbage   garden rose (eudicot)   grapes   elm

Mammalia
This class of chordates is made up of animals that grow hair and feed their young on milk. Humans are mammals.

primate   bat   lion   dolphin   rodent

ORDER
Classes are divided into orders, which may be subdivided into suborders.

Rosales
The Rosales order of eudicots includes many popular flowering plants, as well as nettles, elms, mulberries, and hemp.

garden rose   mulberry   elm

Carnivora
Carnivores are mammals that are specialized in hunting for food. The largest are bears, and the smallest are weasels.

bear   seal   wolf   lion   otter

FAMILY
Orders and suborders are organized into families.

Rosaceae
This family contains many familiar fruits, such as apples, pears, plums, and peaches. Its other members include shrubs, such as rowans.

strawberry   cherry   apple   garden rose   mulberry

Felidae
This is the cat family of mammal carnivores. The family is divided in two: the big cats (Pantherinae) can roar; the small cats (Felinae) cannot.

cheetah   puma   cat   lion

GENUS
A genus is a group of closely related species; some genera contain just one species.

Rosa
Members of the *Rosa* genus are covered in prickles—sharp spikes that grow from the surface of the stem—and produce flowers known as roses.

rosa persica   garden rose

Panthera
The genus of big cats includes lions, tigers, jaguars, and leopards. Mostly these cats hunt alone, killing prey with crushing bites.

lion   tiger   leopard

SPECIES
A species is a group of organisms that look similar and can reproduce with each other.

Rosa centifolia
This species is the main one known as the garden rose. It has been bred into thousands of varieties with desirable colors, scents, and ways of growing, such as climbing.

garden rose

Panthera leo
The lion is the only social member of the cat family, living in groups called prides. Lions are found throughout Africa and in India.

lion

Cell structure

CELLS ARE THE BUILDING BLOCKS OF LIFE.

The cell is the basic unit of living things, with many millions working together to form an individual organism. Each cell is an enclosed sac containing everything it needs to survive and do its job.

Animal cell

The average animal cell grows to about 10 µm across (a 100th of a millimeter) although single cells inside eggs, bones, or muscles can reach several centimeters across. Animal bodies contain a large number of cell types, each specialized to do different jobs. Some kinds of single-celled protists, such as amoebas and protozoans, have a cell body very similar in structure to the cells of animals.

Centrosome
This produces long and thin strands used for hauling objects around the cell.

Cytoplasm
A watery filling of the cell with minerals dissolved in it.

Mitochondrion
The power plant of the cell—it releases energy from sugars.

Rough endoplasmic reticulum (ER)
Networks of ribosome-studded tubes, where proteins are manufactured.

Smooth endoplasmic reticulum
Tubes manufacturing fats and oils, and processing minerals.

Nucleus
This contains the cell's genetic material, DNA—the instructions to build and maintain the cell.

Nucleolus
A dense region of the nucleus, which helps make ribosomes.

Ribosome
Genetic information in DNA is decoded here to make the proteins that build the cell.

Cell membrane
The selectively permeable outer layer through which certain substances pass in and out of the cell.

Golgi apparatus
Where newly made substances are packaged into membrane sacs, or vesicles, for transport around and out of the cell.

▷ **Animal cell construction**
The outer layer of most animal cells is a flexible membrane, which can take on any shape. The cell contains many types of tiny structures called organelles. Each one has a specific role in the cell's metabolism—the chemical processes necessary for the maintenance of life.

Plant cell

The major difference between the cells of plants and animals is that plant cells are surrounded by a cell wall made of a lattice of cellulose strands. The space between the walls of neighboring cells is called the middle lamella. It contains a cement made of pectin, a sugary gel that joins the cells together.

Chloroplast
Folded membranes covered in chlorophyll, a green pigment found in plants.

Nucleus
Contains the nucleolus, which makes ribosomes.

Ribosome
The site where proteins are made.

Vacuole
A container for storing water, which also gives the cell structure.

Druse crystal
A crystal of calcium oxalate, which makes plants less palatable to herbivores.

Amyloplast
This turns sugars into starches.

Vesicles
A membrane sac that can store or transport substances.

Golgi apparatus
This bags up substances into vesicles.

Mitochondrion
This creates the cell's power supply.

Cell wall
A lattice of cellulose, a tough polymer made from chains of glucose.

Cell membrane
The membrane is not attached to the wall, and moves as the cell shrinks and swells.

△ **Plant cell construction**
Plant cells largely contain the same kinds of organelles as animal cells. The main additions are the chloroplasts in the cells of green sections of the plant body. This is where photosynthesis occurs, the process that produces the plant's sugar fuel.

▽ **Membrane structure**
The cell's outer layer, or membrane, is selectively permeable—it allows only some things to enter and leave the cell. The membrane is made from double layers of fat chemicals called lipids. The "head" of a lipid is hydrophilic, meaning it mixes with water and substances on each side of the cell. The "tail" is hydrophobic—it is repelled by water, and forms a barrier that helps keep the cell's contents inside.

Microscopic cells

Most cells are not visible to the naked eye, so microbiologists study them through microscopes. The first person to see cells in this way was 17th-century English scientist Robert Hooke. He named them cells after the small rooms used by monks. Today, microbiologists use dyes and lighting techniques to show a cell's internal structure, such as these human body cells (below).

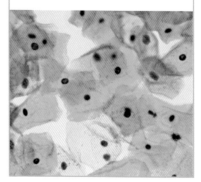

Lysosome
A bag of destructive enzymes that break down unwanted materials in the cell.

Hydrophilic head
The heads floats in the cytoplasm and extracellular liquids.

Hydrophobic tail
The two lipid layers connect by their tails to form a thin, water-repellent film on either side of the membrane.

Cells at work

EACH CELL IS LIKE A MICROSCOPIC FACTORY.

All the processes needed for life, such as releasing energy from food, removing waste materials, and growth, take place inside cells.

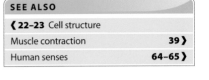
SEE ALSO
⟨ 22–23 Cell structure
Muscle contraction   39 ⟩
Human senses   64–65 ⟩

Cell transport

Cells process a wide range of chemicals. Inside the cell, large molecules such as proteins and even entire organelles are hoisted around by microtubules, which are also used in cell division. Some chemicals must be moved between organelles inside the cell, and others travel in and out through the cell membrane. Here are the main ways substances enter cells.

Bacteria cells can divide every 20 minutes, and one germ can grow to four billion trillion in 24 hours.

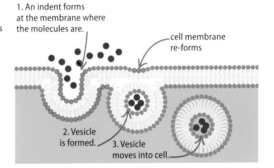

high concentration of molecules outside cell

cell membrane

low concentration of molecules inside cell

△ **Diffusion**
Diffusion happens when a substance spreads out, moving from areas of high concentrations to low.

molecules too big to cross membrane

energy is needed to pump molecules into cell

molecules inside cell

△ **Active transport**
If a molecule is too big or is unable to dissolve in the cell membrane, it is moved across in a process that uses energy.

1. An indent forms at the membrane where the molecules are.

cell membrane re-forms

2. Vesicle is formed.

3. Vesicle moves into cell.

△ **Endocytosis**
If molecules are too big to be pumped into a cell by active transport, a cell uses energy to put them in a sac, called a vesicle. This vesicle is formed from the cell membrane, and breaks open to release its contents once inside. When a cell moves a vesicle of material out, it is called exocytosis.

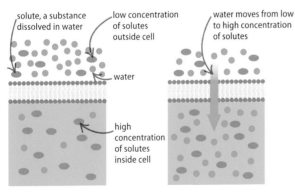

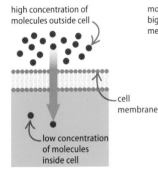

solute, a substance dissolved in water

low concentration of solutes outside cell

water moves from low to high concentration of solutes

water

high concentration of solutes inside cell

△ **Osmosis**
Osmosis is a type of liquid diffusion that takes place when solutions are separated by a membrane. Large dissolved molecules are blocked from diffusing into the cell. Instead, the water balances both sides, by moving from the low concentration side to the high.

Wilted flowers

Osmosis creates a force that moves water in and out of cells. When cut flowers are placed in freshwater, water floods into the plant cells by osmosis, making them full and rigid. When the water has gone, osmosis pulls the water out of the cells. The water evaporates, and the flowers wilt.

Multicellular structures

A living body is made of billions of cells working together. To do that most effectively, the cells are specialized to do certain jobs. A collection of cells that performs a single function—such as producing the mucus in the nose—is called a tissue. Very often, tissues group together to perform a complex set of tasks. They are then described as an organ, such as the nose.

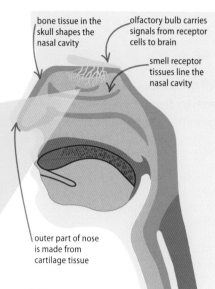

bone tissue in the skull shapes the nasal cavity

olfactory bulb carries signals from receptor cells to brain

smell receptor tissues line the nasal cavity

outer part of nose is made from cartilage tissue

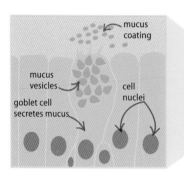

mucus coating

mucus vesicles

cell nuclei

goblet cell secretes mucus

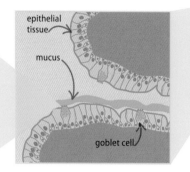

epithelial tissue

mucus

goblet cell

△ **Goblet cell**
This type of cell produces mucus (a mixture of water and a gooey protein called mucin) and other dissolved chemicals.

△ **Epithelial tissue**
Goblet cells form much of the epithelia, the tissue that lines the nose, windpipe, and gut. The mucus they produce protects the cells from chemical attack and dirt.

△ **Nose**
The nose is an organ that carries air in and out of the body. Muscle, cartilage, and bone tissues combine with epithelial tissue to help it do its job.

Cell division

A body grows because the number of its cells increases. This increase in number is achieved by cells dividing in half, to make two identical but fully independent cells. This type of cell division is called mitosis. It involves several stages, in which the cell's contents are split into two groups. That includes doubling the number of chromosomes (which carry the cell's genes).

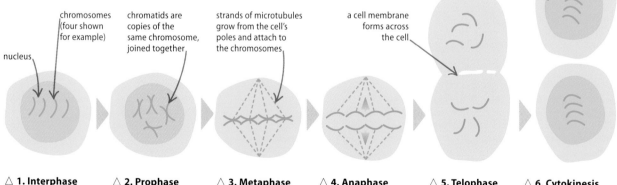

cell splits into two daughter cells, each with a full set of chromosomes

nucleus forms around the chromosomes in each cell

chromosomes (four shown for example)

chromatids are copies of the same chromosome, joined together

strands of microtubules grow from the cell's poles and attach to the chromosomes

a cell membrane forms across the cell

nucleus

△ **1. Interphase**
Cell has usual number of 46 chromosomes inside it.

△ **2. Prophase**
Each chromosome is doubled, forming two chromatids.

△ **3. Metaphase**
The chromosomes line up in the middle of the cell.

△ **4. Anaphase**
The chromatids are pulled apart, to become separate chromosomes.

△ **5. Telophase**
The microtubules disappear, and the cells begin to divide.

△ **6. Cytokinesis**
Two daughter cells are formed, each with 46 chromosomes.

Fungi and single-celled life

LIFE ON EARTH INCLUDES ORGANISMS THAT ARE NEITHER ANIMAL NOR PLANT.

SEE ALSO
❮ **20–21** Variety of life
❮ **22–23** Cell structure
Disease and immunity **50–51** ❯

The life forms within the Bacteria and Archaea domains, and most of the protist kingdom, are single-celled and can be viewed only through a microscope. By contrast, members of the fungi kingdom can grow into the largest organisms in the natural world.

Bacteria

The cells of Bacteria are hundreds of times smaller than those of plants or animals. They do not have a nucleus. Instead, their DNA is stored as a tangled loop called a plasmid. There are no other large organelles bound by a membrane, and all the metabolic reactions occur in the cytoplasm. Many bacteria move by flapping a whiplike flagellum. The hairlike pili are used to attach the bacteria to surfaces.

outer capsule

DNA

ribosomes

cell wall

cytoplasm

plasma membrane

pili

flagellum

△ **Bacterium**
Most bacteria are surrounded by three layers. The plasma membrane is similar to the one in other types of cell. The cell wall is made of proteins and sugars. The starchy outer capsule, which stops the cell from drying out, is missing in some species.

A **honey fungus** in Oregon, USA, is nearly 9 sq km (3.5 sq miles) in area, making it the **largest** single organism on Earth.

Archaea

For many years, these microorganisms were considered to be types of Bacteria, and the two groups were classified together. However, recent DNA analysis suggests that Archaea are a totally separate group. Many archaea are extremophiles—they survive in extreme conditions, such as incredibly hot or cold places. It is likely that their ancestors evolved in the extreme habitats of the young Earth about 3.5 billion years ago.

▷ **Haloquadratum**
This archaea lives in brine pools, where the salt content kills most other life forms. It has a square cell (its name means "salt square") filled with gas bubbles that help it float. No one knows how the cell survives.

▽ **Pyrococcus**
Discovered in the super-hot water that gushes from hydrothermal vents on the deep ocean floor, this archaea's name means "fire sphere." Sunlight never reaches its habitat, and the archaea is sustained by chemicals in the hot water.

flagella

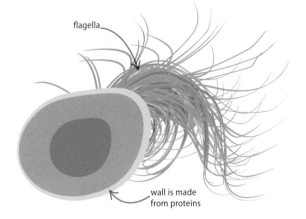

wall is made from proteins

Fungi

The fungal kingdom includes mushrooms, molds, and yeasts. They are saprophytic organisms, which means they grow over a food source and secrete enzymes that digest it externally. Their cells are eukaryotic, with a nucleus and organelles like those of plants and animals. The cells are held inside a rigid cell wall made largely of chitin, the same material that crab shells and beetle wings are made of.

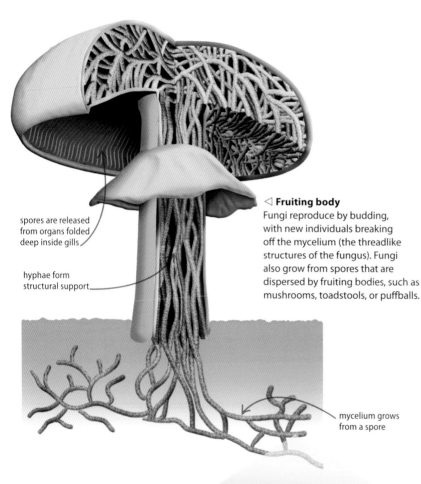

spores are released from organs folded deep inside gills

hyphae form structural support

mycelium grows from a spore

◁ Fruiting body

Fungi reproduce by budding, with new individuals breaking off the mycelium (the threadlike structures of the fungus). Fungi also grow from spores that are dispersed by fruiting bodies, such as mushrooms, toadstools, or puffballs.

▷ Hypha

The main part of a fungus is called the mycelium. This is made up of many strands called hyphae, which are long tubes of cells that extend over food sources. Yeast are single-celled fungi and do not develop hyphae.

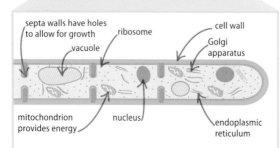

septa walls have holes to allow for growth
ribosome
vacuole
cell wall
Golgi apparatus
mitochondrion provides energy
nucleus
endoplasmic reticulum

Protists

This kingdom includes a wide variety of single-celled organisms. There are at least 30 different phyla and it is likely that at least some of them evolved separately from each other. The protist cell is very diverse, and can resemble that of an animal, plant, or fungus. Some species, such as Euglena, photosynthesize with chloroplasts, but also feed like animals.

▽ Diatom

These single-celled algae live in sunlit waters. They have an ornate cell wall made from silica. In the right conditions, diatoms produce thick blooms in the water. The silica skeletons of dead diatoms are one of the ingredients in clay.

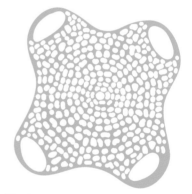

▽ Ciliate

Not every protist is motile (able to move). An amoeba alters the shape of its cell so its contents flow in one direction. Flagellates are powered by tail-like flagella, while ciliates (below) waft hairlike extensions called cilia (singular: cilium) to push themselves along.

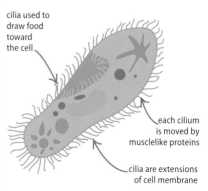

cilia used to draw food toward the cell

each cilium is moved by musclelike proteins

cilia are extensions of cell membrane

Respiration

THE PROCESS OF RESPIRATION SUPPLIES ENERGY FOR LIFE.

SEE ALSO	
Photosynthesis	30–31 >
Combustion	130–131 >
Redox reactions	132–133 >
Energy	170–171 >

All living things are powered by the energy released by a respiration reaction that takes place inside cells. This reaction needs a supply of oxygen taken from the surrounding air or water.

Cellular respiration

Every cell produces its own energy by respiration. The process takes place in tiny power plants called mitochondria. A cell that uses a lot of energy, such as a muscle cell, has a large number of these organelles. Respiration is a chemical reaction in which glucose (a sugar and important source of energy) is oxidized (chemically combined with oxygen). As well as energy, the reaction produces carbon dioxide and water.

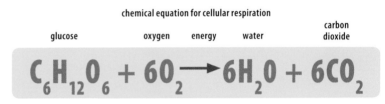

chemical equation for cellular respiration

glucose · · · oxygen · energy · water · carbon dioxide

$$C_6H_{12}O_6 + 6O_2 \longrightarrow 6H_2O + 6CO_2$$

▽ **Storing and releasing energy**
The energy released from respiration is stored by a chemical called adenosine triphosphate (ATP). The energy is used to add a phosphate (P) to adenosine diphosphate (ADP), to store energy. When needed elsewhere in the cell, the phosphate breaks off and releases the energy.

$$ADP + P = ATP$$ energy gained

$$ATP—P = ADP$$ energy released

▽ **Anerobic respiration**
If the cell cannot get enough oxygen to power respiration, it does it anerobically, meaning "without air." This process produces lactic acid as a result, which is what makes hard-working muscles burn with fatigue. Anaerobic respiration releases only part of the energy in glucose, but the rest is released when oxygen is available again.

glucose · · · · · · · · · · · lactic acid

$$C_6H_{12}O_6 = 2C_3H_6O_3$$

Mitochondrion

A mitochondrion is surrounded by an outer membrane, similar to the one around a cell. There is another membrane inside that is folded in on itself. The folded areas are called cristae. The main enzymes that control the production of ATP are bonded to the inner membrane. This is where respiration happens. The cristae increase the surface area of the inner membrane, maximizing the space for the enzymes.

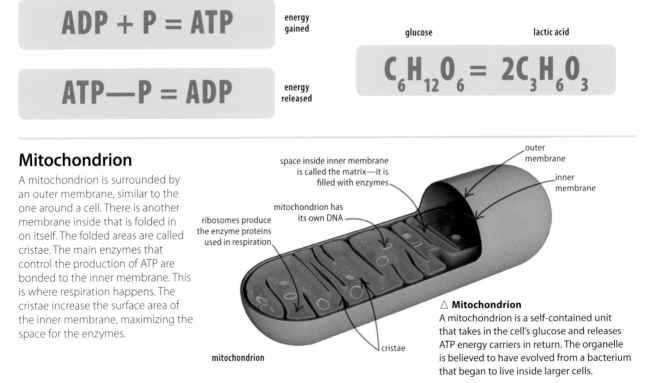

space inside inner membrane is called the matrix—it is filled with enzymes

outer membrane

inner membrane

mitochondrion has its own DNA

ribosomes produce the enzyme proteins used in respiration

cristae

mitochondrion

△ **Mitochondrion**
A mitochondrion is a self-contained unit that takes in the cell's glucose and releases ATP energy carriers in return. The organelle is believed to have evolved from a bacterium that began to live inside larger cells.

Gas exchange

Respiration requires a supply of oxygen, and the body also needs to remove the waste carbon dioxide it produces. The area through which these gases enter and leave the body is called the gas exchange surface. Lungs, gills, and the trachea tubes of insects are lined with these surfaces. A gas exchange surface is thin, moist, and well supplied with blood to take away the oxygen and deliver the waste carbon dioxide. The gases move in and out of the area by diffusion (see page 24).

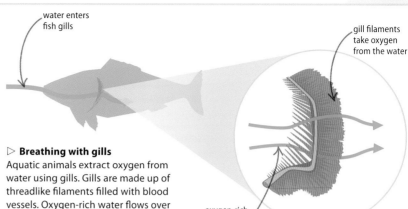

water enters
fish gills

gill filaments
take oxygen
from the water

oxygen-rich
water flow

▷ Breathing with gills
Aquatic animals extract oxygen from water using gills. Gills are made up of threadlike filaments filled with blood vessels. Oxygen-rich water flows over them constantly in one direction.

Breathing with lungs

Most land vertebrates breathe using lungs. The process is called reciprocal breathing: oxygen-rich air is inhaled, gases are exchanged, and then the oxygen-depleted air is exhaled. The lungs of primitive vertebrates, such as salamanders, are simple sacs. The lungs of larger animals are effectively sponges of tissue, with a huge gas exchange surface.

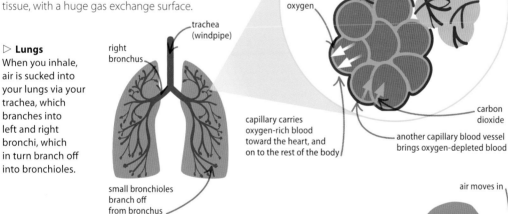

end of
bronchiole

each alveolus is coated
in a thin film of liquid,
which helps with the
diffusion of the gas

oxygen

carbon
dioxide

◁ Alveoli
At the end of each bronchiole are sacs called alveoli (singular: alveolus) where the gases are exchanged.

▷ Lungs
When you inhale, air is sucked into your lungs via your trachea, which branches into left and right bronchi, which in turn branch off into bronchioles.

trachea
(windpipe)

right
bronchus

capillary carries
oxygen-rich blood
toward the heart, and
on to the rest of the body

another capillary blood vessel
brings oxygen-depleted blood

small bronchioles
branch off
from bronchus

▽ Gas mixture
The air we breathe is a mixture of gases. Only about a fifth of it is oxygen, which diffuses into the blood. There is about 100 times more carbon dioxide in exhaled air than in inhaled air.

Gas	Inhaled air %	Exhaled air %
nitrogen	78	78
oxygen	21	17
inert gas	1	1
carbon dioxide	0.04	4
water vapor	little	saturated

▷ Reciprocal breathing
To breathe in, the diaphragm moves down, enlarging the space in the chest. This lowers the pressure in the lungs, forcing in air from outside. To breathe out, the diaphragm goes up, reducing the space in the chest and pushing out the oxygen-depleted air.

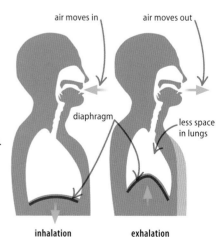

air moves in

air moves out

diaphragm

less space
in lungs

inhalation

exhalation

Photosynthesis

PLANTS MAKE THEIR OWN FOOD FROM
SIMPLE INGREDIENTS AND SUNLIGHT.

**Plants need sunlight to survive. They harness the
energy in light to make food from carbon dioxide
and water in a process called photosynthesis.**

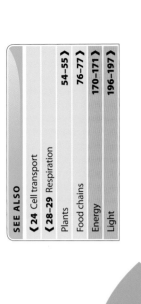

Light reaction

Photosynthesis is a chemical reaction that combines carbon
dioxide gas and water to make a molecule of glucose. The glucose
is the plant's food, and is sent around the plant to provide the
energy it needs. The waste product of the process is oxygen.
Photosynthesis itself is powered by sunlight. A chemical called
chlorophyll in the leaves absorbs some of the light's energy and
uses it to start the reaction.

the Sun's energy is crucial
for photosynthesis

carbon dioxide   water   glucose   oxygen

$$6CO_2 + 6H_2O \xrightarrow{\text{sunlight}} C_6H_{12}O_6 + 6O_2$$

chlorophyll in guard cells
causes them to respond
to light and open the
stomata on the leaf

carbon dioxide from the
air travels into the leaf
through the stomata by
diffusion (see page 24)

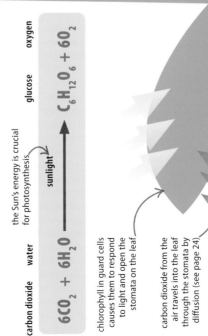

△ **Atmospheric carbon**
During photosynthesis, carbon atoms
are taken from the atmosphere. These
atoms are the building blocks of all
organic (carbon-containing) compounds—
in both plants and the animals that eat them.

Leaf

A leaf is a plant's solar panel. It is flattened to create a larger surface area to catch
as much sunlight as possible. The light shines through the surface of the leaf, and
photosynthesis occurs in the cells inside. Water arrives from the surface of the leaf, and
that runs down the center of the leaf. Carbon dioxide comes into the plant from the
surrounding air through pores called stomata on the underside of the leaf.

Chloroplast
A green structure inside
the cell where the
chlorophyll is located.

Palisade cells
These column-shaped
cells under the upper
surface are where most
of the photosynthesis
takes place.

Vascular bundle
Xylem (blue) brings
water and dissolved
minerals to the leaf.
Phloem (orange)
takes away sucrose
(see page 37).

Lower epidermis
The underside of the leaf is
filled with pores called stomata
(singular: stoma) that let gases
in and out of the plant.

Upper epidermis
A layer of cells
that forms the leaf's
upper surface. These
cells have a waxy
coating to reduce the
amount of water lost
through evaporation.

Spongy mesophyll
Cells with large spaces
between them where
the gases circulate.

Water loss
Leaves lose water
through evaporation and
need a constant supply
so they do not dry out.

Guard cells
A stoma is made of two
guard cells, which move
away from each other to
open the pore when the Sun
is shining, and move together
to close it when it's dark.

Chloroplast

The chloroplast is the organelle where photosynthesis happens. The process has two phases, the light and dark reactions. The light reaction (so-called because it needs light) harnesses the energy in sunlight to create a supply of ATP, an energy-carrying molecule (see page 28). The ATP is used to power the dark reaction, where an enzyme called rubisco combines carbon dioxide and water to make glucose.

oxygen, the waste product of photosynthesis, leaves the leaf through the stomata via diffusion

water moves into the root from the soil due to osmosis (see page 24)

Stroma
The dark spaces between the thylakoids and grana.

Thylakoid
The light reaction happens on membranes called thylakoids when several chlorophyll molecules work together to trap light energy.

Granum
The thylakoids are arranged in stacks called grana (singular: granum).

Stroma lamellae
Single membranes connect the grana.

△ **Inside a chloroplast**
The chlorophyll molecules are attached to membranes called thylakoids. The dark reaction takes place in the stroma, the spaces between the thylakoids and grana. All green parts of a plant contain cells filled with chloroplasts.

REAL WORLD
Fall colors

Deciduous trees drop their leaves in winter, when it is too dark to photosynthesize efficiently. Before they are shed, the leaves change color—turning from green to brown. This change is due to the chlorophyll being absorbed by the plant for use in the next year. The fall colors are formed by pigments called carotenes that are left behind.

Chlorophyll

The chemical pigment chlorophyll is what makes most plants look green. Each chlorophyll molecule absorbs the red and blue light in sunlight, using its energy to power photosynthesis, and reflects the rest back. So what we see is the green light that is not used by photosynthesis reflected back.

△ **Absorption spectrum**
This graph shows the wavelengths, or colors, of light, that are absorbed by chlorophyll. The dip in the middle shows that yellows and greens are absorbed less than reds and blues.

amount of light absorbed

400   500   600   700
wavelength of light (nanometers)

Feeding

THE PROCESS OF COLLECTING AND CONVERTING RAW MATERIALS INTO ENERGY.

Not all living things feed—plants and other photosynthetic organisms make their own food. However animals, fungi, and many single-celled organisms survive by consuming other living things.

What is feeding?

An organism that feeds is called a heterotroph, a name that means "other eater." As the name suggests, heterotrophs collect the nutrients and energy they need by consuming other organisms. Plants are called autotrophs—"self-eaters"—because they generate everything they need to survive themselves. There are several modes of feeding and every organism specializes in getting its food in a specific way.

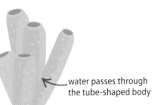

water passes through the tube-shaped body

△ **Absorption**
The simplest feeding method is to absorb food through the surface of the body. The body of a sponge is tube-shaped and food is collected from water flowing through it.

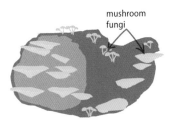

mushroom fungi

△ **External digestion**
A fungus is a saprophyte, meaning it grows over its food source, secreting enzymes that digest the food externally. Nutrients are then absorbed directly into its body.

amoeba   food particle   cell closes around food

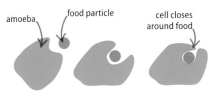

△ **Phagocytosis**
Single-celled organisms such as amoebas engulf their food, moving their cell membrane around it to form a sac in which the food is digested.

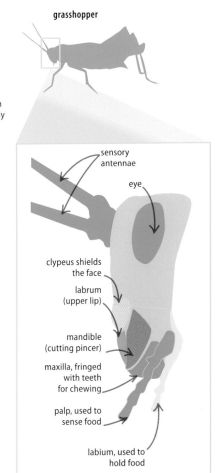

grasshopper

sensory antennae

eye

clypeus shields the face

labrum (upper lip)

mandible (cutting pincer)

maxilla, fringed with teeth for chewing

palp, used to sense food

labium, used to hold food

△ **Mouthparts**
Insects and other arthropods have complex mouthparts. A grasshopper's mouthparts are suited to cutting and chewing, but other insects have mouthparts that can be used for sucking, biting, or soaking up liquids.

cirri

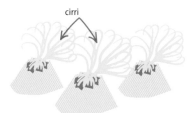

△ **Filter feeding**
Barnacles do not search for food, but sieve it from the water using their long, feathery legs, called cirri. Many shellfish, such as clams, are also filter feeders.

△ **Biting**
Only vertebrates, such as crocodiles, have jaws that open and close in a biting motion. The jaws are lined with teeth, which cut the food into manageable chunks before swallowing.

Teeth

Digestion, the breaking up of food into simpler substances that can be used by the body, follows feeding. The first phase of this is often mechanical digestion, where hard, sharp teeth bite food into small chunks or chew it to a pulp. Some toothless animals, such as birds, grind their food internally in gizzards—muscular stomachs that use stones swallowed by the animals to help break up the food.

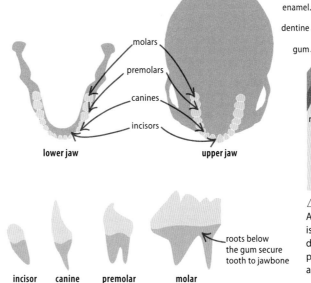

molars

premolars

canines

incisors

lower jaw

upper jaw

enamel

dentine

gum

cementum bonds tooth to gum

nerve

pulp

blood vessel

▷ **Human teeth**
Humans have four types of teeth. Incisors are used to slice and bite, and canines grip and rip. Molars and premolars are flat and are used for grinding food.

incisor canine premolar molar

roots below the gum secure tooth to jawbone

△ **Tooth anatomy**
A hard enamel cover is supported by softer dentine beneath. The pulp contains blood and nerve connections.

Types of consumer

Not all animals eat the same foods, and that difference is reflected in their teeth and jaws. Carnivores eat meat, so their teeth are often structured to help catch prey and rip it to shreds. Plant food is very tough, so herbivores (plant-eaters) use wide, grinding teeth to make it more digestible. Omnivores have teeth suited to a mixed diet of both meat and plants.

▽ **Hunter or hunted?**
Scientists can tell a lot about the way an animal lived by the shape, position, and condition of its teeth.

dolphins have many hooked teeth for gripping slippery fish, so they do not escape

lions have long fangs for gripping prey, while large premolars at the back of the jaw slice meat with a scissor action

the gap in a cow's teeth allows the animal to grab a new mouthful of grass while still chewing the last one

human teeth are adapted to a varied diet of fruits, hard seeds, and flesh

▷ **Rumination**
Chewing food once is not enough for large herbivores, such as cattle or antelopes. They regurgitate food, called cud, from the stomach to chew it a few more times during digestion. Ruminants rely on bacteria living in their complex stomachs to break down the tough cellulose (the main part of plant cell walls) in their food.

4. Finely ground pulp is then churned up in the omasum.

1. Swallowed food goes to the rumen, where it is mixed with digestive bacteria.

6. Nutrients are absorbed in the small intestine.

2. The second stomach chamber, the reticulum, receives cud, a mixture of food and stomach juices, from the rumen.

3. The reticulum pushes cud back up to the mouth for extra chewing.

5. The abomasum digests bacteria, releasing nutrients.

Waste materials

ANIMALS AND PLANTS USE A VARIETY OF METHODS
TO GET RID OF THEIR WASTE MATERIALS.

SEE ALSO	
❮ **32–33** Feeding	
Hormones	**48–49** ❯
Body systems	**62–63** ❯
Human digestion	**66–67** ❯

Excretion is the process of removing the waste produced by living bodies. This process is different to defecation, which is the release of the unused portion of food from the digestive tract.

Waste removal

A waste product is anything that the body cannot use. If they are allowed to build up in the body, they may become toxic. Nitrogen compounds from unneeded proteins form poisons that must be flushed away, and even carbon dioxide from respiration would make the blood dangerously acidic if it were not removed.

▽ **Getting rid of waste**
Organisms tackle their waste in different ways. The methods used to dispose of it safely depend on the nature of the waste and what resources are available. For example, fish flush waste out in water, but this method would dehydrate many animals, so other techniques are used.

REAL WORLD
Crocodile tears

The term "crying crocodile tears," meaning someone acting sad without actually being upset, has a ring of truth to it. Crocodiles do indeed cry, but their tears are not emotional ones. The tears carry away unwanted salts from the body.

Waste product	Organism	Excretory process	Explanation
ammonia	fish	break-down of proteins	ammonia is very poisonous, so it is excreted in very dilute urine by fish and other animals that have plenty of water available around them
urea	mammals	break-down of proteins	to save water, animals chemically convert ammonia into urea, which is soluble and can be excreted in liquid urine
uric acid	birds, reptiles	break-down of proteins	uric acid is a solid form of nitrogen-containing waste excreted as a white paste, which saves water but requires a lot of energy to process
carbon dioxide	all life	respiration of sugars	carbon dioxide, produced as a byproduct of respiration, is released from the body during gas exchange, for example, in the lungs or gills
oxygen	plants and algae	photosynthesis	although oxygen is useful, too much can upset some of the plant's processes, so unwanted oxygen is released through its leaves
feces	most animals	undigested food	unneeded food material, combined with other waste materials (including brown pigments from dead blood cells), is eliminated via the anus
salt	all organisms	balancing concentrations of body fluids	salts help with many body processes, but too much can cause cramps and dehydration, so it is excreted in sweat, urine, or through skin glands

Kidneys and bladder

In humans—and other vertebrates—most waste products are filtered from the blood supply by the kidneys. The liquid produced—known as urine—trickles from each kidney through a long tube called a ureter. Both ureters empty into the bladder, a flexible bag in the pelvic region. When this is about half full, the weight of the liquid creates the urge to urinate. Urine is expelled from the bladder via a channel running through the genital region called the urethra.

▽ **Inside the kidneys**

A renal artery brings waste-filled blood to the kidney. The blood is dispersed to the outer regions, called the cortex, where the filtering happens in thousands of tiny units called nephrons. From there, the clean blood is returned to the body via a renal vein. Drops of the filtered waste are collected by the calyx, a multiheaded funnel that connects to the ureter.

Even **water** can be toxic, because too much in the body causes the brain to swell and can kill.

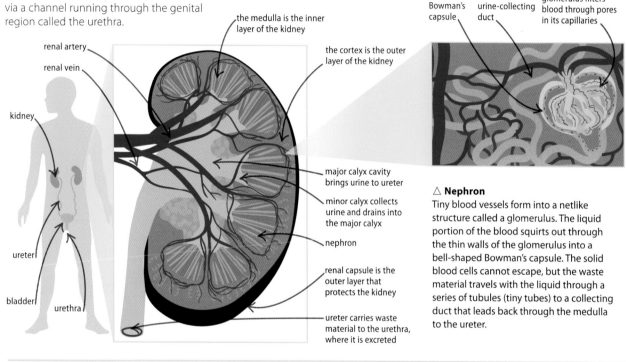

the medulla is the inner layer of the kidney

the cortex is the outer layer of the kidney

renal artery

renal vein

kidney

major calyx cavity brings urine to ureter

minor calyx collects urine and drains into the major calyx

nephron

ureter

renal capsule is the outer layer that protects the kidney

bladder

urethra

ureter carries waste material to the urethra, where it is excreted

Bowman's capsule

urine-collecting duct

glomerulus filters blood through pores in its capillaries

△ **Nephron**

Tiny blood vessels form into a netlike structure called a glomerulus. The liquid portion of the blood squirts out through the thin walls of the glomerulus into a bell-shaped Bowman's capsule. The solid blood cells cannot escape, but the waste material travels with the liquid through a series of tubules (tiny tubes) to a collecting duct that leads back through the medulla to the ureter.

Osmoregulation

The kidneys also carry out osmoregulation, controlling the amount of water in the body. When there is a lack of water, the nephron tubules reabsorb some of it from urine so it is not expelled unnecessarily. Osmoregulation is governed by a hormone called antidiuretic hormone, or ADH, which is produced by the pituitary gland.

▷ **Rising and falling**

The levels of ADH in the blood are constantly adjusting to maintain the right amount of water in the blood in a cycle, shown here.

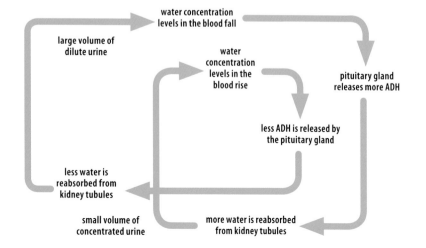

water concentration levels in the blood fall

large volume of dilute urine

water concentration levels in the blood rise

pituitary gland releases more ADH

less ADH is released by the pituitary gland

less water is reabsorbed from kidney tubules

small volume of concentrated urine

more water is reabsorbed from kidney tubules

Transport systems

SUBSTANCES ARE MOVED AROUND INSIDE
LIVING THINGS IN A VARIETY OF WAYS.

The cells in a multicellular organism are specialized into certain roles
and cannot survive on their own. The body's transport system brings
them what they need to stay alive, and takes away their waste materials.

Circulation

Animals transport substances around
their bodies in a liquid. In vertebrates,
this liquid is blood, pumped along by
a heart (or hearts) through a series of
pipes, or vessels. Blood vessels reach
all parts of the body, narrowing to
thin-walled capillaries that deliver
materials to cells by diffusion.

▷ **Arteries and veins**
The vessels that carry blood away from
the heart are called arteries. They pulsate
to push blood along, which can be felt
through the skin in some places. Veins
bring blood back to the heart.

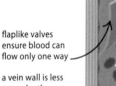

artery

arterial blood is
oxygen-rich
and lighter than
venous blood

arteries have
thick walls made
of layers of
elastic muscle

vein

flaplike valves
ensure blood can
flow only one way

a vein wall is less
muscular than an
artery wall, and
its blue color is
sometimes visible
under skin

venous blood lacks
oxygen and is rich
in carbon dioxide

Composition of blood

Blood contains hundreds of compounds. About
55 percent of blood is a watery mixture known as
plasma. This contains dissolved ions, hormones, and
several proteins, such as the ones that form blood clots
and scabs to seal breaks in vessels. The rest of the blood
is made up of red and white blood cells and platelets.

Blood color ▷
Blood looks red because
most of its cells contain
an iron-rich pigment
called hemoglobin.
This substance bonds
with oxygen arriving
via the lungs and delivers
it to body cells. A few
invertebrates use
copper-rich hemocyanin
to do this, which makes
their blood blue.

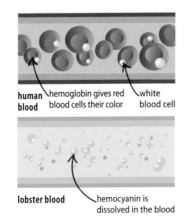

human
blood

hemoglobin gives red
blood cells their color

white
blood cell

lobster blood

hemocyanin is
dissolved in the blood

plasma contains
many substances
dissolved in it, such
as carbon dioxide,
which is produced
as waste by cells

one in 20 blood cells
are white blood cells,
which defend the body
against disease

oxygen-carrying red
blood cells make
up the majority of
blood—there are five
billion in every milliliter

▽ **Red blood cells**
Hemoglobin, the body's
oxygen carrier, is held in red
blood cells. These have a
curved doughnut shape to
maximize their surface area
for collecting oxygen.

Plant vascular system

The transport system of a plant is made up of two sets of vessels—xylem and phloem. Xylem carries water around the plant. Its stiff tubes run from the roots, up the stem, to the leaves. Phloem carries the sugar made in the leaves to the rest of the plant in the form of dissolved sucrose. Both types of vessel are made from columns of cells with openings at either end that form continuous pipes along which liquids can flow.

More than **100 million tons** of sugar are extracted from the sap stored in the phloem tubes of sugar cane **every year**.

xylem tubes are made from the waterproofed cell walls of dead cells

the liquid in phloem tubes is called sap

xylem carries water to the plant

phloem carries sugar from the leaves to the rest of the plant

◁ **Vascular bundle**
The xylem and phloem run together through the plant as a vascular bundle. This structure—especially the xylem—forms a stiff support for the plant. In trees, the wood develops from old xylem tubes.

▽ **Moving sugars and water**
The sugars in phloem diffuse from the leaves, where they are made, to other areas of the plant that lack fuel. Water is essentially pumped up from the roots through xylem tubes by a process called transpiration.

Giant redwood

The largest trees in the world, such as these giant redwoods of California, USA, grow to around 361 ft (110 m) tall. Scientists estimate that this is about the maximum height for a tree, since the pressure needed to pump a continuous column of water any higher would cause the water to pull itself apart inside the tree, and never reach the top.

sunlight is necessary for photosynthesis, and also evaporates water from the leaves

water rises up the stem to replace the water lost higher up

wind blows away moist air, leaving dry air in its place, which increases transpiration, as water is more likely to evaporate in dry air

water is drawn into roots—and up the xylem—by osmosis (see page 24)

root hairs increase the surface area able to suck up water

Movement

ORGANISMS HAVE DEVELOPED DIFFERENT WAYS OF MOVING.

SEE ALSO	
Fish, amphibians, and reptiles	58–59 **〉**
Mammals and birds	60–61 **〉**
Body systems	62–63 **〉**

Organisms move by changing the shape of their body to propel themselves forward. In complex animals these body changes are controlled by muscles, bundles of protein that exert pulling forces on body parts.

Modes of locomotion

Animals move in order to find food, escape a threat, or locate mates. The precise mode of locomotion (movement) used depends heavily on their habitat. Plants and fungi cannot move in the same way—their stiff cell walls make their bodies too rigid. However, many single-celled organisms, such as most protists and algae, can move by using extensions called flagella or cilia in the search for food or better conditions.

△ **Flying**
Wings are modified limbs that create lift and thrust forces to carry birds, bats, and some insects through the air.

△ **Swinging**
Tree-dwellers require a large decision-making brain and nimble limbs to control climbing and jumping.

△ **Walking**
Most land animals walk on four legs (quadrupedal), although humans and flightless birds walk on two (bipedal).

△ **Burrowing**
Burrowers have powerful limbs for digging or are slender enough to be able to wriggle through soft soils.

△ **Floating**
The Portuguese man-of-war cannot move itself, but it is moved by tides, currents, and winds on the water's surface.

△ **Drifting**
Some microscopic plankton can swim, but most float freely in the water and are carried along by ocean currents.

△ **Swimming**
Aquatic animals that can swim strongly enough to control where they move in the water are called nektons.

△ **Staying still**
Some organisms spend their lives anchored in one spot, usually under water, and just move their limbs to catch food.

Snake locomotion

Snakes evolved from four-legged reptiles, with their ancestors losing their limbs over time. Their most common—and fastest—mode of movement is serpentine locomotion, using sideways curves.

muscle contracts on the outside of the curve to pull the body straight

the rear curve is now where the first one was

snake curves around bumps on the ground

the outer edge of curve does the pushing

the straightened front section moves forward

△ **1. Bunching up**
The body is pulled into wide curves so the rear end moves toward the head.

△ **2. Stretching out**
As the body straightens, the curved sections push against the rough ground.

△ **3. Gaining ground**
The head gains ground by moving forward, and then the sequence starts again.

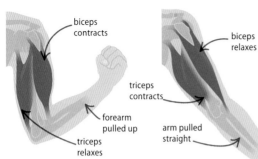

△ **Flex**
The biceps muscle contracts, pulling up on the forearm and causing the whole arm to bend at the elbow.

△ **Extend**
The triceps contracts, and the biceps relaxes, pulling the forearm down and straightening the arm.

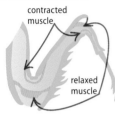
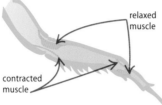

△ **Raised**
Arthropod exoskeletons contain pairs of muscles attached to their jointed inside surfaces.

△ **Extended**
The exoskeleton does not bend when pulled by a muscle. Instead, the force is transferred to the joint, making the whole joint move.

Anchor points

Muscles exert a force by contracting, or pulling, and need a solid anchor point to pull against. This is the main function of a skeleton, with the bones connecting at joints, to allow it to move when muscles pull. Muscles cannot push, so they work in pairs, with each muscle pulling in the opposite direction to the other.

longitudinal muscles running end-to-end can contract to pull body into a ball

circular muscles running around the body can contract to squeeze it into a tube

longitudinal muscles can contract on one side of body to make a crescent shape

△ **Hydrostatic skeleton**
Worms and other soft-bodied animals have a hydrostatic skeleton—made of sacs of liquid surrounded by muscles. These have a fixed volume, but can be changed into different shapes using sets of circular and longitudinal muscles.

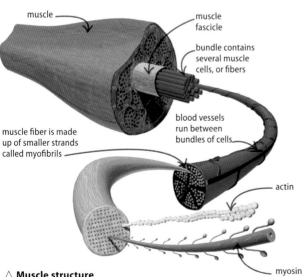

△ **Muscle structure**
Muscles are formed from a hierarchy of bundles. Even the smallest muscle contains several fascicles, which are bundles of muscle cells. In turn, the cells contain bundles of myofibrils that are filled with myosin and actin.

Muscle contraction

A muscle cell takes the form of a long fiber—up to 30 cm (12 in) long in a man's thigh. The cell contains many hundreds of nuclei and several bundles of myofibrils, which are made up of two protein filaments known as myosin and actin. Muscles contract when the two filaments move closer together in the cells. Millions of these tiny movements accumulate into a powerful contraction.

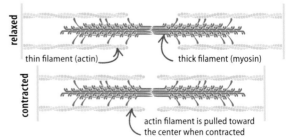

△ **Actin and myosin**
When a muscle receives an electric pulse from a nerve, the signal causes the thick myosin protein to haul itself along two actin strands, pulling them toward the center. When relaxed, the proteins spread apart again, and the muscle lengthens.

Sensitivity

LIVING ORGANISMS SENSE THEIR
ENVIRONMENT IN DIFFERENT WAYS.

**All living things are sensitive to their surroundings, such as changes
in light, sound, or chemistry. This sensitivity allows organisms to
respond, for example to a threat, increasing their chances of survival.**

Tropism

Plants can sense the factors in the environment that
help them maximize their growth. This is called
tropism. A seed is sensitive to gravity (gravitropism),
so its roots grow down into the soil. The roots also
turn toward water in the soil (hydrotropism), while
the stem grows toward sunlight (phototropism).
Phototropism causes a growing point (the meristem)
to face the Sun by growing cells on one side of the
stem longer than those of the other.

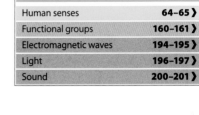

cells on the shady
side grow longer than
on the sunny side

meristem

tip points
toward sunlight

sunlit cells
stay short

◁ **Phototropism**
Sunlight inhibits the
production of growth
hormones or auxins. The
cells on the shady side
of the stem release auxins.
That makes the cells
in shade grow longer,
while the cells on the
sunny side stay short.

▷ **Compound eye**
Many arthropods have
compound eyes, which
have thousands of
individual lenses. Each
lens forms a small dot of
an image, which overlaps
with other dots to form
a larger image.

lenses made
from cone-shaped
crystal can pick up
slightest movements

pigment cells
stop light leaking
to other lenses

rhabdom channels light
toward retina cell

each retina cell
detects light

signal passes
to nerve cell

Animal senses

The human senses of touch,
smell, sight, hearing, and taste
are all used by animals, but not
in the same way. For example, a
grasshopper hears with pressure-
sensitive knees, a housefly tastes
its food by standing on it (its taste
buds are on its feet), while a moth
detects smells with its feathery
antennae. Some animals have
senses that do not compare
to human ones.

electroreceptors (ampullae)
contain gel that carries electrical
current to nerve endings

whiskers are sensory
hair cells like those
in the human ear

detectors run
along the side
of the body

pits are on
the snout

△ **Ampullae of Lorenzini**
Sharks have electroreceptors
that pick up the electric fields
produced by the muscles of
other animals. This allows them
to find prey in the dark water.

△ **Whiskers**
Whiskers are ultra-sensitive hairs
used by mammals to feel their way
in the dark. They are wider than
the head, so the animal knows if
it is heading into a tight spot.

△ **Lateral line**
Fish use a motion sensor, called
the lateral line, running along the
side of the body. It picks up the
swirling water currents created
by other animals moving nearby.

△ **Heat pits**
Pythons and vipers have hollow
pits on their snouts that detect the
body heat of warm-blooded prey.
The pits also warn the snake if it
should avoid the other creature.

Nerve cell

Sensory organs send out signals to the rest of the body as electric pulses that run along nerves. Nerves are made up of bundles of long cells called neurons. The long, wirelike section of the cell is called the axon, and it carries the signal to the next cell in line. Charged ions flood in and out of the axon to create the electric pulse.

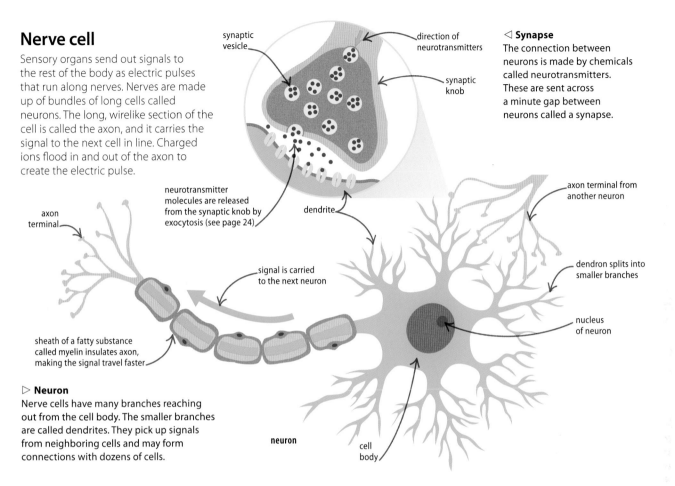

synaptic vesicle

direction of neurotransmitters

◁ **Synapse**
The connection between neurons is made by chemicals called neurotransmitters. These are sent across a minute gap between neurons called a synapse.

synaptic knob

neurotransmitter molecules are released from the synaptic knob by exocytosis (see page 24)

dendrite

axon terminal from another neuron

axon terminal

signal is carried to the next neuron

dendron splits into smaller branches

nucleus of neuron

sheath of a fatty substance called myelin insulates axon, making the signal travel faster

▷ **Neuron**
Nerve cells have many branches reaching out from the cell body. The smaller branches are called dendrites. They pick up signals from neighboring cells and may form connections with dozens of cells.

neuron

cell body

Reflex action

Information from the senses travels toward the brain through sensory neurons. In vertebrates, such as humans, these connect to the spinal cord, and the signal travels up to the brain through the cord. Any immediate response to the stimulus (such as a sharp pin) is sent out to the muscles by motor neurons right away. This means that reflex actions, such as withdrawing the hand from the source of pain, do not involve the brain, but are controlled by the spinal cord alone.

5. The finger moves away from source of pain.

4. The motor neuron signals the muscle to contract.

▷ **Reflex arc**
The nerve pathway controlling a reflex is called the reflex arc. The sensory nerve sends a signal to the spinal cord, where it connects directly to the motor neuron that signals to the muscles, causing them to move.

1. The finger touches source of pain (a sharp pin).

3. The spinal cord connects to a motor neuron.

2. The sensory neuron sends a signal to the spinal cord.

Reproduction I

SPECIES MUST REPRODUCE TO SURVIVE.

Reproduction is the main purpose of the natural world. Living things grow, feed, and survive in order to reproduce and makes copies of themselves.

SEE ALSO	
❰ 22–23 Cell structure	
❰ 25 Cell division	
Human reproduction	**72–73 ❱**
Evolution	**80–81 ❱**
Genetics I	**84–85 ❱**

Asexual reproduction

When a single organism makes an exact copy of itself, the process is called asexual reproduction. The copy is genetically identical, a clone of the parent. Asexual reproduction can be useful for populating new habitats very quickly. However, because all the offspring are identical, a disease or other problem that affects one of them is likely to affect all the others, too.

New Mexico whiptail lizards are all asexual, but all females must "mock mate" with each other before laying eggs.

▽ **Budding**

The most basic form of reproduction is budding, in which a section of the parent breaks off, forming an independent individual. Many single-celled organisms reproduce by budding.

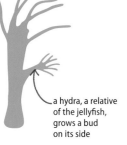
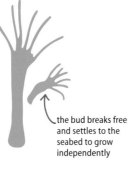

a hydra, a relative of the jellyfish, grows a bud on its side

the bud breaks free and settles to the seabed to grow independently

▽ **Vegetative reproduction**

Some plants send out side roots (called runners) or stems (stolons), that sprout daughter plants nearby. When the daughter plant is established, the connection with the parent breaks.

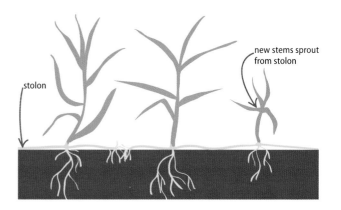

new stems sprout from stolon

stolon

▷ **Sporogenesis**

Fungi, primitive plants (such as ferns and moss), and even some parasitic worms reproduce by releasing hardy spores. These are tiny balls of cells, which can grow into new individuals.

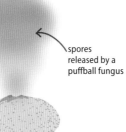

spores released by a puffball fungus

▷ **Parthenogenesis**

Parthenogenesis is a form of reproduction in which animals produce young without mating. Some female aphids give birth to daughters that are identical to themselves in every way except size.

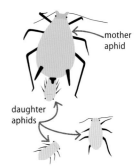

mother aphid

daughter aphids

Sexual reproduction

Sexual reproduction happens when two parents mix up their genes in order to produce offspring with their own unique genetic make-ups. Sexual reproduction requires each parent to produce gametes, or sex cells. While ordinary cells contain two full sets of genes—one from each parent—gametes have just a single set. In a process called fertilization, two gametes—one from each parent—fuse to form a zygote, the first cell of a new individual.

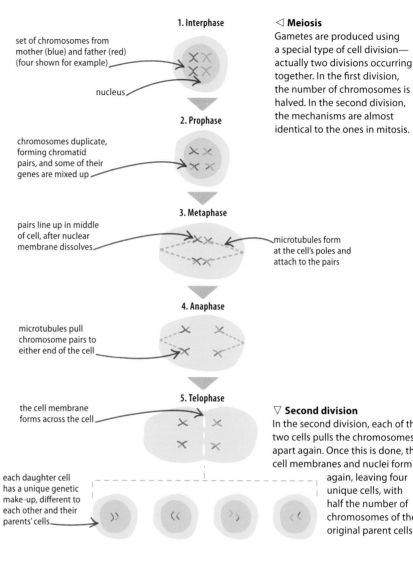

1. Interphase

set of chromosomes from mother (blue) and father (red) (four shown for example)

nucleus

2. Prophase

chromosomes duplicate, forming chromatid pairs, and some of their genes are mixed up

3. Metaphase

pairs line up in middle of cell, after nuclear membrane dissolves

microtubules form at the cell's poles and attach to the pairs

4. Anaphase

microtubules pull chromosome pairs to either end of the cell

5. Telophase

the cell membrane forms across the cell

each daughter cell has a unique genetic make-up, different to each other and their parents' cells

◁ **Meiosis**
Gametes are produced using a special type of cell division—actually two divisions occurring together. In the first division, the number of chromosomes is halved. In the second division, the mechanisms are almost identical to the ones in mitosis.

▽ **Second division**
In the second division, each of the two cells pulls the chromosomes apart again. Once this is done, the cell membranes and nuclei form again, leaving four unique cells, with half the number of chromosomes of the original parent cells.

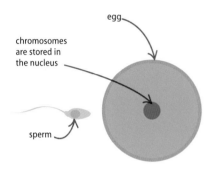

egg

chromosomes are stored in the nucleus

sperm

△ **Sperm and egg**
Male gametes are called sperm, and female ones are called eggs or ova (singular: ovum). Both contain half the usual number of chromosomes. A sperm's purpose is to deliver genes to the egg and it contains nothing else. By contrast, an egg cell needs to be huge to contain the nutrients required to grow a new individual after fertilization.

Animal development

After fertilization, the new individual (embryo) needs to develop and grow until it is ready to feed and live independently. The ways that animals produce their young depends on their habitat and biology.

▷ **Development strategies**
Small creatures, which are under constant threat of predators, produce lots of young quickly. Larger and better protected animals invest in protecting fewer young instead.

Method	Explanation	Example
oviparity	eggs are fertilized after being released by the female	fish, toads
oviparity	eggs are fertilized before release and often protected in a nest	birds
ovoviviparity	fertilized eggs retained in body until after hatching	seahorses
aplacental viviparity	young grow inside mother, feeding on eggs or siblings	some sharks
placental viviparity	young sustained by mother through placenta until birth	mammals

Reproduction II

ANIMALS AND PLANTS EMPLOY A RANGE OF
STRATEGIES TO REPRODUCE.

SEE ALSO	
❰ **22–23** Cell structure	
Life cycles	**46–47** ❱
Plants	**54–55** ❱

Plants and animals employ a number of reproduction strategies to maximize
their breeding potential. This may involve changing from one sex to another,
or relying on other animals to aid in reproduction and dispersal of offspring.

Hermaphrodites

Sex cells are produced by organs
known as gonads. The female gonad
is the ovary; the male one is the testis.
Animals that have both types of
gonads at some point in their lives
are known as hermaphrodites.
Earthworms and land snails are
simultaneous hermaphrodites,
meaning they have both gonads
at the same time. Nevertheless, they
still need to find mates to breed.

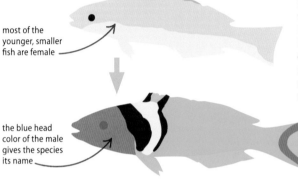

most of the
younger, smaller
fish are female

the blue head
color of the male
gives the species
its name

◁ **Bluehead wrasse**
Like many fish, the bluehead
wrasse is a sequential
hermaphrodite—most are
born female, but they can
change color and become
male later if there are
not enough males for
reproduction to take place.

Marsupials

Most female mammals sustain a developing
fetus inside the uterus or womb using a
placenta. The placenta transfers oxygen
and nutrients into the fetus's blood supply.
The baby is born once it has developed
enough to survive independently. The young
of marsupials are born at an earlier stage of
development than those of other mammals.
Instead of being fed from a placenta in the
uterus, they continue their growth in their
mother's pouch, or marsupium.

the mother sits down and
licks a path through her fur
to make it easier for the
joey to reach her pouch

pouch

the joey hauls itself
forward with its front legs
(its back legs are
not fully developed yet)

▷ **Kangaroo**
Baby kangaroos, or joeys, are born after just
31 days of development inside the mother.
They then make a dangerous journey
from the birth canal, over the mother's
fur to the safety of the mother's pouch.

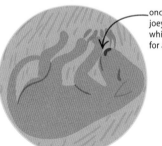

once in the pouch, the
joey finds a teat, from
which it will drink milk
for at least 100 days

when it can feed
independently,
the joey spends
time in and out
of the pouch

Joeys are only about
2 cm (0.8 in) long when
they are born and weigh
less than 1 g (0.4 oz).

Flowering plants

The flower is the reproductive organ of a plant. It has male and female parts. The anthers produce pollen, which contain the male sex cells, while the ovary at the heart of the flower contains the ova (singular: ovum), or eggs. The other structures in the flower are there to aid the pollen from one flower getting to the stigma of another flower, from where the sex cells in the pollen travel to the ovary.

▷ **Animal-pollinated flower**
This flower's bright petals and sweet smell attract insects that come to drink nectar, a sweet liquid produced at the center of the flower. The visiting insects pick up sticky pollen from the anther. When they visit another flower, the pollen transfers to that flower's stigma.

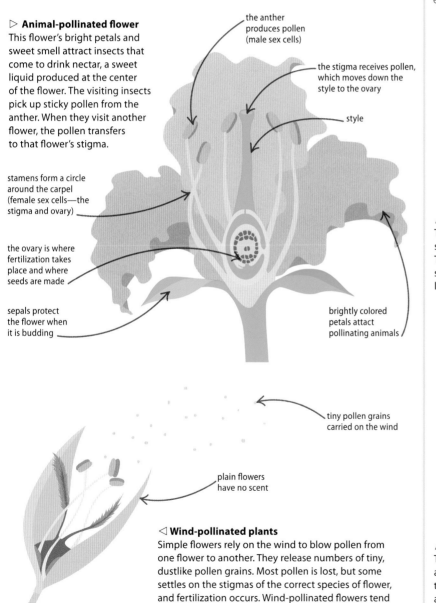

the anther produces pollen (male sex cells)

the stigma receives pollen, which moves down the style to the ovary

style

stamens form a circle around the carpel (female sex cells—the stigma and ovary)

the ovary is where fertilization takes place and where seeds are made

sepals protect the flower when it is budding

brightly colored petals attract pollinating animals

tiny pollen grains carried on the wind

plain flowers have no scent

◁ **Wind-pollinated plants**
Simple flowers rely on the wind to blow pollen from one flower to another. They release numbers of tiny, dustlike pollen grains. Most pollen is lost, but some settles on the stigmas of the correct species of flower, and fertilization occurs. Wind-pollinated flowers tend to be dull in color, because they do not need to attract animals for pollination.

Fruit and seeds

When a plant ovum is fertilized, the ovary develops into a seed. The seed is an embryonic form of the adult plant, with a root, stem, and food store. A fruit is the coating around the seed, developed from the wall of the ovary. Fruits have evolved to have many functions.

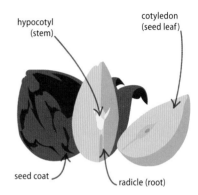

hypocotyl (stem)

cotyledon (seed leaf)

seed coat

radicle (root)

△ **Seeds**
The embryonic root and stem are ready to sprout from the seed during germination. They get their energy from a cotyledon—some seeds have two—which is an embryonic leaf structure packed with starch fuel.

berry

maple key

coconut

△ **Different fruits**
The main job of a fruit is to protect the seed and help it move far away from the parent tree. Sweet fruits, such as berries, are eaten by animals and the seed is deposited later. A maple key is a wind-borne fruit, while the coconut is able to float vast distances across the ocean.

Life cycles

DIFFERENT PLANTS AND ANIMALS GROW
TO MATURITY IN DIFFERENT WAYS.

The early, or juvenile, phase of a multicellular organism's life is devoted to growth. Organisms use a range of systems to reach an adult size, only then developing sexual organs and reproducing.

Germination

A seed is a plant embryo. It already has a root (radicle) and a tiny stem (plumule) inside. The embryonic leaf, called a cotyledon, is a food store that powers the first stage of growth, known as germination. Germination is stimulated by environmental conditions. Longer days—indicating the approach of spring—are a common cue. Some seeds require other cues, such as temperature changes, being soaked in water for a long period, or even heat from fire.

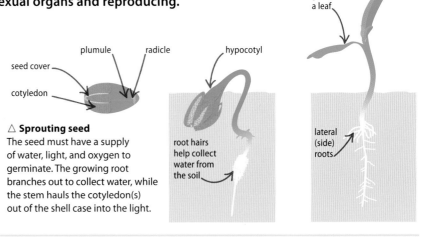

△ **Sprouting seed**
The seed must have a supply of water, light, and oxygen to germinate. The growing root branches out to collect water, while the stem hauls the cotyledon(s) out of the shell case into the light.

Plant life cycles

All flowering plants produce seeds but they do it according to one of three life cycles. Annual plants, such as grasses, sprout, seed, and then die all within one year. Biennials spend the first year growing a storage root, such as a carrot, which then resprouts and flowers in the second year. Perennials live for more than two years and produce repeated batches of seeds.

▽ **Annual (grass)**
The grass seed stays in the soil during winter, grows rapidly, and flowers within a few months. The plant drops new seeds onto the fresh soil before it dies.

▽ **Biennial (carrot)**
In the first year, leaves above ground fuel the creation of a carrot root, which remains even when the leaves and shoots die off over winter. The next spring, the carrot root's stored sugar fuels new shoots, rapid flowering, and seed production.

▽ **Perennial (oak tree)**
The oak tree grows for several years before flowering for the first time. Its seeds are dispersed by animals. During winters, the plant becomes dormant, before growing more and flowering again the following year.

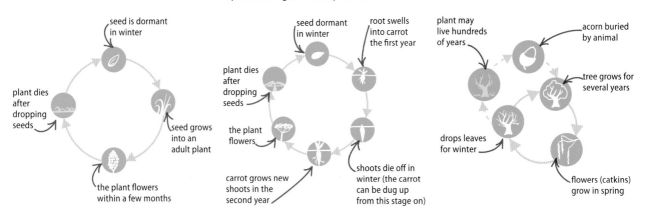

Metamorphosis

Animals that produce large numbers of young can find they are in direct competition for food with their own offspring. Many insects avoid this problem by having larval stages, which look and live in very different ways to the adults. A larva must undergo a complete metamorphosis, where its body rebuilds itself in the adult, sexually mature form. Other young insects are nymphs, which, unlike larvae, resemble their adult form.

Woolly bear caterpillar

The larvae of tiger moths are called woolly bears, and the species that live in the Arctic take years to reach adulthood. The woolly bears freeze solid during the long winters, and can only manage one molt during the short Arctic summer. After 14 molts and 14 years, the caterpillars finally pupate into tiger moths.

▷ **Incomplete metamorphosis**
The cicada nymph looks like its adult parent, but lacks wings. After several molts, the nymph reaches its largest size, called the final instar. During the next molt, it develops wings and sex organs and emerges as an adult, ready to reproduce.

▷ **Complete metamorphosis**
After hatching, the caterpillar (larva) is an eating machine and undergoes several molts as it outgrows its inflexible exoskeleton. Then it becomes a pupa, a dormant phase inside a protective case, where metamorphosis takes place. The insect emerges as an adult butterfly.

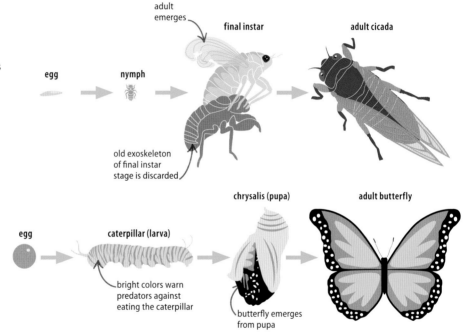

adult emerges

final instar

adult cicada

egg

nymph

old exoskeleton of final instar stage is discarded

chrysalis (pupa)

adult butterfly

egg

caterpillar (larva)

bright colors warn predators against eating the caterpillar

butterfly emerges from pupa

Reproductive strategies

Animals employ different strategies to ensure their offspring survive until they can reproduce. There are two main options: producing huge numbers of young, but leaving their survival to chance, or protecting just a few young, and giving them parental care and protection.

▷ **Pros and cons**
All reproductive strategies have advantages and disadvantages. An animal's place in the food chain and its habitat are the two factors that influence its reproductive strategy.

Animal	Type of care	Benefits	Costs
salmon	many thousands of eggs are laid each year	young can populate a new habitat quickly, and at least a few will always survive	effort kills the parents, and most young die before they reproduce
lion	one or two young are produced every few years; mother looks after them until adulthood	young are more likely to survive until adulthood, and help raise and protect younger siblings	investing energy into just a few young over many years is risky

Hormones

CHEMICAL MESSAGES CALLED HORMONES
CONTROL DAY-TO-DAY BODY PROCESSES.

Complex life forms use hormones to control growth, metabolic rate, and to prepare the body for activity or sleep. Hormones are produced in special organs called glands throughout the body.

Glands

Any body part that secretes a substance is called a gland. Exocrine glands send chemicals out of the body. They include sweat, salivary, and the seminal gland, which releases semen. Hormones are produced by endocrine glands, which release substances into the blood and internal body fluids. From there, hormones are carried to the parts of the body that they influence.

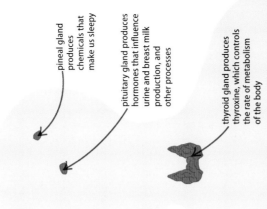

pineal gland produces chemicals that make us sleepy

pituitary gland produces hormones that influence urine and breast milk production, and other processes

thyroid gland produces thyroxine, which controls the rate of metabolism of the body

adrenal glands produce epinephrine to combat stress

pancreas produces insulin to process sugar in food

▽ **Human hormones**
The glands shown here secrete hormones used in a variety of processes in the human body.

▽ Melatonin

This hormone is released by the pineal gland underneath the brain. Its production is linked to the time of the day. In humans, it is released in the evening to prepare the body for night time, making us sleepy. In nocturnal animals, it is activated to wake them up.

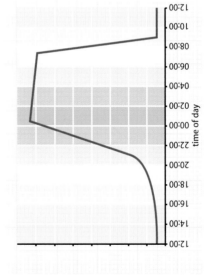

time of day

▽ Epinephrine

This powerful hormone is released by the adrenal glands. It triggers the body's response to stress (the "fight-or-flight" response). When released, epinephrine (also known as adrenaline) gives an immediate energy boost to prepare the body to act. Some of the common signs of this are listed here.

Effect	Explanation	Purpose
skin goes pale	blood vessels in the skin contract	blood directed to muscles for movement
heart rate goes up	heart pumps more blood	oxygen reaches muscles faster
heavy breathing	lungs take bigger breaths	boost to oxygen supply

Thermoregulation

The human body maintains a constant body temperature so its metabolism runs at a constant rate. As a result, the body must conserve its heat in cold conditions and shed excess heat in warm ones. Its processes for keeping its body temperature steady are called thermoregulation.

▽ Hot and cold

Thermoregulation makes use of the basic principles of heat transfer. Heat is lost via skin flooded with warm blood, and also by the evaporation of sweat secreted on the skin. When cold, the body curls up to decrease its surface area and so reduce heat loss.

Hot conditions	Cold conditions
Vasodilation The blood vessels in the skin widen to allow warm blood to radiate heat into the air.	**Vasocontraction** The blood vessels contract, so less blood reaches the skin, reducing heat transfer.
Sweating As sweat evaporates, it takes some of the body's heat with it.	**Shivering** The rapid movement of muscles when shivering generates heat.
Pilorelaxation Body hair lies flat, allowing cool breezes to get close to the skin.	**Piloerection (goosebumps)** Body hair stands up, to keep a layer of warm air next to the body.
Stretching out Moving around allows heat to be lost from a larger surface area.	**Curling up** Curling tight reduces the surface area losing heat.

In **very cold water**, a human's heart rate slows and blood is sent only to the brain and vital organs to conserve oxygen—so the person can survive for several minutes without **breathing.**

▽ Insulin

This hormone is produced by the pancreas. Its main job is to convert the sugar that has entered the blood following a meal into a starchy fuel store called glycogen. This takes place in the liver. If blood sugar drops, another hormone, called glucagon, reverses the process, turning the glycogen back into sugar to fuel the body.

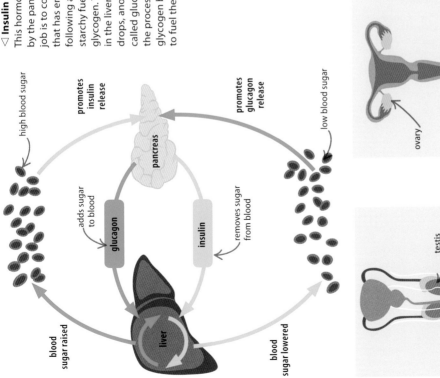

high blood sugar

promotes insulin release

blood sugar raised

adds sugar to blood

glucagon

pancreas

promotes glucagon release

low blood sugar

insulin

removes sugar from blood

liver

blood sugar lowered

△ Testosterone

The male hormone is produced by the testes (singular: testis), the male reproductive organs. As well as controlling the production of sperm, testosterone makes the body develop male characteristics, such as increased body hair and larger muscles. Testosterone also increases the willingness to fight (although it does not make you any better at it).

testis

△ Estrogen

This is a female hormone produced by the ovaries. It is involved in the production of eggs, making them ready for reproduction on a monthly cycle. Estrogen is also responsible for making the body develop secondary sexual features, such as mammary glands and pubic hair during puberty.

ovary

Disease and immunity

WHEN THE BODY IS ATTACKED BY DISEASE-CAUSING
ORGANISMS, IT HAS A RANGE OF RESPONSES.

SEE ALSO	
⟨ 24–25 Cells at work	
⟨ 26–27 Fungi and single-celled life	
Body systems	62–63 ⟩

The immune system is a highly complex defense system that looks for, and
then destroys, foreign bodies that get inside the body. These foreign bodies
use the body as a place to live and reproduce, which can cause illness.

Pathogens

The agents that cause disease are called
pathogens. Most are living organisms, such as
bacteria (often called germs), but illnesses are
also caused by viruses, which are not generally
considered to be living. The pathogens infect
body tissues, and cause symptoms by killing cells,
or by releasing poisons as they grow and spread.

Infection name	Type of pathogen	What it does	Symptom
streptococcus	bacterium	lives on skin and throat	sore throat
plasmodium	protist	kills body cells	malaria
threadworm	nematode worm	lives in intestines	itchy anus
H1N1	virus	invades body cells	fever and joint pain

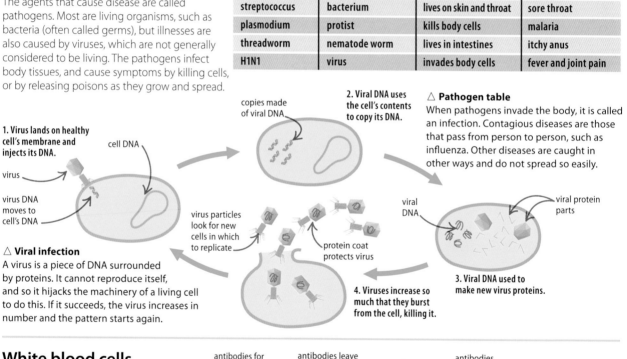

△ **Pathogen table**
When pathogens invade the body, it is called
an infection. Contagious diseases are those
that pass from person to person, such as
influenza. Other diseases are caught in
other ways and do not spread so easily.

1. Virus lands on healthy
cell's membrane and
injects its DNA.

cell DNA

virus

virus DNA
moves to
cell's DNA

copies made
of viral DNA

2. Viral DNA uses
the cell's contents
to copy its DNA.

virus particles
look for new
cells in which
to replicate

protein coat
protects virus

viral
DNA

viral protein
parts

4. Viruses increase so
much that they burst
from the cell, killing it.

3. Viral DNA used to
make new virus proteins.

△ **Viral infection**
A virus is a piece of DNA surrounded
by proteins. It cannot reproduce itself,
and so it hijacks the machinery of a living cell
to do this. If it succeeds, the virus increases in
number and the pattern starts again.

White blood cells

White blood cells are the detectives of
the body, patrolling the bloodstream
looking for invaders. When they find
one, the white blood cells copy the
invading pathogen's antigen—the
chemical marker on its surface. Then,
the blood cells generate a protein,
called an antibody, that flags the attacker
for removal from the body. Amazingly,
the immune system remembers the
antibodies for all past attacks, and
so can only be infected once by
the same pathogen.

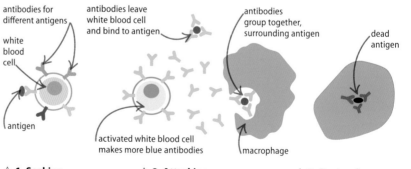

antibodies for
different antigens

antibodies leave
white blood cell
and bind to antigen

antibodies
group together,
surrounding antigen

dead
antigen

white
blood
cell

antigen

activated white blood cell
makes more blue antibodies

macrophage

△ **1. Seeking**
The white blood cell
recognizes the antigen
on an object in the
body as being foreign.

△ **2. Attacking**
Antibodies are released,
and large white blood
cells called macrophages
engulf the antigen.

△ **3. Destroying**
Destructive enzymes
called lysosomes finally
kill the antigen inside
the macrophage.

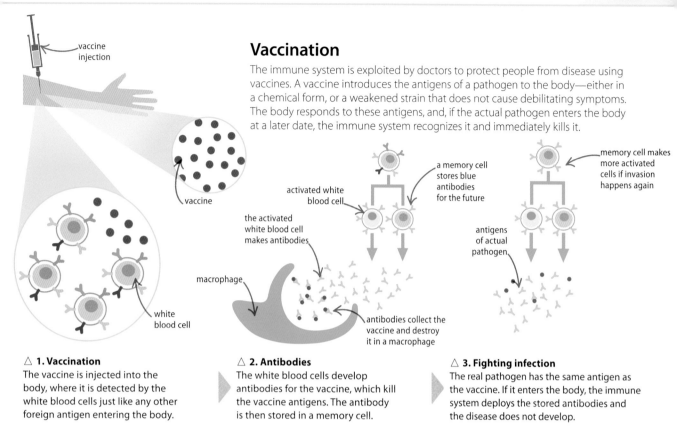

vaccine injection

Vaccination

The immune system is exploited by doctors to protect people from disease using vaccines. A vaccine introduces the antigens of a pathogen to the body—either in a chemical form, or a weakened strain that does not cause debilitating symptoms. The body responds to these antigens, and, if the actual pathogen enters the body at a later date, the immune system recognizes it and immediately kills it.

vaccine

activated white blood cell

a memory cell stores blue antibodies for the future

memory cell makes more activated cells if invasion happens again

the activated white blood cell makes antibodies

antigens of actual pathogen

macrophage

white blood cell

antibodies collect the vaccine and destroy it in a macrophage

△ **1. Vaccination**
The vaccine is injected into the body, where it is detected by the white blood cells just like any other foreign antigen entering the body.

△ **2. Antibodies**
The white blood cells develop antibodies for the vaccine, which kill the vaccine antigens. The antibody is then stored in a memory cell.

△ **3. Fighting infection**
The real pathogen has the same antigen as the vaccine. If it enters the body, the immune system deploys the stored antibodies and the disease does not develop.

Healing skin

The body's first line of defense against attack is the skin. When the skin is broken by a cut or scrape, bacteria and other germs can get into the body. Therefore, blood rushes to the area, making it swell and helping seal the gap. The liquid blood quickly coagulates (thickens) into a solid scab that forms a temporary seal while the skin grows back.

Sufferers of **hemophilia** have a reduced ability to form blood clots, which can result in a small scrape causing them to bleed to death.

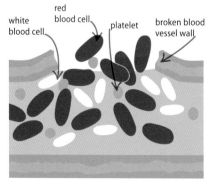

white blood cell

red blood cell

platelet

broken blood vessel wall

scab

fibrin mesh

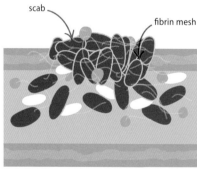

healed skin

△ **1. Broken skin**
The blood floods into the gap in the skin. Tiny cells called platelets react to the skin proteins. They trigger swelling, which brings white blood cells to mop up any invaders.

△ **2. Coagulation**
The platelets release the enzyme thrombin, which converts a soluble protein called fibrinogen into an insoluble one, fibrin. The fibrin forms a solid mesh across the gap.

△ **3. Healing**
The temporary seal, or scab, lasts until the skin has grown back. When this happens, the inflammation reduces, and the fibrin dissolves back into the blood.

Animal relationships

ANIMALS LIVE TOGETHER IN DIFFERENT WAYS.

Competition for resources is central to animal life. Many species go it alone, but others team up to make life easier. This teaming-up may be between members of the same species, or involve completely different animals working together.

SEE ALSO	
❰ 44–45 Reproduction II	
Ecosystems	74–75 ❱
Evolution	80–81 ❱
Co-evolution	83 ❱

Social groups

The strongest competition for survival is between members of the same species. Solitary animals avoid each other, so that competition is at a minimum. The animals that live in groups must strike a balance between the benefits of sticking together, and the increased competition for food and mates. Social groups range from simple ones that provide safety in numbers, to complex societies, where members hunt together and protect each other's young.

△ **Lion pride**
Prides feature one top male, who protects the rest and fathers all the cubs.

△ **Wolf pack**
Wolves work together to hunt animals much larger than themselves.

△ **Fish school**
Within a school, an individual has less chance of being picked off by a predator.

△ **Sheep flock**
Together, a flock is more likely to spot a threat before it gets too close.

△ **Baboon troop**
Baboons work together to defend their young and secure food supplies.

△ **Okapi**
Living alone is best in a dense forest habitat where food is widely available.

A single **super colony** of Argentine ants runs 6,000 km (3,700 miles) along the southern European coast.

Eusocial colony

The most highly social animals are ants, wasps, and bees. They are eusocial, which means there is division of labour, with different members of the colony performing specific jobs for the good of the whole. The colony works for their mother, the queen, to raise huge numbers of yet more sisters. All work is done by females. Only a few males are produced every year to mate with the next generation of queens.

woodcutter ants feed on fungus grown on a compost of cut leaf fragments

eggs develop into more workers to help out in the colony

queen is considerably larger than other ants

wings used to fly to find a mate

△ **Forager**
Foragers collect food from the surroundings and bring it back to the nest for the colony to eat.

△ **Worker**
Small workers feed and clean the eggs, larvae, and pupae, and help build the nest.

△ **Queen**
A large female controls the colony, and uses chemicals to stop other females from laying eggs.

△ **Male**
Male ants are produced at the end of summer. They die after they have mated with a queen.

Symbiosis

When animals of two species cooperate with each other, the relationship is known as symbiosis. There are two types. In mutualistic relationships, both partners benefit from the actions of the other. Commensal relationships are rarer. They involve one animal benefiting from the association, while the other receives no benefit, but is not harmed either.

shark gets cleaned by the pilot fish

△ **Pilot fish and shark**
The small fish follow a large predator and snap up the leftovers from its meals, keeping the predator clean in the process.

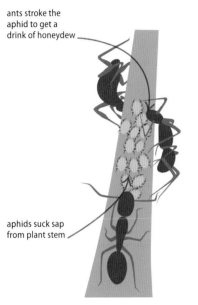

ants stroke the aphid to get a drink of honeydew

aphids suck sap from plant stem

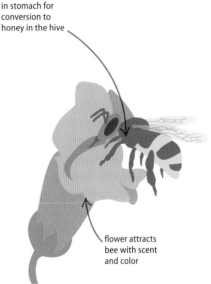

bee stores nectar in stomach for conversion to honey in the hive

flower attracts bee with scent and color

impala tolerates the oxpecker because the bird keeps its coat clean of parasites

△ **Ants and aphids**
Aphids produce a sweet urine called honeydew. Ants protect a herd of aphids from predators, in order to feed on the aphids' honeydew.

△ **Honeybee and flower**
Flowers rely on honeybees to transfer their pollen to another plant. In return, the bees feed on nectar supplied by the plant.

△ **Oxpecker and impala**
An oxpecker bird lives on the back of a large herbivore, such as an impala. The bird feeds on ticks and insects living on the larger animal's hair and skin.

Parasites

A parasite is an organism that lives on or inside another, known as the host. The parasite either eats the body of the host or consumes some of its food. The host is disadvantaged by the relationship, but is not killed—if it was, the parasite would soon die as well in many cases. A parasitoid is an animal that does kill its host, generally as a larva eating it alive. Once the host is dead, the parasitoid takes on an independent mode of life (see page 91).

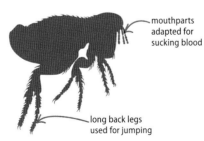

mouthparts adapted for sucking blood

long back legs used for jumping

△ **Flea**
This insect is an ectoparasite, meaning it lives on the outside of the host. Fleas suck the blood of their hosts, moving to new ones in great leaps.

hooks and suckers on the tapeworm's head hold it in place in its host's intestines

tapeworm has no mouth so it absorbs food through its skin

suckers

segmented body

△ **Tapeworm**
This flatworm is an endoparasite, meaning it lives inside its host. Egg-carrying segments break off the worm and end up in the host's droppings, where they hatch and spread.

Plants

THE ORGANISMS THAT MAKE UP THE PLANT KINGDOM RANGE FROM SIMPLE MOSSES TO COMPLEX FLOWERING PLANTS.

The study of plants is called botany. Plants reach such huge sizes and are so widespread that botanists estimate there are 1,000 tons of living plant material for every tonne of animal.

The plant kingdom

There are thought to be around 300,000 species of plants, far fewer than animals. Unlike animals, plants are restricted to sunlit habitats to power photosynthesis, so they cannot grow in deep water or underground. They fall into three main groups: seaweeds (including algae), nonvascular land plants, and vascular plants (plants with xylem and phloem vessels). This last group makes up 90 percent of plant species.

plant kingdom

— **water**
  — **seaweeds**

  anchored to rocks by structures called holdfasts, seaweeds are hardy plants that use leaflike structures called laminae to catch light for photosynthesis

— **land**
  — **non-vascular**
    — **mosses**

    simple stems are flattened to catch light

    moss

    the plant is held in place by extensions called rhizoids

    ▽ **Mosses and liverworts**
    The most simple land plants are the mosses and liverworts. They do not have true leaves or roots. Without xylem or phloem to transport material, they are restricted in size and require a damp habitat. Mosses and liverworts reproduce using eggs and sperm that swim between plants.

  — **vascular**
    — **ferns**

    structures on the fronds called sporangia disperse spores—which form new ferns.

    fern

    a young bud unfurls into a frond

    ▽ **Ferns**
    Ferns are primitive vascular plants that do not produce seeds. They have roots and stems with bundles of xylem and phloem. These not only transport water and sugars, but also provide support for the plants to reach large sizes. Ferns include the first "trees", about 350 million years ago.

    — **gymnosperms**

      pine cone

      pollen develops into a seed inside the cone, before the cone lets the seed fall to the ground

      opening in top of female cone receives pollen from smaller male cones

      ▽ **Gymnosperms**
      The first plants to reproduce using seeds were the gymnosperms, which include today's conifer trees, such as pines, cycads, and firs. Their scientific name means "naked seed," which refers to the way the seeds are not enclosed in a coat or fruit, as in the flowering plants (angiosperms) that evolved later.

    — **angiosperms (see below)**

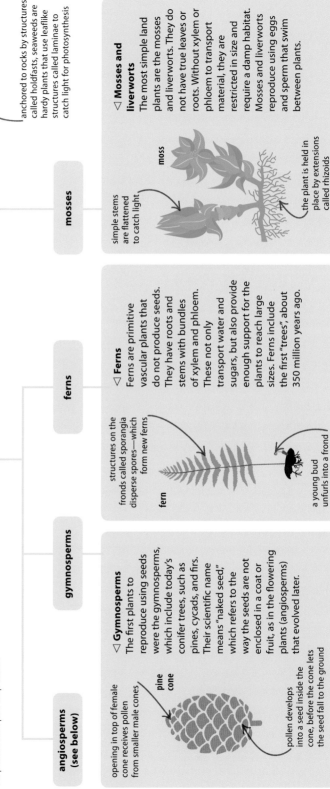

Angiosperms

Plants that reproduce using flowers are called angiosperms. Their seeds develop with a protective coat. They evolved from gymnosperms about 200 million years ago, and are the most common plant group, at least on land. Unlike the seeds of more primitive plants, angiosperm seeds include a starchy endosperm as a source of nutrition for the growing plant. Wheat, rice, and corn all come from endosperm seeds, and form much of the staple diet of humans.

The **largest flower** belongs to the corpse flower of Southeast Asia—it is 1 m (3¼ ft) wide and smells of rotting meat.

△ **Fruit**

Only angiosperms produce fruits. These fruits develop from the outer layers of the ovary after seeds have formed inside. The seeds are often spread by animals, who eat the fruit but cannot digest the seeds.

△ **Wood**

Tall trees are supported by dead xylem tubes that have been strengthened with a waterproof compound called lignin. The xylem grows out from the center. What remains of the original stem forms the bark.

petal

△ **Flower**

The flower is the reproductive organ of a flowering plant. Most produce both pollen (male sex cells) and ova, or eggs (female sex cells). The flower is structured to disperse and collect pollen.

vein

petiole

△ **Leaf**

Most angiosperms are broad-leaved. However, plants that live in extreme conditions—such as cacti—have spiked leaves to save water loss. Pine needles have the same function.

Dropping leaves

Botanists call the way plants drop their leaves abscission. Deciduous plants drop their leaves all in one go, generally in fall, because there will not be enough sunlight in winter to photosynthesize, and the leaves will be damaged by frost. New leaves grow in spring. An evergreen plant also drops its leaves, but evergreen abscission occurs continually throughout the year, along with new growth.

Climate	Conditions	Which?	Why?
tropical	wet and hot	evergreen	growth possible all year around
monsoon	rainy season	deciduous	avoid water loss through leaves
temperate	cold winter	deciduous	avoid frost damage to leaves
polar	short summer	evergreen	no time to grow new leaves for summer

▽ **Evergreen or deciduous?**

Evergreen plants live in places that are warm or cold all year, while deciduous species are adapted to habitats with changing seasons.

▽ **Abscission**

Leaf loss is triggered by changing conditions, such as shortening day lengths. The area at the base of the petiole has thin cell walls. These are broken when spongy bark expands underneath, breaking the water supply to the leaf, so the leaf falls away.

green chlorophyll is transported back to the tree before leaf is shed

base of petiole

weight of leaf pulls on abscission zone

weak cells in abscission zone

bud will grow into a leaf next year

spongy bark fills gap

Many conifer seeds need to be frozen **over winter** before they will sprout.

Invertebrates

AN INVERTEBRATE IS AN ANIMAL WITHOUT A BACKBONE.

The invertebrates are made up of dozens of phyla, many as distantly related to each other as they are to vertebrate animals. They range from microscopic to some of the largest creatures on Earth.

Arthropods

By far the largest group of invertebrates, the Arthropoda phylum includes insects (which make up 90 percent of the group), arachnids, and crustaceans. They all have a stiff exoskeleton made from chitin, a protein-based substance. All arthropods have legs made from several jointed sections; the phylum's name means "jointed foot." Insects are the only invertebrates that are able to fly.

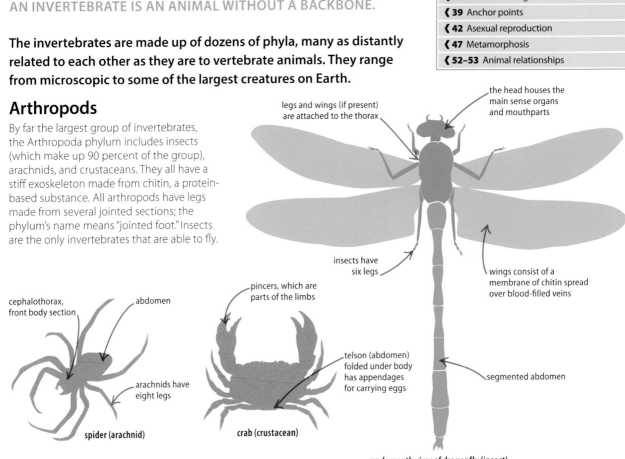

legs and wings (if present) are attached to the thorax

the head houses the main sense organs and mouthparts

insects have six legs

wings consist of a membrane of chitin spread over blood-filled veins

cephalothorax, front body section

abdomen

arachnids have eight legs

spider (arachnid)

pincers, which are parts of the limbs

telson (abdomen) folded under body has appendages for carrying eggs

segmented abdomen

crab (crustacean)

underneath view of dragonfly (insect)

Radiata

Most animals are bilaterally symmetrical, which means they can be divided into two halves that mirror each other. Radiata is a subkingdom (a group of phyla) made up of simple animals with round bodies. Radiata have both radial (symmetry around a fixed point, called the center) and bilateral symmetry. They do not have a mouth as such, but one body opening through which both food and waste pass. The main phylum is the Cnidaria, which includes corals, jellyfish, and anemones. Cnidarians have two types of body form, the polyp and the medusa.

▽ **Polyp**
The polyp is the upright form used by corals or sea anemones. They sit on the seabed with feeding tentacles facing upward, sifting food from the water.

▽ **Medusa**
Adult jellyfish are medusae, the bell-shaped form of cnidarians. Medusae are free swimming, and have stinging tentacles that hang down.

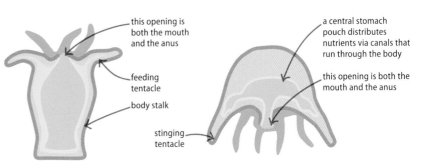

this opening is both the mouth and the anus

feeding tentacle

body stalk

stinging tentacle

a central stomach pouch distributes nutrients via canals that run through the body

this opening is both the mouth and the anus

Mollusks

The second-largest phylum of invertebrates is the mollusks. Mollusks range from filter-feeding bivalves and grazing gastropods to highly intelligent cephalopods. All mollusks share a common body plan. The main muscle is the "foot," which is used for locomotion in snails. In cephalopods, the foot is divided into tentacles, while bivalves use it to move and dig.

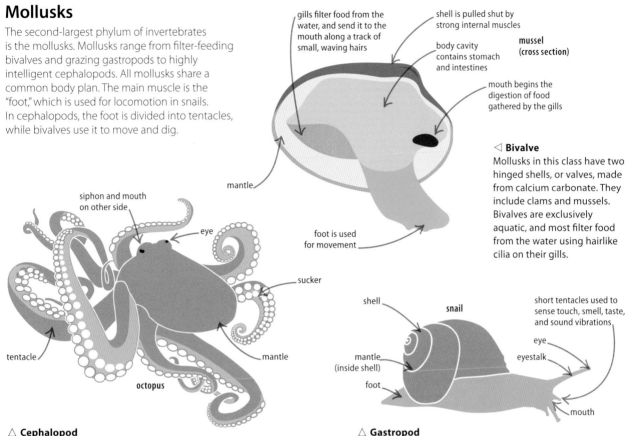

gills filter food from the water, and send it to the mouth along a track of small, waving hairs

shell is pulled shut by strong internal muscles

mussel (cross section)

body cavity contains stomach and intestines

mouth begins the digestion of food gathered by the gills

mantle

foot is used for movement

◁ **Bivalve**

Mollusks in this class have two hinged shells, or valves, made from calcium carbonate. They include clams and mussels. Bivalves are exclusively aquatic, and most filter food from the water using hairlike cilia on their gills.

siphon and mouth on other side

eye

sucker

mantle

tentacle

octopus

shell

snail

short tentacles used to sense touch, smell, taste, and sound vibrations

eye

eyestalk

mantle (inside shell)

foot

mouth

△ **Cephalopod**

This class includes octopuses, squid, and the nautilus. All but the latter have evolved out of their shells. They catch food with suckered—and in many cases clawed—tentacles that surround a beaklike mouth. They squirt a jet of water from a funnel near their mouth, called the siphon, to move.

△ **Gastropod**

This class of mollusks includes snails, slugs, winkles, and limpets. They have one shell—although this can be either reduced in size or absent completely in slugs. Snails and slugs are the only mollusks to live on land, although they require damp habitats. Snails breathe using a lunglike cavity in the mantle.

Worms

Worms are simple animals. They all lack legs, but can live in a wide range of habitats from the deep sea to inside the bodies of other animals. About half of the nematodes, also known as roundworms, are intestinal parasites, while the rest live in soil. The platyhelminthes, or flatworms, are parasitic or aquatic. They do not have intestines, and absorb food through their skin.

▷ **Annelid**

Also known as segmented worms, the Annelid phylum includes ragworms living in the ocean, oligochaetes such as earthworms on land, and leeches, which can live in freshwater or on land. Small, hairlike structures called setae help earthworms to burrow and sense their environment, while a series of pseudohearts pumps blood around their bodies.

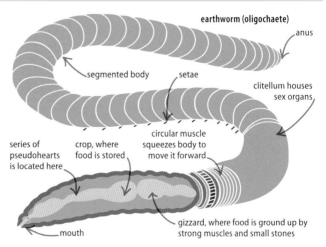

earthworm (oligochaete)

anus

segmented body

setae

clitellum houses sex organs

series of pseudohearts is located here

crop, where food is stored

circular muscle squeezes body to move it forward

mouth

gizzard, where food is ground up by strong muscles and small stones

Fish, amphibians, and reptiles

THESE GROUPS ARE THE MOST PRIMITIVE
VERTEBRATES (ANIMALS WITH BACKBONES).

Fish, amphibians, and reptiles are
three classes of vertebrates, the
group to which birds and mammals—
including humans—belong.

What is a vertebrate?

Vertebrates make up most of the phylum
Chordata. "Chordata" refers to a flexible
supporting rod, called the notochord, that is
present at some point in the life of all chordates.
In most cases, it develops into a vertebral
column—a chain of interlinked bones that
form the spine, or backbone. This protects
a spinal cord, a thick nerve bundle that
connects the brain to the rest of the body.

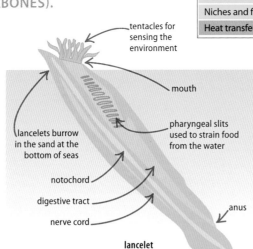

tentacles for
sensing the
environment

mouth

pharyngeal slits
used to strain food
from the water

lancelets burrow
in the sand at the
bottom of seas

notochord

digestive tract

nerve cord

anus

lancelet

◁ **No skull**
The first vertebrates are
thought to have looked
like today's lancelets,
simple aquatic animals
that live on the seabed.
Lancelets have no skull,
unlike true vertebrates,
but they share other
features, including
a notochord and
pharyngeal slits (which
form gills in fish).

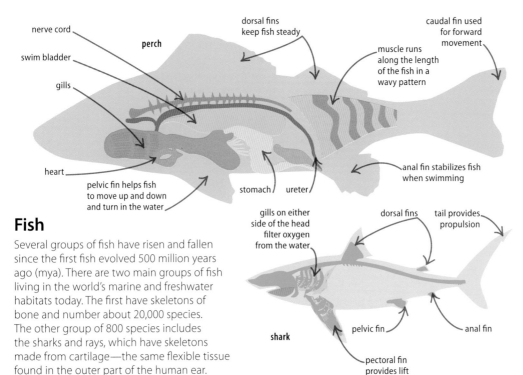

nerve cord

swim bladder

gills

heart

pelvic fin helps fish
to move up and down
and turn in the water

perch

dorsal fins
keep fish steady

muscle runs
along the length
of the fish in a
wavy pattern

caudal fin used
for forward
movement

anal fin stabilizes fish
when swimming

stomach   ureter

◁ **Bony fish**
Bony fish, unlike
cartilaginous fish, can
control their buoyancy
by altering the levels of
gas in an internal float
called the swim bladder.

gills on either
side of the head
filter oxygen
from the water

dorsal fins

tail provides
propulsion

pelvic fin

anal fin

shark

pectoral fin
provides lift

◁ **Cartilaginous fish**
A shark's cartilaginous
skeleton (in dark blue)
and streamlined body
shape help it move quickly
through water. Flexible rods
of cartilage stiffen the flat
fins and tail lobes. The
dorsal fins keep the shark
from rolling over as swishes
of its long tail power it
through the water.

Fish

Several groups of fish have risen and fallen
since the first fish evolved 500 million years
ago (mya). There are two main groups of fish
living in the world's marine and freshwater
habitats today. The first have skeletons of
bone and number about 20,000 species.
The other group of 800 species includes
the sharks and rays, which have skeletons
made from cartilage—the same flexible tissue
found in the outer part of the human ear.

Reptiles

Reptiles were the first vertebrates to make the break from living in water completely. They became the ancestors of birds and mammals as a result. They are a varied group with several distinct branches, but all share two common features. They all have waterproofed keratin scales covering their skin, and their eggs all have waterproof shells to keep in their moisture, so they won't shrivel up out of water.

The **Johnstone river turtle** breathes underwater by absorbing oxygen through its anus.

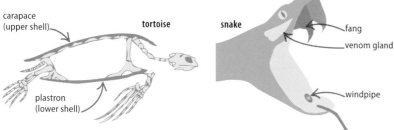

carapace (upper shell)

tortoise

plastron (lower shell)

snake

fang

venom gland

windpipe

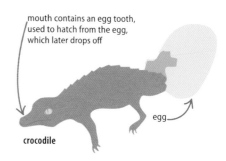

mouth contains an egg tooth, used to hatch from the egg, which later drops off

crocodile

egg

△ **Turtles and tortoises**
Turtles and tortoises evolved separately from dinosaurs and other reptiles. They have a defensive bony shell covered in giant horny scales (called scutes) attached to the ribs.

△ **Squamates**
Most of today's reptiles belong to this order, which includes lizards and snakes. Many snakes and a few lizards have venom glands, which are modified salivary glands, used for attacking prey.

△ **Crocodilians**
The crocodilians are archosaurs, a group of large reptiles that also included the dinosaurs. They are predatory hunters, waiting for prey to come close before snapping with powerful jaws.

Amphibians

Amphibians were the first creatures to live part of their life on land, evolving about 400 mya. They must return to water or moist habitats to lay eggs. After hatching, most amphibians spend their early growth phase in water, breathing with gills. They then transform—in a process called metamorphosis—into an air-breathing adult form that feeds on land.

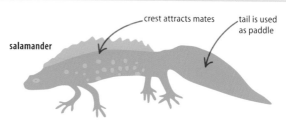

salamander

crest attracts mates

tail is used as paddle

△ **Newts and salamanders**
These amphibians were the first vertebrates to evolve a neck. Their neck lets them move their head from side to side, which is different from frogs and toads, who must move their whole body to look left or right.

Ectothermy

Fish, amphibians, and reptiles are ectothermic (cold-blooded), meaning their bodies are the same temperature as their surroundings. Ectotherms become more active in warm weather. Reptiles and amphibians influence their temperature by basking in the sun to heat up, or diving into water to cool down.

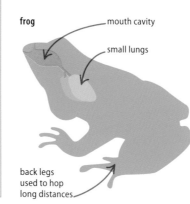

frog

mouth cavity

small lungs

back legs used to hop long distances

◁ **Frogs and toads**
Frogs are hunters that ambush prey using a sticky tongue and huge mouth. They have small lungs, and absorb much of their oxygen through their skin. Toads tend to have warty skin and legs designed for walking, while frogs have smoother skin and legs suited to hopping.

Mammals and birds

THESE GROUPS ARE WARM-BLOODED VERTEBRATES.

The vertebrate classes Aves (birds) and Mammalia (mammals) are among the most widespread groups of animal. They live on all continents and in almost all aquatic habitats.

SEE ALSO	
❮ 20–21 Variety of life	
❮ 32–33 Feeding	
❮ 38–39 Movement	
❮ 40 Animal senses	
❮ 42–43 Reproduction I	
❮ 58–59 Fish, amphibians, and reptiles	
Adaptations	82–83 ❯
Heat transfer	188–189 ❯

Endothermy

Birds and mammals are endothermic (warm-blooded) animals, meaning they maintain a constant body temperature. This requires energy to warm or cool the body, but it ensures that the animal's metabolism runs at a constant rate. As a result, its body systems function fully—even in colder habitats where ectothermic (cold-blooded) animals cannot survive. Endotherms have anatomical features to help them manage their body heat.

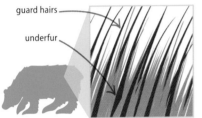

△ **Fur layers**
The hairs of many mammals are in two layers. The short underfur traps an insulating blanket of air. The longer, oily guard hairs keep out water, which would reduce the effectiveness of the underfur.

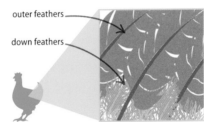

△ **Down insulation**
Birds prevent heat loss using fluffy down feathers that grow close to the body, under their outer feathers. Down traps air in pockets, insulating the body, and preventing valuable body heat from escaping.

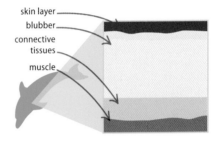

△ **Blubber**
In water, wet fur is a hindrance, so marine mammals have a thick insulating layer of blubber. This is a layer of soft fat, which has blood vessels running through it to help keep the animal warm.

Mammals

The largest vertebrates around today are mammals. The group gets its name from their mammary glands—modified sweat glands that produce milk. They are used by female mammals to suckle their young after birth. All mammals have at least a few hairs on their skin—although they are lost soon after birth in whales and dolphins. The hairs are made from keratin, the same waxy protein that builds reptile scales and bird feathers.

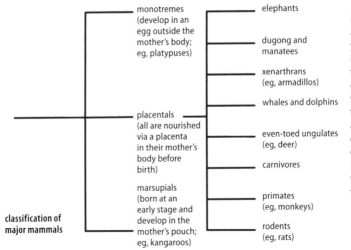

classification of major mammals

- monotremes (develop in an egg outside the mother's body; eg, platypuses)
- placentals (all are nourished via a placenta in their mother's body before birth)
  - elephants
  - dugong and manatees
  - xenarthrans (eg, armadillos)
  - whales and dolphins
  - even-toed ungulates (eg, deer)
  - carnivores
  - primates (eg, monkeys)
  - rodents (eg, rats)
- marsupials (born at an early stage and develop in the mother's pouch; eg, kangaroos)

◁ **Mammal variety**
Mammals appeared around 200 million years ago (mya) as small insect-eaters similar to today's shrews. The great variety of species we see today evolved from these primitive ancestors after the dinosaurs became extinct 65 mya. By 30 mya, mammals were the dominant vertebrate group.

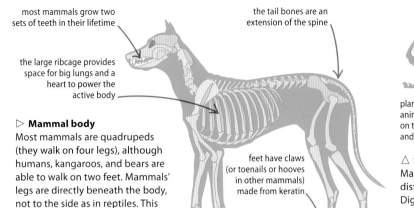

most mammals grow two sets of teeth in their lifetime

the tail bones are an extension of the spine

the large ribcage provides space for big lungs and a heart to power the active body

feet have claws (or toenails or hooves in other mammals) made from keratin

dog

▷ **Mammal body**
Most mammals are quadrupeds (they walk on four legs), although humans, kangaroos, and bears are able to walk on two feet. Mammals' legs are directly beneath the body, not to the side as in reptiles. This allows them to walk long distances and run at high speeds.

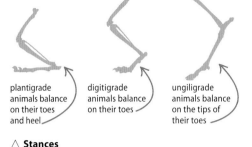

plantigrade animals balance on their toes and heel

digitigrade animals balance on their toes

ungiligrade animals balance on the tips of their toes

△ **Stances**
Mammals stand in three ways. Animals that walk long distances, such as humans and bears, are plantigrade. Digitigrade feet are used by agile animals that run and jump, such as dogs. Ungiligrade animals, such as horses, are suited to high-speed running.

Birds

The first birds evolved from forest-living dinosaurs about 150 mya. With about 10,000 species known, birds are the dominant flying vertebrates. Their wings are formed from long feathers attached to the bones of the forelimb. Feathers are stiff but lightweight, making them ideal for forming a rigid flight surface.

▽ **Bird skeleton**
Birds evolved from bipedal (upright-walking) dinosaurs. The wing is formed from the forelimb with thickened finger bones extending from the end to increase the length.

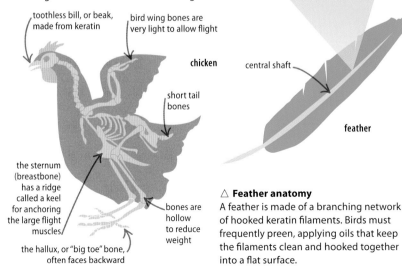

toothless bill, or beak, made from keratin

bird wing bones are very light to allow flight

chicken

short tail bones

the sternum (breastbone) has a ridge called a keel for anchoring the large flight muscles

bones are hollow to reduce weight

the hallux, or "big toe" bone, often faces backward

barb extends from the central shaft and then divides again into barbules, which hook together

barbule

central shaft

feather

△ **Feather anatomy**
A feather is made of a branching network of hooked keratin filaments. Birds must frequently preen, applying oils that keep the filaments clean and hooked together into a flat surface.

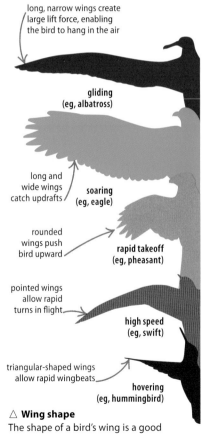

long, narrow wings create large lift force, enabling the bird to hang in the air

gliding (eg, albatross)

long and wide wings catch updrafts

soaring (eg, eagle)

rounded wings push bird upward

rapid takeoff (eg, pheasant)

pointed wings allow rapid turns in flight

high speed (eg, swift)

triangular-shaped wings allow rapid wingbeats

hovering (eg, hummingbird)

△ **Wing shape**
The shape of a bird's wing is a good indicator of how it flies. Scavenging birds need long, curved wings to glide, while ground birds need short wings to take off and get away from predators quickly.

Body systems

THE HUMAN BODY SYSTEMS THAT PERFORM SPECIFIC JOBS.

SEE ALSO	
❮ **38–39** Movement	
❮ **40–41** Sensitivity	
❮ **60–61** Mammals and birds	
Human senses	**64–65** ❯

The human body takes between 18 and 23 years to develop to full size. Medical science divides the body into several body systems—each featuring a set of organs that work together to perform certain jobs.

Skeletal and muscular systems

Bones give strength and support to the body, and are the main tissue in the skeleton. An adult human skeleton is made up of about 200 bones. These are covered by about 640 skeletal muscles, each one connected by a stiff, cordlike tendon to a specific joint. As the muscle contracts (tightens), it pulls on that joint, creating movement in a variety of ways. Bones are joined to each other by bands of cartilage, called ligaments.

▽ **Muscular system**
There are two main sets of muscles in the human body. The skeletal muscles work in pairs to move the body, while smooth muscles produce rippling pulses in the digestive system and arteries, to push material along tubes.

▷ **Synovial joints**
Most joints are synovial joints—the bone ends have a covering of smooth cartilage and the space between them contains lubricating synovial fluid. Different kinds of joints allow different types of movement.

Pivot
The head rotates from side to side using a pivot joint in the neck.

Ball-and-socket
The shoulder can move in all directions thanks to a circular bone connected to a round socket.

Hinge
Like the hinge on a door, the elbow can move only in one plane—it cannot twist like other joints.

Saddle
Made of two curved bones, a saddle joint allows the thumb to move in two planes.

muscles in the head control facial expressions

intercostal muscles at the ribs help control breathing

Ellipsoidal
The wrist has an oval bone sitting in a socket, allowing it to move in two planes—up-and-down, and side-to-side.

hand muscles allow us to grip and use objects

cartilage stops bones grinding into each other

blood vessels bring oxygen and food to the bone

matrix

marrow

peroneus longus muscle pulls the foot up and outward

sartorius muscles enable many movements, such as bending the knee

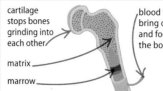

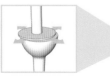

Gliding
Gliding joints occur in many places in the skeleton, and are usually very small. They feature bones that are almost flat that can glide over each other.

△ **Bone structure**
Bones are made of living cells that secrete a matrix of flexible calcium phosphate. In the core, or marrow, of a bone red blood cells are manufactured.

Other systems

The human body can be divided into a total of ten internal body systems (the skin and other outer body coverings can be counted as an external system). The organs and tissues in each system work closely together to perform the vital tasks that keep the body alive. If any one system fails, the other body systems cannot replace its function and are unable to work properly themselves.

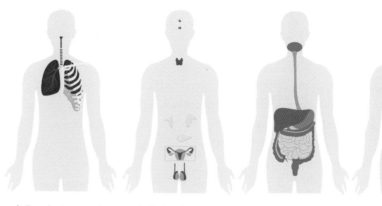

△ **Respiratory system**
Centered on the lungs, this system takes oxygen, needed by the body, from the air and puts it into the blood.

△ **Endocrine system**
The glands that make up this system produce hormones and other secretions that control other body systems.

△ **Digestive system**
This system processes food to extract its nutrients, which are taken into the bloodstream.

△ **Urinary system**
The kidneys filter waste materials from the blood, which are then flushed away in urine.

the brain and spinal cord control the activity of the other nerves

arteries (red) carry blood from the heart

veins (blue) carry blood to the heart

nerves branch out to all parts of the body

When compressed, human bone is **four times** stronger than concrete.

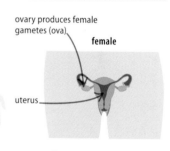

ovary produces female gametes (ova)

female

uterus

testes produce male gametes (sperm)

male

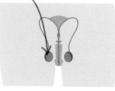

△ **Reproductive system**
The reproductive systems of males and females produce gametes, or sex cells. When these fuse, they form the first cell of a new person, which develops in the uterus.

△ **Nervous system**
A network of nerves carries signals around the body as electric pulses. The brain and spinal cord form the central nervous system.

△ **Circulatory system**
This system takes blood pumped by the heart around the body. The blood delivers oxygen and other materials to body tissues.

△ **Lymphatic system**
Body cells leak slightly, so this system collects waste liquids that build up in tissues, and empties them into the circulatory system.

Human senses

THE WAY WE GATHER INFORMATION ABOUT OUR SURROUNDINGS.

Our senses of hearing, vision, smell, taste, and touch constantly relay information to our brain about the world around us. The brain can then respond if necessary—such as moving us away from danger.

Hearing

The ear is a hypersensitive touch organ that picks up pressure waves moving through air, which make the eardrum vibrate. This vibration travels along three tiny bones, called the auditory ossicles, to the fluid-filled labyrinth, made up of the cochlea and semicircular canals. The sound waves become ripples in this labyrinth fluid, wafting hairlike sensory nerve endings that send signals on to the brain along the auditory nerve.

semicircular canal helps us balance by detecting movements of the head in all directions

auditory nerve sends signals from the cochlea to the brain

auditory ossicles transmit sound vibrations

sound waves

cochlea and semicircular canals form the inner ear

Eustachian tube connects the ear to the throat

pinna

eardrum separates the outer ear from the middle ear

◁ **Collecting sounds**
The trumpet-shaped pinna of the outer ear directs sound into the ear canal.

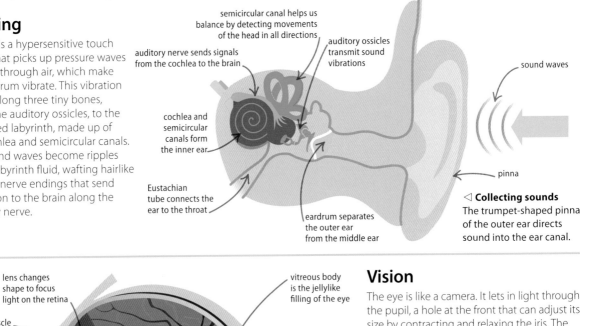

Vision

The eye is like a camera. It lets in light through the pupil, a hole at the front that can adjust its size by contracting and relaxing the iris. The cornea and lens work together to focus the light onto the retina, a lining of light-sensitive cells at the back of the eye. The cells pick up the pattern of light falling across them, which is transmitted to the brain by the optic nerve and made into an image.

lens changes shape to focus light on the retina

ciliary muscle

anterior chamber is filled with a clear fluid

pupil

cornea

iris gives the eye its color

vitreous body is the jellylike filling of the eye

sclera is the white outer layer of the eye

retina

optic nerve

blood vessels

choroid contains blood vessels

△ **Focus on the eye**
The ciliary muscles control the shape of the lens to focus light from near and far onto the retina. The retina then sends the image to the brain along the optic nerve.

rods outnumber cones about 17 to one

light

choroid

cone

◁ **Rods and cones**
The cells in the retina have light-sensitive pigments that produce electric nerve pulses when light hits them. Rod cells are used in night vision and cannot detect color. There are three types of cone cells, each sensitive to light within a different range of colors, and are used to produce color images by day.

Smell and taste

Our senses of smell and taste both involve collecting chemicals and analysing them. The nose collects chemicals carried in the air. Inside the nose, scent chemicals dissolve in the mucus lining of the nasal cavity (which also helps clean the air). The chemicals are detected by hairlike nerve endings that send signals to the brain.The tongue detects similar chemicals in food.

olfactory bulb takes signals from the nose to the brain

nasal conchae force inhaled air to flow steadily

the nasal cavity is shaped to increase the size of the odor-sensitive layer

tastebuds on tongue can detect five distinct tastes: sweet, sour, bitter, salty, and umami (savory)

nerves under the tongue's surface take taste signals to the brain

▷ **Taste bud**

Taste buds are located on the tongue, gums, and throat. They have nerve endings covered in proteins that can detect specific chemicals associated with certain foods, such as sweet sugar or sour acids.

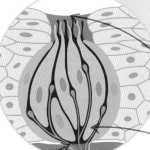

the nerve endings send signals from the taste receptor cells to the brain

openings in the the tongue called taste pores allow food dissolved in saliva to reach the taste receptors

each taste receptor cell picks up certain chemicals such as sugars (sweet) and acids (sour)

▷ **Skin**

Mechanoreceptors pick up pressure, touch, stretching, vibration, and sharp pains. Heat and cold are picked up by thermoreceptors.

hair shaft

pain receptors

sweat pore

pressure receptor

heat receptor

touch receptor

cold receptor

nerves send messages to the nervous system

sebaceous gland secretes oily sebum to lubricate skin and hair

sweat gland secretes sweat to cool the skin

Touch

The sense of touch relies on several types of receptors, mainly located in the skin, but also found in muscles, joints, and internal organs. There are about 50 touch receptors for every square inch of skin, although more sensitive body parts, such as the fingertips and tongue, have more, while the back has fewer.

Braille

Fingertips (touch), and not eyes (vision), are used to read Braille. Letters are represented by patterns of between one and five small bumps, or dots, arranged in a grid. Skilled Braille readers can read about 200 words per minute.

Human digestion

THE DIGESTIVE SYSTEM PROCESSES THE FOOD WE EAT.

SEE ALSO

❮ 62–63 Body systems

Human health   **70–71** ❯

Catalysts   **138–139** ❯

Digestion is a complex process that breaks down food into simple substances. These fats, sugars, proteins, and other nutrients are then absorbed, leaving unwanted waste to be expelled.

The digestive tract

Food is digested in the digestive tract—the passage food takes from the mouth to the anus. Nutrients are absorbed in the intestines (also known as the gut). The material that cannot be digested is mixed with other waste products from the body, such as the brown pigments from old blood cells, and pushed out of the body through the anus.

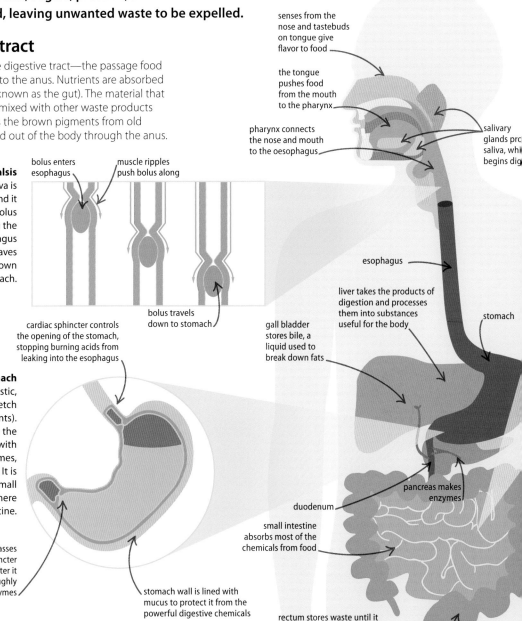

▷ **Peristalsis**
In the mouth, saliva is mixed with the food and it is chewed to form a bolus (ball). Muscles along the walls of the esophagus (throat) contract in waves to push the bolus down to the stomach.

bolus enters esophagus

muscle ripples push bolus along

bolus travels down to stomach

cardiac sphincter controls the opening of the stomach, stopping burning acids from leaking into the esophagus

▷ **Stomach**
The stomach is an elastic, muscular sac that can stretch to hold up to 4 liters (8 pints). The stomach churns up the food and mixes it with powerful acids and enzymes, turning it into a liquid. It is then sent to the small intestine, and from there to the large intestine.

food in the stomach passes through the pyloric sphincter to the small intestine after it has been thoroughly mixed with enzymes

stomach wall is lined with mucus to protect it from the powerful digestive chemicals

senses from the nose and tastebuds on tongue give flavor to food

the tongue pushes food from the mouth to the pharynx

pharynx connects the nose and mouth to the oesophagus

salivary glands pro saliva, whi begins dig

esophagus

liver takes the products of digestion and processes them into substances useful for the body

stomach

gall bladder stores bile, a liquid used to break down fats

pancreas makes enzymes

duodenum

small intestine absorbs most of the chemicals from food

rectum stores waste until it can be ejected from the anus

anus

large intestine absorbs water from food

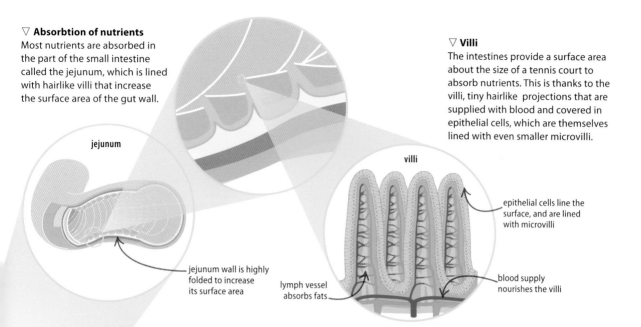

▽ **Absorbtion of nutrients**
Most nutrients are absorbed in the part of the small intestine called the jejunum, which is lined with hairlike villi that increase the surface area of the gut wall.

jejunum

jejunum wall is highly folded to increase its surface area

lymph vessel absorbs fats

▽ **Villi**
The intestines provide a surface area about the size of a tennis court to absorb nutrients. This is thanks to the villi, tiny hairlike projections that are supplied with blood and covered in epithelial cells, which are themselves lined with even smaller microvilli.

villi

epithelial cells line the surface, and are lined with microvilli

blood supply nourishes the villi

Digestive chemicals

Digestion is both a physical and a chemical process. It starts in the mouth, when the teeth mechanically grind up the food. This pulp is mixed with saliva, which contains enzymes that work on the food. Enzymes target specific foods, dividing complex foods, such as starches and proteins, into smaller, simpler ingredients—sugar and amino acids respectively—that can absorbed more easily.

▽ **Chemical chart**
A range of digestive chemicals work on the food at each stage of its journey through the gut. The chemicals each have a specific role to play in breaking down the food, and are produced by glands and organs along the alimentary canal.

	Enzyme or other chemical	Function	Produced by
Mouth	lipase (enzyme)	digests fats	salivary gland
	amylase (enzyme)	digests starch	salivary gland
	mucin	lubricates food	salivary gland and gut lining
	bicarbonate (enzyme)	kills bacteria, neutralizes acids	salivary gland
Stomach	pepsin (enzyme)	digests proteins	stomach cells
	hydrochloric acid	kills bacteria	stomach cells
	rennin (enzyme)	digests milk	stomach cells
Small intestine	bile	aids digestion of fats	liver, via gall bladder
	trypsin (enzyme)	digest proteins	pancreas
	nuclease (enzyme)	digest nucleic acids	pancreas
	phospholipase (enzyme)	digests fats	pancreas
	amylase (enzyme)	digests starches	pancreas
	sucrase (enzyme)	digests sucrose	duodenum
	lactase (enzyme)	digests lactose (sugar found in milk)	duodenum
	maltase (enzyme)	digests maltose (sugar found in starch)	duodenum

Brain and heart

THE BODY'S MOST VITAL ORGANS
ARE THE BRAIN AND THE HEART.

The brain and the heart are the most important parts of
the body. While the heart is the engine that keeps the body
supplied with nutrients, the brain is the control center.

SEE ALSO

❮ **36** Circulation
❮ **36** Composition of blood
❮ **39** Muscle contraction
❮ **62–63** Body systems

Brain

The brain forms the main part of the central
nervous system (CNS), which receives signals
from every part of the body, and sends out
responses if necessary. The brain is split
into two halves, or hemispheres, made of
masses of nerve cells that have thousands
of high-speed connections with their
neighbors. The outer layer of the brain is
called the cerebral cortex, or gray matter,
and the inner layer is called white matter.

REAL WORLD

Magnetic resonance imaging (MRI)

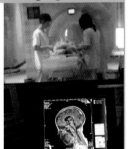

An MRI scanner causes
soft body tissues, such as
the brain, to release radio
waves for a split second.
These are used to build
a detailed picture of
internal tissues, and
help doctors diagnose
and treat illnesses.

the cerebrum is highly
folded, which increases
the brain's surface area

corpus callosum
connects the two
halves of the brain

thalamus relays
sensory signals to
the cerebral cortex

hypothalamus
controls the
endocrine system

pons is responsible
for motor control
and analyzing senses

cerebellum
controls learned
movements and balance

brain stem controls
involuntary functions, such
as breathing and heart rate

▷ **Human brain**
The human brain consists of the
hindbrain, midbrain, and forebrain
(cerebrum). The hindbrain (made up of
the brainstem, pons, and cerebellum)
and midbrain control basic functions,
such as breathing and balance, while the
forebrain—especially large in humans—
is used for thinking and making decisions.

Brain functions

Neuroscience, the study of the
brain, has found that different areas
of the cerebrum are devoted to specific
functions. If one of the areas—often
known as a cortex—is damaged, that
function, such as speech or sight,
ceases while the others continue
unaffected. Neuroscientists have
learned a lot about the human brain
in recent years. For example, we now
know that each cortex has more
connections between its cells than
there are stars in the Milky Way Galaxy.

▷ **Mapping the brain**
The functional areas are mapped
on the outside of the brain. Different
parts of the brain cooperate and
interact with each other to produce
other functions, such as planning
or operating machinery.

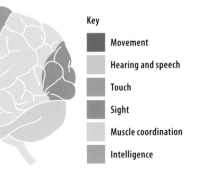

Key

Movement

Hearing and speech

Touch

Sight

Muscle coordination

Intelligence

Circulatory system

The human circulatory system is a double loop of vessels. The pulmonary loop carries deoxygenated blood to the lungs, where it picks up oxygen and releases carbon dioxide. The reoxygenated blood then goes back to the heart, where it enters the second loop, the systemic loop, which takes it around the body.

▷ Vessel types

The arteries (in red) take oxygenated blood to the tissues. The system of veins (in blue) then brings back the used, deoxygenated blood—which is then returned to the lungs. Capillary vessels run between the arteries and veins, carrying blood through the tissues.

The **total length** of your circulatory system stretches an amazing 96,600 km (60,000 miles)—more than **twice** the distance around Earth.

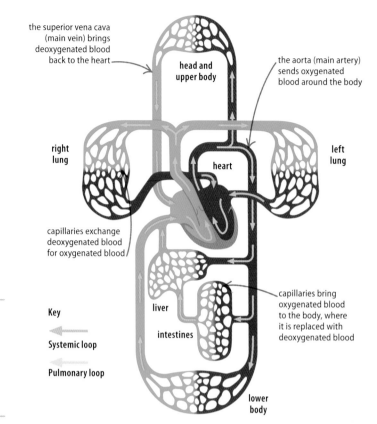

the superior vena cava (main vein) brings deoxygenated blood back to the heart

head and upper body

the aorta (main artery) sends oxygenated blood around the body

right lung

left lung

heart

capillaries exchange deoxygenated blood for oxygenated blood

capillaries bring oxygenated blood to the body, where it is replaced with deoxygenated blood

liver

intestines

Key

◁ Systemic loop

◁ Pulmonary loop

lower body

In a heartbeat

The human heart is a powerful pump made from a type of muscle (cardiac) that never needs to rest—so a heart can keep working throughout a person's life. The heart has two sides, each one divided into an upper chamber called the atrium, and a lower chamber (the ventricles). The right side receives deoxygenated blood from the body. Reoxygenated blood is pumped out again from the left side.

The heart beats around **three billion times** in the average person's life.

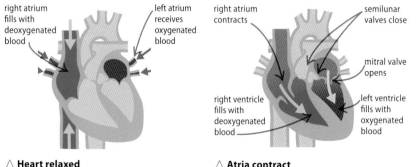

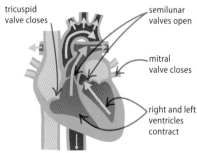

right atrium fills with deoxygenated blood

left atrium receives oxygenated blood

right atrium contracts

semilunar valves close

tricuspid valve closes

semilunar valves open

mitral valve opens

mitral valve closes

right ventricle fills with deoxygenated blood

left ventricle fills with oxygenated blood

right and left ventricles contract

△ Heart relaxed

When the cardiac muscles are relaxed, deoxygenated blood flows into the right atrium from the vena cava, the main vein. Oxygenated blood flows into the left atrium.

△ Atria contract

The contraction of the heart starts at the top, squeezing the atria, so the blood moves down into the ventricles. One-way valves prevent the blood from moving back into the atria.

△ Ventricles contract

The lower part of the heart contracts, squeezing the ventricle. The right ventricle pumps the blood toward the lungs. The left ventricle pushes blood into the aorta (main artery).

Human health

DIET, EXERCISE, AND AVOIDING DANGEROUS
SUBSTANCES HELP TO MAINTAIN A HEALTHY BODY.

SEE ALSO
❮ 32–33 Feeding
❮ 62–63 Body systems
❮ 66–67 Human digestion

Medical science and improved living conditions have resulted in human life
expectancies being twice, if not three times, those of prehistoric people. However,
some aspects of a modern lifestyle are at odds with maintaining a healthy body.

Healthy eating

Food is made up of four groups of
substances: carbohydrates, fats, proteins,
and fiber. All four are essential for a
nutritious diet. Carbohydrates are found
in simple form in sugary food and in
complex form in starchy food. Fiber is
an indigestible form of carbohydrate
that keeps the digestive tract healthy.
Fats and oils are concentrated energy
stores, and too much of them can lead
to weight problems. Finally, protein,
needed for muscles and digestion, is
mainly found in animal-based foods, such
as meat and dairy products, but is also
found in beans, chickpeas, and lentils.

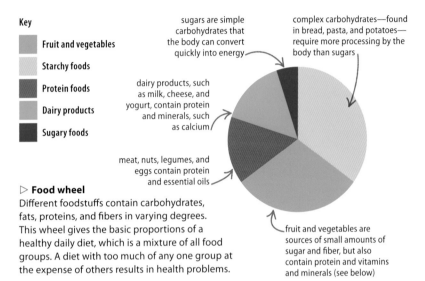

Key

- Fruit and vegetables
- Starchy foods
- Protein foods
- Dairy products
- Sugary foods

sugars are simple
carbohydrates that
the body can convert
quickly into energy

complex carbohydrates—found
in bread, pasta, and potatoes—
require more processing by the
body than sugars

dairy products, such
as milk, cheese, and
yogurt, contain protein
and minerals, such
as calcium

meat, nuts, legumes, and
eggs contain protein
and essential oils

fruit and vegetables are
sources of small amounts of
sugar and fiber, but also
contain protein and vitamins
and minerals (see below)

▷ **Food wheel**
Different foodstuffs contain carbohydrates,
fats, proteins, and fibers in varying degrees.
This wheel gives the basic proportions of a
healthy daily diet, which is a mixture of all food
groups. A diet with too much of any one group at
the expense of others results in health problems.

Vitamins and minerals

A healthy diet contains a series
of nutrients called vitamins.
These are chemicals the body
cannot make itself, which
are essential for important
metabolic processes. The
health problems that vitamin
deficiencies produce can
usually be reversed by eating
a balanced diet. The body also
requires a supply of minerals,
which are metals that are
important for maintenance.

▷ **Required nutrients**
Humans require the following
vitamins and minerals in small
amounts in their diets.

Name	Beneficial for	Sources	Deficiency results in
vitamin A	good eyesight	liver, carrots, green vegetables	night blindness
vitamin B1	healthy nerves and muscles	eggs, red meat, and cereal	loss of appetite
vitamin B2	healthy skin and nails	milk, cheese, and fish	itchy eyes
vitamin B6	healthy skin and digestion	fish, bananas, and beans	inflamed skin
vitamin B12	healthy blood and nerves	shellfish, poultry, and milk	fatigue
vitamin C	healthy immune system	citrus fruits, kiwi, and fruits	scurvy
vitamin D	strong bones and teeth	sunlight and oily fish	rickets
vitamin E	removing toxins	nuts, green vegetables	weakness
folic acid	red blood cell formation	carrots, yeast	anemia
calcium	strong bones and healthy muscles	dairy products	bad teeth
iron	healthy blood and body cells	red meat and cereals	anemia
magnesium	healthy bones	nuts and green vegetables	insomnia
zinc	normal growth and immune system	meat and fish	growth retardation

Body weight

The human body is primed to survive long periods of starvation. When food is available, the body lays down stores of fat to fuel the body during the lean times. In developed countries, food is always available, so people may become overweight, taking in more food than their body uses each day. This can lead to a variety of illnesses.

Key

- Obese
- Normal
- Overweight
- Underweight

▷ **Body mass index**
This chart is used to work out the healthiness of a person's weight-to-height ratio. Being overweight causes problems for the body, especially the circulatory system. People who are underweight may have a weaker immune system.

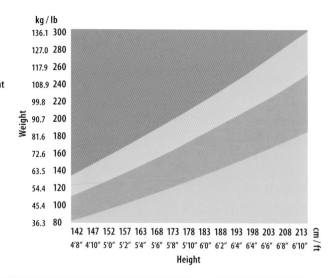

Exercise

The human body is built for walking long distances and having short bursts of activity. Modern working practices require people to sit still for long periods, so it is necessary these days to do regular exercise to keep the body in good condition. Exercise helps to burn the energy in food (measured in kilocalories, or calories for short), reducing weight gain due to overeating.

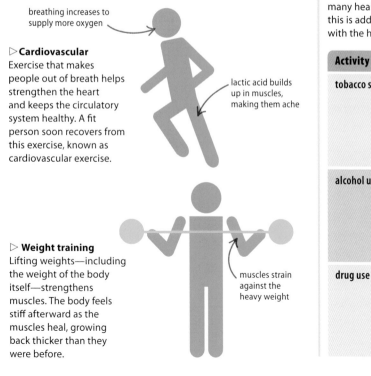

breathing increases to supply more oxygen

▷ **Cardiovascular**
Exercise that makes people out of breath helps strengthen the heart and keeps the circulatory system healthy. A fit person soon recovers from this exercise, known as cardiovascular exercise.

lactic acid builds up in muscles, making them ache

▷ **Weight training**
Lifting weights—including the weight of the body itself—strengthens muscles. The body feels stiff afterward as the muscles heal, growing back thicker than they were before.

muscles strain against the heavy weight

Dangerous substances

Alcohol and tobacco products are sold legally to adults because they have a long history of use across the world, but they cause serious illness. Other substances—often just called drugs—are illegal, and cause many health and social problems.

▽ **Threats**
Misuse of alcohol, over-the-counter drugs, and smoking can lead to many health issues, both physical and mental. The main reason for this is addiction and dependency, which means the addict continues with the harmful behaviour and finds it hard to break away.

Activity	Associated health problems
tobacco smoking	cancer of lungs, mouth, esophagus, and pancreas; heart disease; lung problems, specifically emphysema, bronchitis, and scarring of lung tissue; addiction
alcohol use	physical damage to liver (cirrhosis); mental instability; poor judgment; dangerous behavior; increased risk of heart attack; inflammation of digestive tract and pancreas; addiction
drug use	mental and physical problems, depending on drug; severe addiction and dependency; risk of various cancers; addict may resort to crime to pay for drugs

Human reproduction

EVERY HUMAN BEING STARTS LIFE AS A TINY FERTILIZED EGG.

SEE ALSO

❮ **43** Sexual reproduction
❮ **62–63** Body systems
Genetics I   **84–85** ❯
Genetics II   **86–87** ❯

Human reproduction begins with a sperm from a man combining with an egg inside a woman's uterus to produce an embryo. The baby develops for nine months inside the mother, sustained by a temporary organ called the placenta.

Sex organs

Gametes, or sex cells, are produced in sex organs or gonads. They carry a half set of chromosomes. The man produces sperm cells in organs called testes, while a woman produces egg cells (ova) in two ovaries. The ovaries release about 400 eggs in a woman's lifetime, at a rate of one every 28 days or so, while the testes produce many millions of sperm each day. Sperm cells are delivered to the cervix during sexual intercourse, and from there they swim into the oviduct (fallopian tube) to reach the single egg.

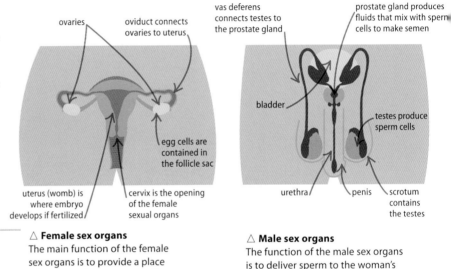

ovaries

oviduct connects ovaries to uterus

egg cells are contained in the follicle sac

uterus (womb) is where embryo develops if fertilized

cervix is the opening of the female sexual organs

vas deferens connects testes to the prostate gland

prostate gland produces fluids that mix with sperm cells to make semen

bladder

testes produce sperm cells

urethra

penis

scrotum contains the testes

△ **Female sex organs**
The main function of the female sex organs is to provide a place where an embryo can grow. Once it has developed enough to survive independently, the baby is born.

△ **Male sex organs**
The function of the male sex organs is to deliver sperm to the woman's uterus in a liquid called semen. The sperm cells make up about five percent of this mixture.

The record for the **most children** with the same mother is 69, born in Russia in the 18th century.

Ovulation

The process of producing and releasing an egg cell, known as ovulation, is controlled by hormones. The amount of oestrogen rises, causing one follicle in one ovary to prepare an egg cell. The ripe egg bursts from the ovary and travels into the oviduct ready to meet a sperm. The rest of the follicle then releases another hormone, progesterone, which causes the lining of the uterus to thicken, ready to receive an embryo.

▽ **Hormones and ovulation**
The egg follicle produces estrogen around day ten. A few days later, the hormone progesterone causes the lining of the uterus to thicken, so it is ready to receive a fertilized egg cell. If fertilization does not happen, the progesterone level drops, and the thickened lining of the uterus is shed as menstrual blood. The process then repeats.

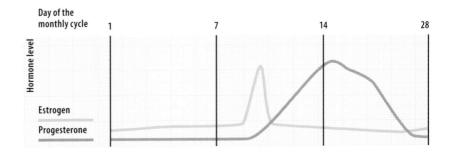

Day of the monthly cycle

Hormone level

1   7   14   28

Estrogen

Progesterone

Fertilization

After ovulation, the egg travels toward the uterus along the oviduct. It lives for about 18 hours, during which time it is ready for a sperm to fertilize it. During fertilization, the half sets of DNA from both sex cells combine to make a full set. At this point the cell becomes a zygote, the first cell of a genetically unique individual. The zygote divides into a ball of cells, called a blastocyst.

▷ **Fusing cells**
A single sperm burrows into the much larger egg cell. It drops its tail-like flagella, and any sperm arriving afterward are blocked from getting in.

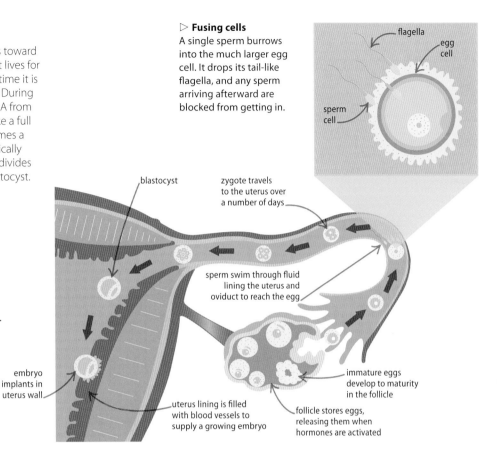

flagella

egg cell

sperm cell

blastocyst

zygote travels to the uterus over a number of days

▷ **Implantation**
The blastocyst can survive only for a few days on its own. It must implant in the uterus wall within about ten days in order to receive oxygen and nutrients. Once it does this, it continues to divide into new cells, and is then called an embryo.

sperm swim through fluid lining the uterus and oviduct to reach the egg

embryo implants in uterus wall

uterus lining is filled with blood vessels to supply a growing embryo

follicle stores eggs, releasing them when hormones are activated

immature eggs develop to maturity in the follicle

Gestation

From fertilization, the human embryo takes between 38 and 42 weeks to develop to the point where it can survive outside the uterus. The embryo is nourished by the placenta. Both develop from the same single blastocyst. Birth is triggered by a hormone released by the growing baby. This makes the cervix and vagina soften, and contractions of the uterus push the baby out.

placenta nourishes the baby via the umbilical cord

Fetal development

After about eight weeks of growth, the baby has all of its primary organs and recognizable human features. From this point it is known as a fetus. The development of a fetus can be monitored by scanning the womb with ultrasound to produce an image (below).

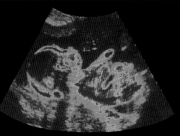

▷ **Early development**
Every baby grows from a single cell called a zygote. This divides rapidly into a growing ball of cells. The cells differentiate into those that form the placenta, the membranes around the embryo, and the embryo itself. The cells that form the baby are identifiable after eight days of growth.

embryo

umbilical cord brings nutrients to baby

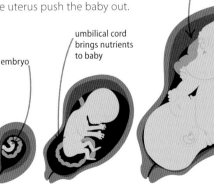

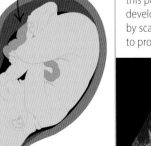

two months   five months   nine months

Ecosystems

THE SCIENCE OF ECOLOGY STUDIES HOW ORGANISMS
FORM COMMUNITIES CALLED ECOSYSTEMS.

An ecosystem is a complex set of relationships between the plants, animals, and other life forms that live in a habitat. These living things are also affected by other factors, such as weather and climate.

Niches and factors

Each species in an ecosystem occupies a niche—which means both the place and roles it carries out in the habitat. The mode of survival in any niche depends on the activity of other species in the ecosystem, such as predators looking for prey, or fast-growing algae using up the available resources. In a stable ecosystem, these influences, or factors, are in balance. If one factor changes, the rest of the ecosystem rebalances.

▽ **Wildlife community**
This freshwater ecosystem, like all ecosystems, is affected by physical factors, such as sunlight, climate, and fire, while its living members depend on each other for food.

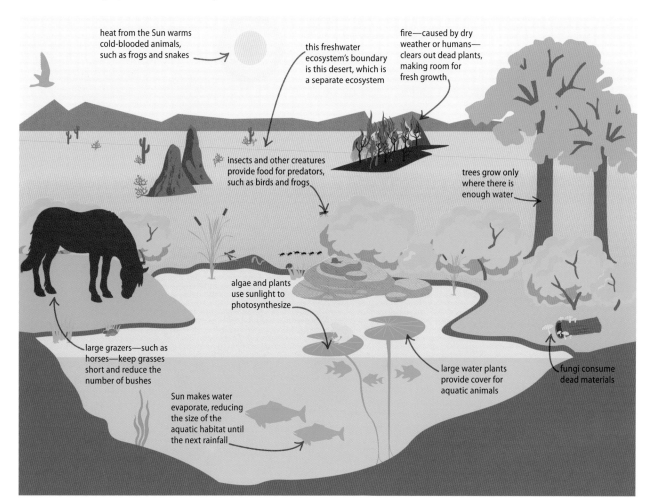

heat from the Sun warms cold-blooded animals, such as frogs and snakes

this freshwater ecosystem's boundary is this desert, which is a separate ecosystem

fire—caused by dry weather or humans—clears out dead plants, making room for fresh growth

insects and other creatures provide food for predators, such as birds and frogs

trees grow only where there is enough water

algae and plants use sunlight to photosynthesize

large grazers—such as horses—keep grasses short and reduce the number of bushes

large water plants provide cover for aquatic animals

fungi consume dead materials

Sun makes water evaporate, reducing the size of the aquatic habitat until the next rainfall

Predators and prey

Within an ecosystem, hunters and the hunted are closely linked. Their populations rise and fall in a repeating pattern. When there are a lot of prey animals, predators also increase in number, as there is more food to sustain them. However, more predators soon results in fewer prey, and the number of predators drops as there is less food available. Without many predators, the prey population rises again, and the cycle repeats.

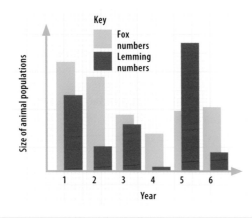

◁ **Foxes and lemmings**
This graph shows that, in years with high numbers of lemmings, Arctic foxes do well and have large numbers of pups. The next year, the lemming population falls as a result of this and the population of foxes decreases, too.

Biomes

The land habitats on Earth are grouped into ten climate zones, also known as biomes. Each biome is home to a particular set of animals and plants, which are adapted to the challenges of surviving in the different conditions. Desert animals must conserve water, while polar ones contend with long periods of extreme cold. Aquatic habitats are divided into marine and freshwater biomes.

Temperate forest
Trees grow in summer, before dropping their leaves and becoming dormant in winter.

Taiga
Conifer forests dominate the far north, where the cold and short summers are the main factors.

Polar
The temperature around the poles is below freezing for most of the year.

Temperate grassland
When there is too little rainfall for trees to grow, huge expanses of grass cover the land.

Savannah
In these warm, grass-covered regions, there is low rainfall and few trees.

Tropical forest
High rainfall and warm conditions all year result in thick jungles around the tropics.

Mountains
At high altitude, the air is thin (lacking oxygen) and temperatures are low.

Tundra
All but the upper layer of soil is permanently frozen, making it hard for plants to grow.

Chaparral
Also known as the Mediterranean biome, this region is filled with dry woodlands.

Desert
The driest parts of Earth have hardly any rainfall and very little vegetation.

Food chains

ENERGY PASSES ALONG FOOD CHAINS
FROM PLANTS TO TOP CARNIVORES.

Living things require a supply of energy and nutrients to power,
maintain, and grow their bodies. Scientists track how energy and
nutrients move from one organism to another using food chains.

Producers and consumers

Food chains always begin with plants and other photosynthetic
organisms, which are known as the producers. Animals and other
heterotrophs (organisms that eat others to survive) are known as
the consumers. The nutrients and energy gathered by the producers
passes up the food chain via a series of consumers.

△ **Producer**
Green plants harness
the energy of sunlight to
power themselves, and
are called producers.

△ **Primary consumer**
Herbivores, such as cows,
eat only producers, and
form the second step in
the food chain.

△ **Secondary consumer**
Omnivores, such as
raccoons, eat both
producers and small
primary consumers.

△ **Top predator**
The food chain ends
with a powerful
predator, such as
an eagle or shark.

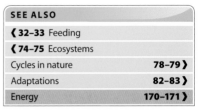

△ **Detritivore**
Worms, vultures, and
most fungi recycle the
dead remains and waste
of other organisms.

Energy pyramid

Most of the energy consumed by
organisms is given off as heat, becoming
unavailable to the rest of the food chain,
so less energy is passed onto the next
level. As a result, the total quantity of
organisms—the biomass—also decreases.
This gives the food chain a pyramid
structure—with many producers at the
base, and fewer and fewer consumers
at each stage above.

▷ **Trophic levels**
Scientists call each level of a food chain a
trophic level—from the Greek word for
food. As a rough estimate, only about
10 percent of the energy in one trophic
level passes to the one above.

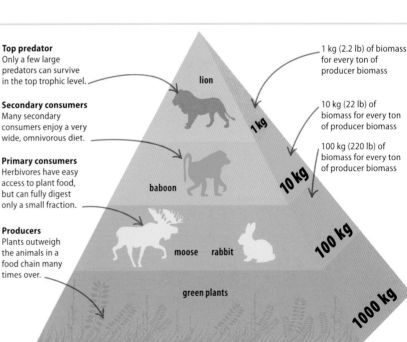

Top predator
Only a few large
predators can survive
in the top trophic level.

Secondary consumers
Many secondary
consumers enjoy a very
wide, omnivorous diet.

Primary consumers
Herbivores have easy
access to plant food,
but can fully digest
only a small fraction.

Producers
Plants outweigh
the animals in a
food chain many
times over.

lion

baboon

moose   rabbit

green plants

1 kg

10 kg

100 kg

1000 kg

1 kg (2.2 lb) of biomass
for every ton of
producer biomass

10 kg (22 lb) of
biomass for every ton
of producer biomass

100 kg (220 lb) of
biomass for every ton
of producer biomass

Food webs

No food chain exists on its own. In real wildlife communities, the chains interlink to make a food web—a representation of an ecosystem. Food webs vary a great deal between habitats. They may contain a keystone species, through which a lot of the nutrients pass, or on which many other species in the web rely for food.

▽ **Arctic Ocean**
Despite being one of the coldest places on Earth, the Arctic has a rich food web. Minute algae called phytoplankton are the producers. Arctic cod is a keystone species, since many of the predators would die out without it.

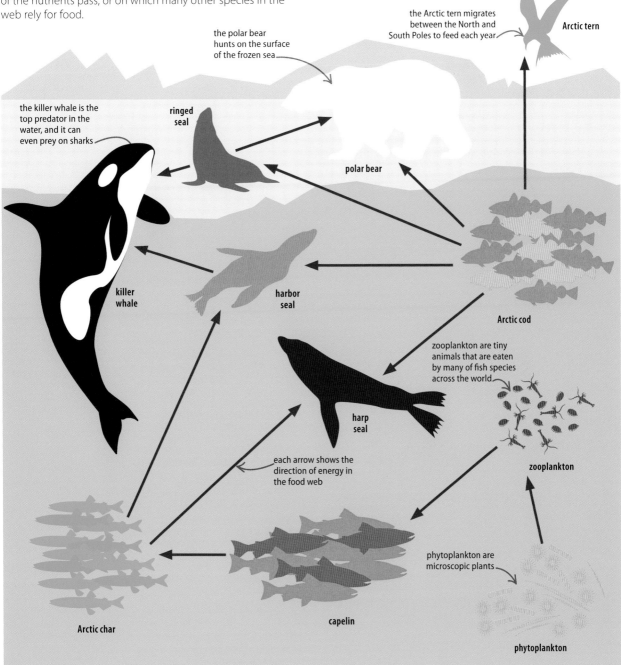

the polar bear hunts on the surface of the frozen sea

the Arctic tern migrates between the North and South Poles to feed each year

Arctic tern

the killer whale is the top predator in the water, and it can even prey on sharks

ringed seal

polar bear

killer whale

harbor seal

Arctic cod

zooplankton are tiny animals that are eaten by many of fish species across the world

harp seal

each arrow shows the direction of energy in the food web

zooplankton

phytoplankton are microscopic plants

Arctic char

capelin

phytoplankton

Cycles in nature

**NUTRIENTS AND OTHER SUBSTANCES
ARE RECYCLED IN THE ENVIRONMENT.**

Living things require many nutrients—substances used to build their
bodies. There is a finite supply of these in the environment, so they are
recycled through the environment by biological, chemical, and physical
processes—and also by human activities.

The carbon cycle

Carbon is essential to life. It is one of the
most abundant elements in a living body and
its atoms are in just about every chemical in
cells. During photosynthesis, plants fix (collect)
carbon dioxide from the atmosphere and turn
it into sugars and other nutrients. These then
pass to animals and other organisms that eat
the plants. Eventually, the carbon in them
is returned to the atmosphere as a waste
product of respiration.

▷ **Nonbiological factors**
Carbon is not only cycled between the
atmosphere and organisms. Carbonates, a
combination of carbon and oxygen found
in rocks and fossil fuels, are locked away
underground for millions of years. Burning
fossil fuels releases this carbon dioxide
back into the atmosphere.

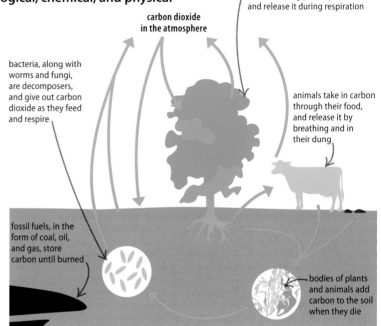

green plants take in carbon
dioxide during photosynthesis,
and release it during respiration

carbon dioxide
in the atmosphere

bacteria, along with
worms and fungi,
are decomposers,
and give out carbon
dioxide as they feed
and respire

animals take in carbon
through their food,
and release it by
breathing and in
their dung

fossil fuels, in the
form of coal, oil,
and gas, store
carbon until burned

bodies of plants
and animals add
carbon to the soil
when they die

The oxygen cycle

Almost all organisms require a supply of oxygen,
which is used in respiration to release energy from
sugar. Organisms take in the oxygen and give out
carbon dioxide (a waste product of respiration).
However, oxygen does not run out, because it is
constantly being replaced by the photosynthesis
of plants. In this process carbon dioxide is taken
in as a raw ingredient of glucose, and oxygen is
given out as a waste material.

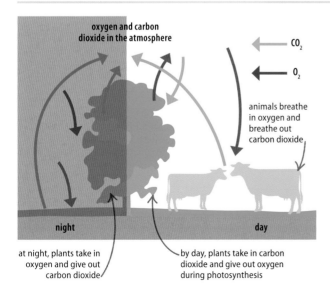

oxygen and carbon
dioxide in the atmosphere

CO_2

O_2

animals breathe
in oxygen and
breathe out
carbon dioxide

night

day

at night, plants take in
oxygen and give out
carbon dioxide

by day, plants take in carbon
dioxide and give out oxygen
during photosynthesis

◁ **Night and day**
Plants take in carbon dioxide as a raw material for
photosynthesis, and give out oxygen as a waste material
of the process. Plants photosynthesize only during daylight,
and this is when oxygen is released into the atmosphere. By
night, plants take in some oxygen to power their respiration,
but they use less than they produce.

The nitrogen cycle

Nitrogen is an essential component in amino acids, the basic units of protein, which all living creatures need. Therefore, all life needs a supply of nitrogen compounds. Animals cannot manufacture most amino acids themselves, so they obtain them from plant foods. Plants make amino acids from nitrates (a combination of nitrogen and oxygen) absorbed from the soil. The nitrates are added to the soil by bacteria that fix nitrogen from the air.

REAL WORLD

Carnivorous plant

The Venus flytrap grows in soils that lack nitrates, so the plant collects it from prey instead. It traps insects in its pressure-sensitive, pincer-shaped leaves. The leaves shut to form a stomachlike space, where enzymes digest the insect to release its nutrients.

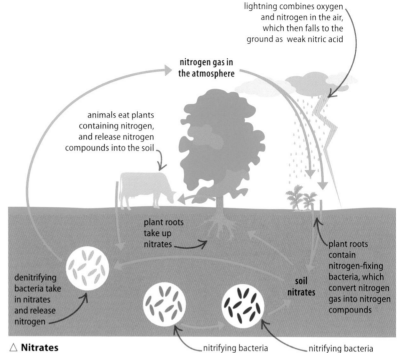

lightning combines oxygen and nitrogen in the air, which then falls to the ground as weak nitric acid

nitrogen gas in the atmosphere

animals eat plants containing nitrogen, and release nitrogen compounds into the soil

plant roots take up nitrates

denitrifying bacteria take in nitrates and release nitrogen

soil nitrates

plant roots contain nitrogen-fixing bacteria, which convert nitrogen gas into nitrogen compounds

nitrifying bacteria in the soil convert nitrogen compounds to nitrites, which add oxygen

nitrifying bacteria in the soil convert nitrites into nitrates

△ Nitrates

Nitrogen is not a very reactive element, mostly staying unchanged in the atmosphere. However, the enzymes in certain bacteria and the high energy of lightning can convert nitrogen into nitrates, a form that can be used by all life.

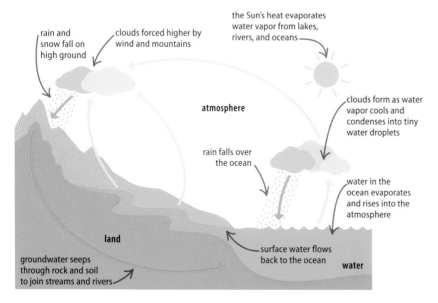

the Sun's heat evaporates water vapor from lakes, rivers, and oceans

rain and snow fall on high ground

clouds forced higher by wind and mountains

atmosphere

clouds form as water vapor cools and condenses into tiny water droplets

rain falls over the ocean

water in the ocean evaporates and rises into the atmosphere

land

surface water flows back to the ocean

water

groundwater seeps through rock and soil to join streams and rivers

The water cycle

Earth's water is always on the move, collecting in vast quantities in the ocean, but rarely finding its way to deserts. Life cannot exist without water. It is one of the ingredients in the production of glucose in photosynthesis, and water is also the medium in which the metabolic process takes place inside cells. Most living bodies are mainly water—about 60 percent in the case of humans— and, where water is rare, so is life.

◁ Movement of water

Most of Earth's water is in the oceans, but it also moves constantly into the atmosphere, falling as rain to form freshwater running over and into the ground or freezing as ice on high mountains and in the polar regions.

Evolution

THE ORGANISMS MOST SUITED TO SURVIVE IN AN
ENVIRONMENT ARE MOST LIKELY TO PASS ON THEIR GENES.

SEE ALSO	
❰ 20–21 Variety of life	
❰ 42–43 Reproduction I	
Adaptations	82–83 ❱
Genetics I	84–85 ❱

Set out by English naturalist Charles Darwin in 1859, the theory of evolution
by natural selection was one of the most controversial scientific theories ever. It
has become accepted since, and has been updated to include the role of genes.

The drive to breed

Everything an organism does is meant to increase
the chance of it producing as many surviving
offspring as possible. These offspring compete with
each other and other species for limited resources,
such as food, water, and a place to live. Those best
able to survive are the ones that pass on their
genes to the next generation. The individuals that
cannot compete die without producing young,
so their genes are not passed on.

▷ **Increasing genes**
Rabbits have a very high reproduction rate, with one female able to
produce 70 young in just one year. The following year, her offspring
could potentially produce almost 5,000 more. In reality, competition
between all these rabbits is so fierce that far fewer than this survive.

1st generation	2nd generation	3rd generation

2 rabbits → 5 rabbits → 15 rabbits → 35 rabbits

Natural selection

The most successful, or "fit," offspring are the ones with genes
that allow them to out-compete their rivals. When they mate,
their fit genes are passed on to the next generation. This is called
natural selection. Eventually every animal in the species has the
fit genes—meaning the species gradually evolves over time.

peppered moth gives
rise to both dark and
light colored offspring

△ **Adding variety**
Most variation between animals is the result of sexual
reproduction. Every offspring inherits a slightly different
mixture of genes from both parents. The variation in color
of these moths ensures that at least some of the offspring
will survive if the habitat begins to change.

pale peppered moth is harder
for a predator to see

△ **Pale moths hidden**
Before the Industrial Revolution,
most peppered moths were pale
and could hide in the lichens
growing on tree trunks, while
the darker moths stood out.

pale peppered moths are easier to
see for predators, so experience a
decline in numbers

△ **Pale moths stand out**
Then soot from factories killed
the lichens, making tree trunks
darker. The pale moths then
became more preyed upon,
which made the dark moths
more common.

A mass extinction known as the **Great Dying** wiped out 90 percent of all species on Earth about **252 million years ago**.

How new species evolve

A species is a group of organisms that look the same and survive in the same way, and can breed together to produce viable young in the wild. Some species of bat look very similar to one another and live in the same areas, but attract mates using different calls and cannot breed with each other. Speciation is the formation of a new species. Species can evolve sympatrically (from one ancestor) or allopatrically (when populations are isolated).

REAL WORLD
Extinction

Most of the knowledge of evolution comes from fossils of extinct species. An extinct species is one that has no members left alive. Fossils, such as this primitive bird, form when body parts are replaced with rocky minerals over a long period of time. Fossils show us what the ancestors of today's species looked like. If the lost species died out after it evolved into another species, experts call this pseudoextinction.

sympatric speciation

fish feed in all parts of the lake

△ 1. One species of fish
One species of fish lives in the lake. The fish feed mainly on small animals in the water and on the lakebed.

these fish specialize as plankton feeders

these fish prey on crustaceans that live on the lakebed

△ 2. Specialist feeders
Gradually, the fish split into two groups as they specialize on catching certain animals. The groups evolve in different ways to exploit the different food sources.

plankton-eaters

lakebed-feeders

△ 3. Two species
The groups rarely mix, and eventually they evolve into two distinct species that cannot breed with each other any more.

allopatric speciation

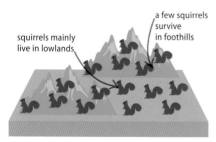

a few squirrels survive in foothills

squirrels mainly live in lowlands

△ 1. One species of squirrel
A species of squirrel has spread across a wide range. Although they do not meet, squirrels from either end of the range could breed with one another.

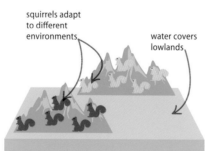

squirrels adapt to different environments

water covers lowlands

△ 2. Geographic isolation
Rising sea levels turns the range into isolated islands. Squirrels on both islands adapt to life in their own mountain habitat, forming two new species.

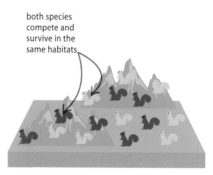

both species compete and survive in the same habitats

△ 3. Sharing habitat
The sea level falls again, creating a new lowland habitat for the squirrels to exploit. Both species mix throughout the original range, but are no longer able to breed.

Adaptations

ORGANISMS CHANGE OVER TIME IN ORDER TO SURVIVE.

Adaptations are the visible results of evolution. Natural selection alters the anatomy and behavior of organisms so they become adapted to new ways of living.

Adaptive radiation

When several different adaptations evolve from a single ancestor, it is known as adaptive radiation. The result is a group of species that share many features, but differ in ways that adapt them to a specific way of life. For example, rodents all have long, sharp incisors inherited from their common ancestor. However, in gophers these teeth are adapted to digging burrows, in beavers they fell trees, while squirrels use them to nibble through hard seed casings.

▷ **Darwin's finches**
Adaptive radiation can be seen in Darwin's finches—named after the discoverer of evolution. Most finches are seed-eaters, but the songbirds that live on the Galápagos Islands, Ecuador, have adapted to tackle other foods too.

ancestor cracked seeds with thick bill

probing bill pulls soft seeds from cactus flowers

hooked bill slices into soft fruits and buds

pointed bill is used to peck insects from leaves

overbite is useful for digging up grubs

woodpecker finch digs out prey from under bark using a stick

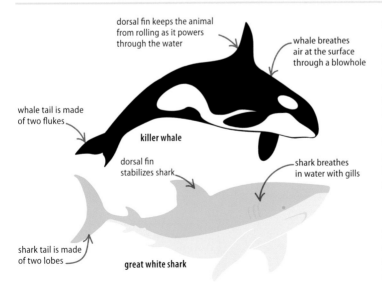

dorsal fin keeps the animal from rolling as it powers through the water

whale breathes air at the surface through a blowhole

whale tail is made of two flukes

killer whale

dorsal fin stabilizes shark

shark breathes in water with gills

shark tail is made of two lobes

great white shark

Convergent evolution

A lot of evolution is divergent, with groups of related animals becoming less alike as they adapt to different environments. However, evolution can also be convergent, where unrelated species adapt to the same environment in the same way. For example, both birds and bats have evolved wings to enable them to fly. The shape, structure, and function of both kinds of wings are very similar, but birds and bats are only distantly related to each other, and their common ancestor did not have wings.

◁ **Marine hunters**
Sharks and toothed whales, such as dolphins and orcas, are all fast-swimming hunters. They look similar, but have very different body systems, because sharks are fish and whales are mammals.

Coevolution

Sometimes, two species evolve together, adapting to ways of life by relying on each other for survival. Each organism affects the other in small ways, so the two become better adapted to each other and surviving together. Many animals and flowering plants have undergone this coevolution.

Corals have coevolved with **microscopic algae**—the algae live inside them, and provide food in return.

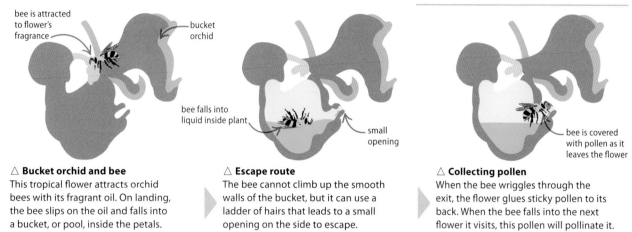

bee is attracted to flower's fragrance

bucket orchid

bee falls into liquid inside plant

small opening

bee is covered with pollen as it leaves the flower

△ **Bucket orchid and bee**
This tropical flower attracts orchid bees with its fragrant oil. On landing, the bee slips on the oil and falls into a bucket, or pool, inside the petals.

△ **Escape route**
The bee cannot climb up the smooth walls of the bucket, but it can use a ladder of hairs that leads to a small opening on the side to escape.

△ **Collecting pollen**
When the bee wriggles through the exit, the flower glues sticky pollen to its back. When the bee falls into the next flower it visits, this pollen will pollinate it.

Sexual selection

Not all adaptations increase an ability to survive in competition with others. Sexual selection can produce traits that can be a hindrance, such as unwieldy antlers or long, ornate tail feathers that make flying difficult. This type of selection happens because the female chooses a particular trait in the male. The female will select the mate with the best features, and, because of this, males with that trait pass on their genes, increasing the size of the trait in the next generation.

▽ **Bird tail tale**
To attract a mate, a male pheasant displays its tail feathers. The females prefer long, clean feathers because they show the male is a strong specimen. Sexual selection results in larger, more ornate tail feathers. The process stops only when the tail size hinders the male and so weakens it.

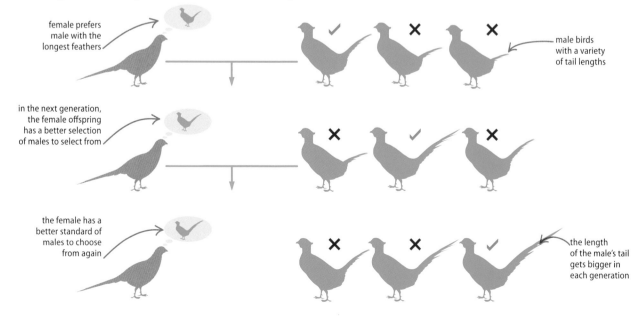

female prefers male with the longest feathers

male birds with a variety of tail lengths

in the next generation, the female offspring has a better selection of males to select from

the female has a better standard of males to choose from again

the length of the male's tail gets bigger in each generation

Genetics I

THE FIELD OF BIOLOGY THAT INVESTIGATES INHERITANCE OF
CHARACTERISTICS FROM PARENTS IS CALLED GENETICS.

SEE ALSO

❮ **22–23** Cell structure
❮ **44–45** Reproduction II
❮ **80–81** Evolution
Polymers   **162–163** ❯

The instructions for making a living body are called genes. Each gene relates
to a specific characteristic, such as eye color or height. A full set of genes is inherited
from both parents, so a child shares many of his or her parents' characteristics.

Chromosomes

Genes are carried on long chemical
chains of deoxyribonucleic acid (DNA).
DNA is stored inside a cell's nucleus
on chromosomes, which are the
vehicles that carry the genes as they
pass from one generation to the next.
The number of chromosomes in a cell
is called the diploid number. Sperm
and eggs contain a half-set, or haploid
number, of chromosomes.

Genes and alleles

Each gene has a specific position on its chromosome. Everyone has two
versions of each chromosome, one from each parent. That means they have
two versions, known as alleles, of each gene. The two alleles form a person's
genotype. One allele is often dominant over the other, which is recessive,
so just one characteristic (known as a person's phenotype) is expressed.

▽ **Genetic probability**
In this example, we see two parents and their possible offspring.
Both parents give one allele to their children. The allele for brown
eyes (B) is dominant, and the allele for blue eyes (b) is recessive.
The mother has a recessive allele, so it is equally likely that they
have a brown or a blue-eyed child.

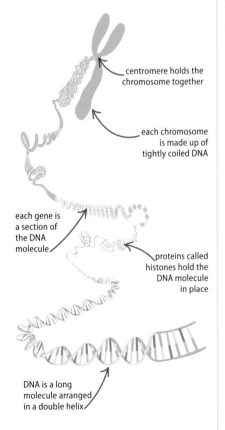

centromere holds the
chromosome together

each chromosome
is made up of
tightly coiled DNA

each gene is
a section of
the DNA
molecule

proteins called
histones hold the
DNA molecule
in place

DNA is a long
molecule arranged
in a double helix

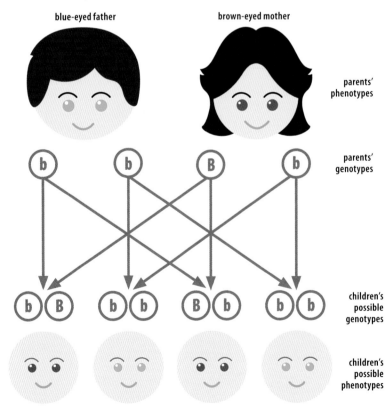

blue-eyed father

brown-eyed mother

parents'
phenotypes

parents'
genotypes

children's
possible
genotypes

children's
possible
phenotypes

Codominance

Not all alleles are dominant or recessive. Sometimes, both alleles are expressed at once in a system called codominance, or incomplete dominance. In the example below, the red parent flower has two red alleles (R), while the white parent flower has two white alleles (r). When the pair breed, all the offspring have the genotype Rr and codominance makes all the flowers pink. However, breeding two pink flowers produces red and white, as well as pink, blooms.

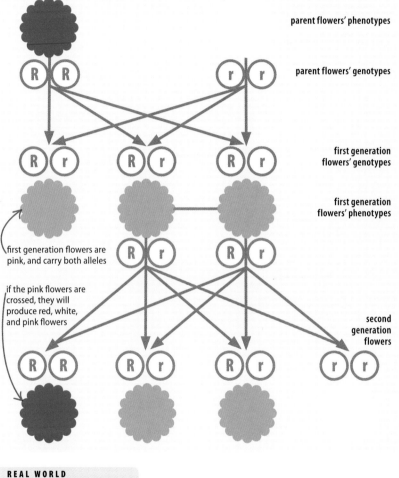

parent flowers' phenotypes

parent flowers' genotypes

first generation flowers' genotypes

first generation flowers' phenotypes

first generation flowers are pink, and carry both alleles

if the pink flowers are crossed, they will produce red, white, and pink flowers

second generation flowers

Sex chromosomes

A person's sex is determined by inheriting particular chromosomes from the mother and father. Females have two X chromosomes, while males have an X and a Y chromosome—the Y is much smaller, and has fewer genes, than the X. A mother's gametes always contain an X, while a sperm can have an X or Y.

▽ **Determining gender**
Because the mother always gives an X chromosome, it is the father's gamete that determines the sex of the baby. If an X sperm fertilizes the egg, the baby will be female, and male if a Y sperm achieves it.

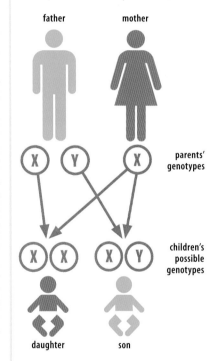

father

mother

parents' genotypes

children's possible genotypes

daughter

son

X-linked diseases

Males are more likely to be color blind, because they have a defective gene on their X chromosome and their shorter Y has no alternative allele for the problem. Females can carry the same defective gene, but have normal vision thanks to a healthy allele from their other X chromosome.

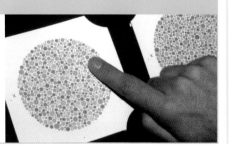

Humans have **46 chromosomes**, which is less than some rats, at **92**, but more than kangaroos, which have only **16**.

Genetics II

GENETIC CODES ARE USED TO MAKE
PROTEINS NECESSARY FOR THE BODY.

Genetic information is held as a code stored on DNA molecules. This code
is translated into the many proteins that do the work in a cell. When errors
occur in this process, genetic illnesses are possible as a result.

Double helix

A DNA (deoxyribonucleic acid) molecule is a double helix,
a ladder-shaped spiral. The "sides" are chiefly ribose sugars,
while the "rungs" are made up of four chemical compounds
called nucleotides, or bases. The bases are called thymine (T),
adenine (A), cytosine (C), and guanine (G). The sequence of
these bases is a code that adds up to the instructions for
a particular gene.

Transcription

The first step in turning a gene into a protein involves making
a copy of the gene's DNA stored in the nucleus. This involves
transcribing the DNA's code onto ribonucleic acid (RNA). The DNA
double helix is unzipped into two unwound strands, and an RNA
strand forms next to one of them. The RNA has bases too, but
instead of thymine it has a base called uracil (U). The RNA copies
the DNA strand and then travels to a ribosome.

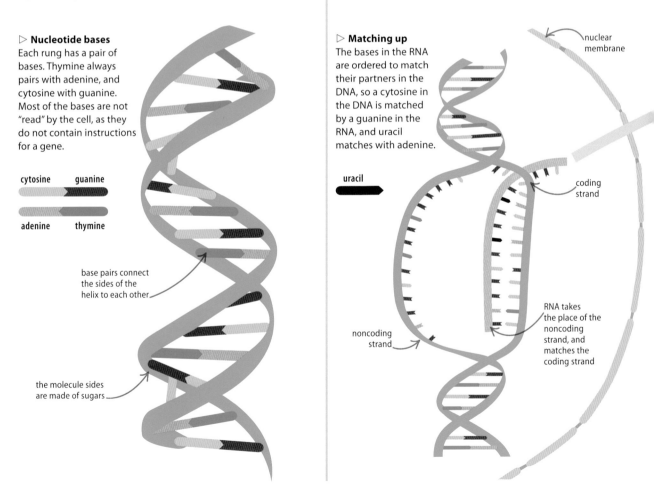

▷ **Nucleotide bases**
Each rung has a pair of
bases. Thymine always
pairs with adenine, and
cytosine with guanine.
Most of the bases are not
"read" by the cell, as they
do not contain instructions
for a gene.

cytosine   guanine

adenine   thymine

base pairs connect
the sides of the
helix to each other

the molecule sides
are made of sugars

▷ **Matching up**
The bases in the RNA
are ordered to match
their partners in the
DNA, so a cytosine in
the DNA is matched
by a guanine in the
RNA, and uracil
matches with adenine.

uracil

nuclear
membrane

coding
strand

RNA takes
the place of the
noncoding
strand, and
matches the
coding strand

noncoding
strand

Translation

The genetic code is translated into a protein in the cell's ribosome. This tiny organelle pulls the RNA through itself three bases at a time. Every three bases form a triple-character sequence called a codon, which is specific to a certain amino acid. Proteins are composed of chains of these acids, and the codons on the RNA tell the ribosome in which order to arrange them.

▽ **Making proteins**
Anticodons carry specific amino acids to the ribosome to add to the chain. When the anticodon matches the codon on the RNA, the anticodon adds its amino acid to the chain, and the next codon is pulled into the ribosome.

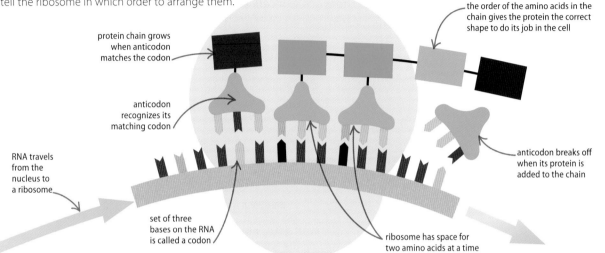

the order of the amino acids in the chain gives the protein the correct shape to do its job in the cell

protein chain grows when anticodon matches the codon

anticodon recognizes its matching codon

anticodon breaks off when its protein is added to the chain

RNA travels from the nucleus to a ribosome

set of three bases on the RNA is called a codon

ribosome has space for two amino acids at a time

Mutations

When DNA is copied, mistakes can occur. These are called mutations. A mutation may be made in the unread part of DNA, and so have no effect. If it happens in the read section, the result can make the cell die. However, occasionally a mutation improves the way the cell and the body works. These useful mistakes are spread by natural selection and drive evolution.

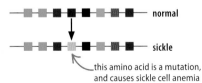

normal

sickle

this amino acid is a mutation, and causes sickle cell anemia

△ **Genetic disease**
Some mutations are not deadly, but cause diseases. For example, sickle cell anemia is caused by one different amino acid in the structure of hemoglobin, the chemical that carries oxygen in the blood. The mutant hemoglobin forms long chains, which makes a sufferer's blood cells sickle-shaped.

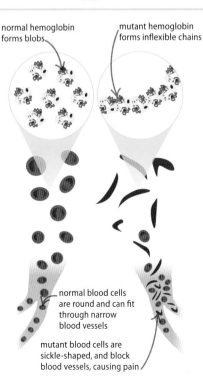

normal hemoglobin forms blobs

mutant hemoglobin forms inflexible chains

normal blood cells are round and can fit through narrow blood vessels

mutant blood cells are sickle-shaped, and block blood vessels, causing pain

REAL WORLD

Human genome project

A genome is the complete collection of a species' genes. In 2003, scientists finished a complete record of the human genome. They identified about 25,000 genes and sequenced three billion base pairs. Below is a section of the genome, with a color for each base. However, geneticists have still to figure out what most of the genes do and record their many different versions, or alleles.

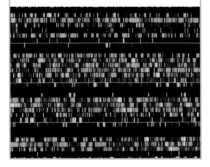

Pollution

**CHEMICALS FROM HUMAN ACTIVITIES AFFECT
THE ECOSYSTEMS AND FUTURE OF THE EARTH.**

SEE ALSO	
❰ 74–75 Ecosystems	
Human impact	**90–91 ❱**
Acids and bases	**144–145 ❱**
Electromagnetic waves	**194–195 ❱**

**Pollution is anything that is added to the environment in amounts large
enough to have a harmful effect. Sound, light, and heat can be pollution,
but the most damaging pollutants are chemicals in Earth's soil, water, and air.**

Ozone hole

Ozone is a type of oxygen in the
atmosphere that blocks dangerous
ultraviolet light (UV) coming from the
Sun. Large amounts of chlorofluorocarbon
(CFC) gases, used in aerosols and
refrigerators and thought to be inactive,
were released in the 1980s. The CFCs
reacted with ozone, and, over the years,
have depleted the ozone layer in places,
especially above the North and South
Poles. CFCs are now banned and the
ozone holes are shrinking.

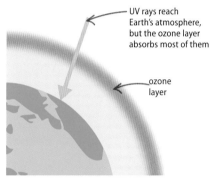

UV rays reach
Earth's atmosphere,
but the ozone layer
absorbs most of them

ozone
layer

△ **Safe levels**
While some does hit the Earth's surface,
the ozone layer, 25 km (15.5 miles)
above Earth, deflects much of the
harmful UV light back into space.

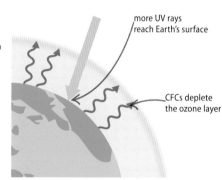

more UV rays
reach Earth's surface

CFCs deplete
the ozone layer

△ **High layer**
A chemical reaction between the CFCs and
the ozone layer turns the latter into oxygen,
which does not shield Earth from UV light,
meaning more of it reaches the surface.

REAL WORLD

Global dimming

Burning fossil fuels releases carbon dioxide,
which contributes to global warming.
However, the soot released may also keep
temperatures down in a process called
global dimming. The tiny dark particles
in the air reflect the Sun's light, reducing
the amount that heats Earth's surface.

Greenhouse effect

The "greenhouse gases" of water vapor,
carbon dioxide, and methane in the
atmosphere stop heat being lost to space.
Without this process, Earth's average
surface temperature would be below
freezing. However, human activities, such
as burning fossil fuels and intensive
farming, are increasing the amount of
greenhouse gas. This greenhouse effect is
gradually increasing Earth's surface
temperature, resulting in more extreme
weather, such as flooding and drought.

Venus, warm enough
to melt lead, is the
hottest planet,
due to an **extreme**
greenhouse effect.

▽ **Trapped heat**
Sunlight absorbed by the Earth's surface
warms it up, and the surface sends out
heat in the form of infrared radiation.
Some infrared is absorbed in turn by
the atmospheric greenhouse gases.

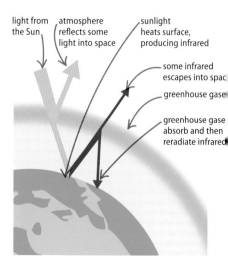

light from
the Sun

atmosphere
reflects some
light into space

sunlight
heats surface,
producing infrared

some infrared
escapes into spac

greenhouse gase

greenhouse gase
absorb and then
reradiate infrared

Acid rain

All rainwater is slightly acidic. This is because carbon dioxide in the air dissolves in it, making weak carbonic acid (as found in carbonated drinks). However, sometimes oxides of sulfur and nitrogen are released into the atmosphere as industrial waste. When these dissolve in water, they form much more potent acids, which have a damaging effect on wildlife when they fall as rain.

▷ **Gases released**
Coal-burning power plants and engines fueled by oil or gasoline release gases that can form acid rain. The rain often falls far from its source. It has many effects, including killing animal and plant life and damaging buildings.

acid-forming pollution rises high into the atmosphere and can be carried far away from its source

pollution dissolves in water droplets in clouds, forming acid rain

rain increases the acidity of freshwater habitats, killing animals such as fish

acid rain damages stone buildings

acid rain damages the bark of forest trees

Eutrophication

Fertilizers provide crops with nutrients such as nitrates and phosphates. These compounds boost the growth of any plant, and cause thick blooms of algae if they are washed by heavy rains into lakes and rivers. This leads to a process of eutrophication, where these blooms choke out life in the water.

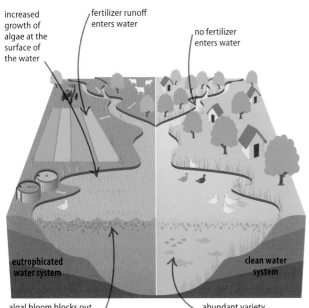

increased growth of algae at the surface of the water

fertilizer runoff enters water

no fertilizer enters water

eutrophicated water system

clean water system

algal bloom blocks out light for plants and animals deeper down

abundant variety of life at all levels of the system

Biomagnification

Even when present in tiny amounts, some pollutants can have an impact through a process called biomagnification. When an animal cannot break down a chemical, it is stored in its body, and is passed on to any predator that eats it. The concentration of this pollutant in animal tissues increases at each stage of the food chain, reaching damaging levels in top predators.

▷ **DDT disaster**
The biomagnification of an insecticide called DDT nearly wiped out many birds of prey in the USA in the 1940s and 1950s. DDT was thought harmless to vertebrates, but it built up in the bodies of fish and other animals in their environments. DDT poisoning resulted in birds, such as ospreys and bald eagles, laying eggs with very thin shells, so many eggs smashed in the nest.

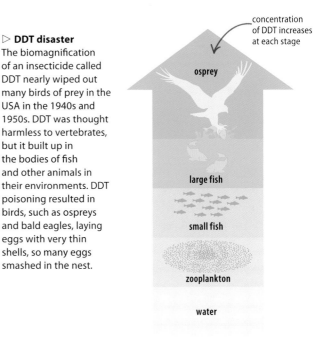

concentration of DDT increases at each stage

osprey

large fish

small fish

zooplankton

water

Human impact

ACTIVITIES BY HUMANS CAN CHANGE ECOSYSTEMS
AND THE PLANTS AND ANIMALS WITHIN THEM.

SEE ALSO

❮ 74–75 Ecosystems
❮ 82–83 Adaptations
❮ 84–85 Genetics I
❮ 88–89 Pollution

Scientists know there have been five mass extinctions in Earth's history,
all with natural causes. Many of today's species are becoming extinct due to human
activities. Some experts think we are living through a sixth mass extinction right now.

Habitat loss

Humans have the ability to alter a habitat to suit their needs, turning natural landscapes
into artificial ones, such as farmland or urban developments. The wildlife from the original
habitat has been evolving for millions of years, and its species are specialized to a life within
that community. They cannot survive in other habitats, and sometimes face extinction as a result.

▽ **Climax habitat**
Climax habitats carry the
maximum number of life forms
possible. This patch of tropical
forest has a unique community
of species.

▽ **Slash and burn**
Humans need places to grow
crops, so they cut down the
forest and burn the logs. The
ash makes a nutrient-rich soil
for the first crops.

▽ **Fertile soil**
For a few years the ashy soil
makes good farmland. However,
the orginal jungle soil beneath
does not hold nutrients for long,
so eventually the crop yields fall.

▽ **Secondary forest**
The farmers abandon this plot,
and move on, leaving a new
forest habitat to develop.
It will never recover to its
original climax state.

Fragmentation

Many forest animals never leave their
habitat. For example, the gibbons of
Southeast Asia can walk only short
distances on the ground—they move
about by swinging from branch to branch.
Even a narrow gap in the forest, such
as a road, is enough to divide forest
communities permanently. Fragmentation
breaks forest animals into small groups,
making it harder for them to survive.

▷ **Living with relatives**
The biggest problems caused by fragmentation
are loss of diversity and inbreeding. In a small
group, every member is closely related to each
other. Relatives share the same genes and any
offspring tend to be weak.

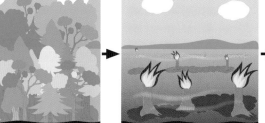
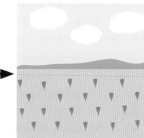
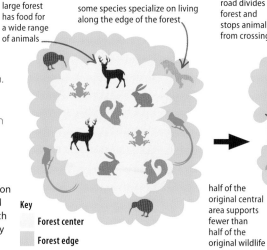
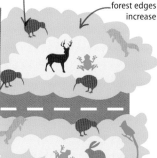

large forest
has food for
a wide range
of animals

some species specialize on living
along the edge of the forest

road divides
forest and
stops animals
from crossing

species on habitat edge benefit
most from fragmentation, due
to fewer predators

forest edges
increase

half of the
original central
area supports
fewer than
half of the
original wildlife

Key

Forest center

Forest edge

Controlling pests

Pests are animals that impede human activities. They tend to be able to live in a wide variety of habitats, and can damage crops, spread disease, or infest homes. Pest control usually involves the use of chemicals, but this can cause pollution. By contrast, biological control makes use of natural predators and parasites to control pests.

▽ **Fly pests**
Flies that lay their eggs in animal waste and rotting food are a serious pest in stables, farms, and sewage treatment plants. Their maggot larvae eat the waste before pupating into adults that can spread diseases.

▽ **Parasitic wasp**
The spalangia wasp is native to Australia, but humans use it to kill flies naturally around the world. The wasp lays its eggs inside a fly pupa. The wasp larva hatches and eats the fly pupa from the inside out.

▽ **Becoming an adult**
The wasp then pupates itself inside the empty fly case. When fully grown, the adult wasp bites a hole in the case and flies away. After mating, the female wasps will lay more eggs on fly pupae, until all the flies are dead.

Introduced species

As humans move around the world, they take animals and plants with them. Introducing species to new habitats can upset the balance of an ecosystem. There have been several disastrous introductions, such as the introduction of the cane toad to Australia (below), or the 60 starlings that were released in New York City, in 1890. There are now 200 million starlings in North America, which have pushed many native birds close to extinction.

Key
Spread of cane toads today

△ **Beetle pest**
The grubs of a native beetle were damaging valuable plantations of sugar cane across tropical parts of eastern Australia. Farmers looked for a small predator to control the beetle numbers and protect the crops.

△ **Marine toad**
A large toad from South America was introduced in 1936 to tackle the beetles. It was known as the marine toad because it was tough enough to survive almost anywhere, even along the seashore.

△ **Spread across Australia**
The toads ate almost everything except the cane beetle, upsetting delicate ecosystems. They spread and became a major pest, renamed the cane toad. There are more than 200 million living in Australia today.

Genetic modification

Humans have been altering the genetic makeup of animals and plants for thousands of years by selectively breeding animals with desired characteristics. However, in recent years, genetic engineers have been adding completely new genes to animals in the laboratory. These fish glow in the dark because of a gene added from a bioluminescent (light-emitting) deep sea jellyfish.

Humans are the only species to live on all **seven continents**, including a permanent settlement in the Antarctic since 1956.

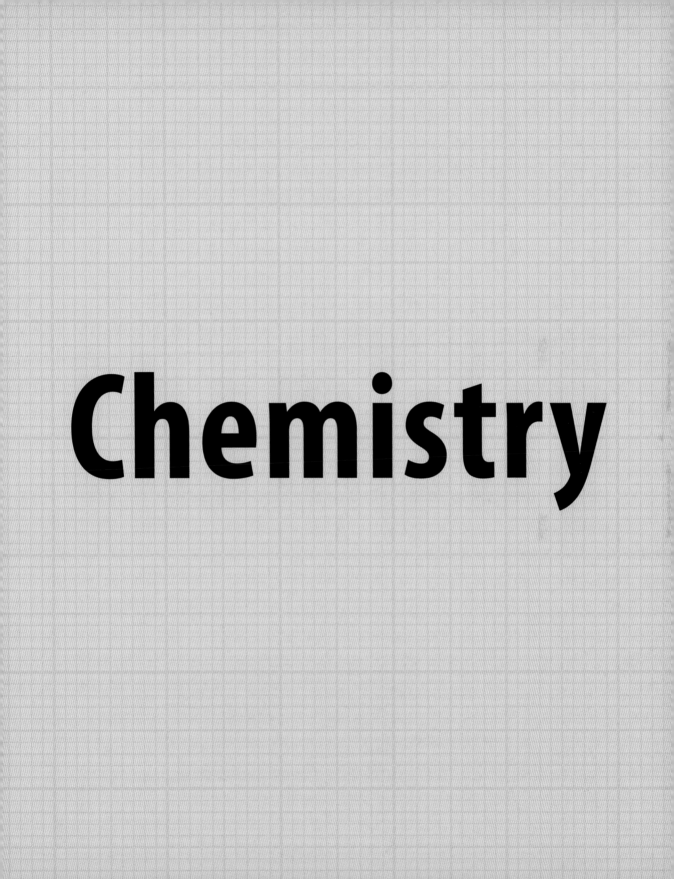

Chemistry

What is chemistry?

THE SCIENCE THAT DEALS WITH THE PROPERTIES OF SUBSTANCES AND LOOKS AT HOW THEY CAN CHANGE FROM ONE TYPE TO ANOTHER.

Chemistry is sometimes called the central science, because it forms the link between physics and biology. Chemistry builds on the knowledge of physics and then, in turn, is used to provide the basis for much of biology.

Understanding substances

Chemists seek to understand how the characteristics and structure of a substance, natural or artificial, can be described. What is it about water that makes it a flowing liquid, while the plastic bucket used to carry it is a rigid solid? A chemist finds the answer at the very smallest of scales. Every substance contains atoms, and the way they are arranged dictates how a substance behaves.

▽ **Describing materials**
Chemists have many ways of describing substances. They include the substance's state—solid, liquid, or gas—or whether it is metallic like the screw, or nonmetallic like the seashell.

water—liquid, nonmetallic

helium in balloon—gas, nonmetallic

seashell—solid, nonmetallic

screw—solid, metallic

Elements

Everything in the Universe is made out of raw materials called elements. There are about 91 naturally occurring elements. Most, such as gold or mercury, are pretty rare, while others, such as carbon, chlorine, and iron, are found in great quantities. Few elements are found pure in nature; they are usually combined with other elements to form entirely different materials called compounds. (Water is a compound of hydrogen and oxygen, for example.) Compounds can be separated into their elements, but an element cannot be broken down into anything simpler.

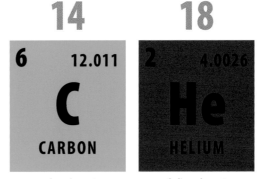

14

6   12.011

C

CARBON

carbon element

18

2   4.0026

He

HELIUM

helium element

◁ **Defining elements**
Chemists arrange the elements in the periodic table (see pages 116–117) according to the structure of their atoms. For example, the number of electrons (one of the parts of an atom) at a certain part of the atom means carbon is in group 14 whereas helium is in group 18.

Atoms

Atoms are the building blocks of all material on Earth and out in space (as far as we know). They are not all the same. In fact, every element is made up of its own type of atom. All atoms have positively charged protons in the central nucleus. These are surrounded by negatively charged electrons. The number of electrons and protons varies from element to element and this is what gives each element its properties.

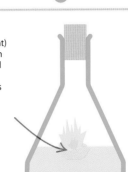

nitrogen atom

seven electrons

seven protons

most nitrogen atoms have seven neutrons, too

◁ **Balanced charge**
The number of protons in an atom always equals the number of electrons, so overall the atom has no charge. Neutrons are particles with no, or a neutral, charge.

Reactions

Chemists investigate how elements and compounds behave in reactions. During a chemical reaction, substances known as reactants are transformed into new substances called products. The reaction rearranges the atoms, breaking up the reactants and combining them in new ways to make the products. Most products are different compounds, but some may be pure elements.

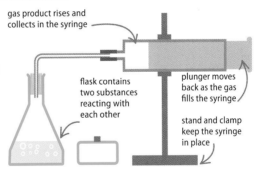

sodium (a reactant) combusts when in water (the second reactant)—this reaction produces a liquid (sodium hydroxide) and a gas (hydrogen)

◁ **Chemical energy**
Reactions take in and release energy, and can be very violent events. Explosions and combustion (burning) are among the most energetic reactions.

Analysis

One role of a chemist is to use knowledge of the physical and chemical properties of different elements and compounds to figure out the content of an unknown substance. This process is called analysis. It involves using a number of tests, such as burning substance (the flame's color gives clues to its contents) or reacting it with a known compound, to see the products created.

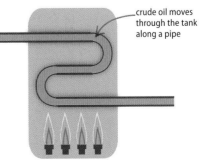

gas product rises and collects in the syringe

flask contains two substances reacting with each other

plunger moves back as the gas fills the syringe

stand and clamp keep the syringe in place

◁ **Laboratory apparatus**
Chemical reactions are carried out in laboratories. These science workshops contain a range of apparatus for containing and heating reactants, and collecting and measuring products.

Chemical industry

Chemistry is also used to manufacture useful substances. Manufacturing chemicals on an industrial scale is very different to making them in a laboratory. Scientists use their knowledge of what controls the speed of a reaction to come up with the best possible manufacturing process—making the most product for the least expenditure on heat and raw materials.

crude oil moves through the tank along a pipe

Heated crude oil

◁ **Petrochemicals**
The many hundreds of chemicals in crude oil are used as raw materials for making fuels, plastics, waxes, and medicines. The oil is heated to separate it into different materials (see page 157).

Properties of materials

SUBSTANCES CAN BE UNDERSTOOD BY OBSERVING THEIR PROPERTIES.

Every substance has its own unique set of properties—color, density, smell, and flammability. Chemists try to understand why the substances in nature have such varied properties.

Mass and density

All objects have a mass: a measure of how much matter they contain. Mass is not an indicator of size. A piece of lead has more mass than an identically sized piece of polystyrene, for example. The difference in mass is due to a property known as density: a measure of how tightly packed matter is inside a substance. Density is calculated as mass divided by volume, and expressed with the units kg/m^3—or often g/cm^3. Lead is one of the most dense elements of all, which is why it is used in weights—a small, manageable lead object contains a lot of matter and so weighs a large amount.

Buoyancy

The density of a substance can be tested by putting it in water. If an object has a density higher than water, it will sink; if it is less dense than water, it will float.

REAL WORLD
Physical versus chemical

The spokes of this bike are bent. The bending is due to the physical properties of the metal. Physical properties do not change the substance (in this case, metal). Some parts of the bike have rusted. The rusting is due to the chemical properties of the metal. Chemical properties relate to how the substance changes into other materials (rust) when it reacts with other substances (air and water).

units of matter in the cube are shown as spheres

matter is equally spread out, as in the smaller cube on the left

more matter is packed into the cube

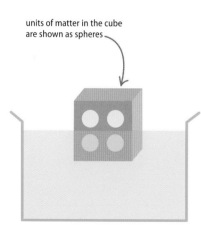

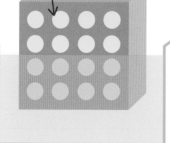

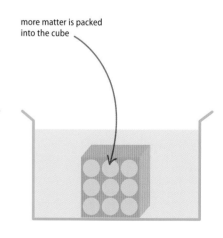

△ **Low-density object**
This cube is less dense than water. The matter in the cube is spread out more than the matter in water, so it weighs less than an equal volume of water.

△ **Larger object**
This cube is made of the same material as the first, only it has four times the volume—and weighs four times as much. So it has the same density as the first and floats.

△ **High-density object**
This cube is the same size as the first cube, but has a higher density. In this case, the cube weighs more than the same volume of water, so it sinks.

Comparing properties

Substances can be described and identified in terms of their properties. Chemists compare the properties of materials to find similarities and differences between them. Then they can start to investigate why these similarities and differences exist.

Substance	Floats in water?	Color	Transparency	Luster	Solubility	Conductivity	Texture
copper	no	red	opaque	shiny	in acid	conductor	smooth
natural chalk	no	white	opaque	dull	in acid	insulator	powdery
pencil lead (graphite)	no	black	opaque	shiny	no	conductor	slippery
pine wood	yes	brown	opaque	dull	only in special solvents	insulator	fibrous
salt crystals	no	white	translucent	shiny	in water	insulator when solid	gritty
glass	no	various	varies	shiny	only in special solvents	insulator	smooth
talc	no	various	opaque	waxy	in acid	insulator	greasy
diamond	no	various	transparent	sparkling when cut	no	insulator	smooth

Hardness

The hardness of a substance is normally measured on the Mohs scale, named after its inventor Friedrich Mohs. The scale is based on ten "guide" minerals, which all occur naturally in rocks. The hardness of a substance is measured in comparison with these guides. A material is harder than another when it can leave a scratch on it. For example, a piece of ordinary glass can scratch apatite but not orthoclase, and so its hardness is somewhere between 5 and 6.

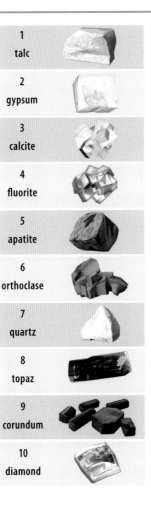

1 talc

2 gypsum

3 calcite

4 fluorite

5 apatite

6 orthoclase

7 quartz

8 topaz

9 corundum

10 diamond

▷ **Mohs scale**
The Mohs scale is only a comparative measure of hardness. In reality, a diamond is not ten times harder than talc. However, the Mohs scale is the preferred measure because it gives meaningful results using a quick and simple method.

Chemical properties

A substance can be described in terms of its chemical properties. It could be an element (a pure substance that cannot be reduced into simpler constituents), a compound (a combination of two or more elements), or be described as a metal, nonmetal, or semimetal. Chemists also look at a substance's chemical behavior, cataloguing its reactions and analyzing the products. A full set of properties—chemical and physical—can belong only to one substance.

▷ **Reactivity series**
Every element has a certain reactivity, which is part of its chemical behavior. Common metals are often ordered by how reactive they are. This is called the reactivity series. Metals at the top are most reactive. Potassium is so reactive that it is rarely found on its own. If two metals are competing to bond with another element, the one higher up the scale would win.

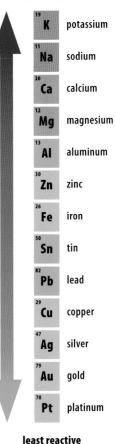

most reactive

19 **K** potassium

11 **Na** sodium

20 **Ca** calcium

12 **Mg** magnesium

13 **Al** aluminum

30 **Zn** zinc

26 **Fe** iron

50 **Sn** tin

82 **Pb** lead

29 **Cu** copper

47 **Ag** silver

79 **Au** gold

78 **Pt** platinum

least reactive

States of matter

THERE ARE THREE MAIN STATES OF MATTER: SOLID, LIQUID, AND GAS.

What sets each state apart is how the atoms and molecules (groups of atoms) are bonded together. This bonding is determined by factors such as temperature and pressure.

Physical difference

A solid that is melting into a liquid or boiling into a gas is changing physically. However, all three states share the same chemical formula.

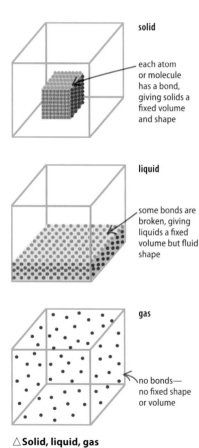

solid

each atom or molecule has a bond, giving solids a fixed volume and shape

liquid

some bonds are broken, giving liquids a fixed volume but fluid shape

gas

no bonds—no fixed shape or volume

△ **Solid, liquid, gas**
As a substance gets warmer, its molecules break bonds. The substance's structure becomes more chaotic, and changes state, from a solid to a liquid to a gas.

Solids

A solid is the most ordered state of matter, with every atom or molecule connected to its neighbors, forming a fixed shape with a fixed volume. Solids are either crystalline, with their units built up in repeating units, or amorphous, with the units grouped together randomly.

△ **Crystalline halite**
Large crystals of common salt are called halite. The crystal is made up of sodium and chlorine atoms arranged in a cube.

△ **Amorphous silica**
Glass is silica, the same material found in sand. It has an amorphous structure, with the units arranged randomly.

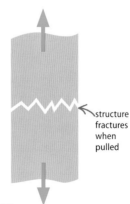

structure fractures when pulled

structure stretches when pulled

△ **Brittle solid**
In a brittle solid, the particles are held in a crystalline structure. Small forces do not alter the solid, but a force stronger than the bonds between the molecules can break it.

△ **Ductile solid**
Metals and some other solids can be pulled into a wire without breaking. This is because their molecules are held in an amorphous structure and can slide past each other.

Liquids

In a liquid, most of the atoms and molecules are still bonded together, but about one in ten of the links between them is broken. As a result, a liquid still has a more or less fixed volume and density—squeezing it does not really reduce its volume much. However, the constituents of the liquid are freer to move around than in a solid. Liquid can flow down slopes under the force of gravity, and take on the shape of any container it is poured into.

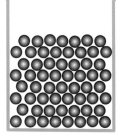

◁ **Liquid metal**
Mercury is the only metal that is liquid at room temperature. This is because its atoms form only weak bonds with each other.

◁ **Viscosity**
How a liquid flows is called its viscosity. When molecules are often blocked from moving past each other, the liquid is viscous (thick) and flows slowly. In low-viscosity liquids, molecules move around with little resistance.

honey is very viscous and flows slowly

oil is less viscous, but does not splash much

water has low viscosity, and drips and splashes easily

Plasma

The aurora, or Northern Lights and Southern Lights, are an example of a fourth state of matter: plasma. Plasma is a mixture of high-energy charged atoms and smaller subatomic particles. The aurora is formed by plasma streaming from the Sun being trapped in the Earth's magnetic field. It crashes into the atmosphere over the polar regions, creating the amazing light show.

Gases

In a gas, there are no bonds at all between the atoms or molecules. The units are free to move independently of each other in any direction. As a result, a gas has no fixed shape or volume and can be squeezed into a small space or spread out to fill a container of any shape. Like a liquid, it can also be made to flow from one place to another.

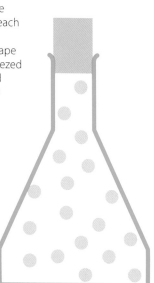

▷ **Helium**
Helium is made up of just single atoms. As they move around, the atoms bounce off each other and the sides of the container.

Changing states

MATTER CHANGES FROM ONE STATE TO ANOTHER
ACCORDING TO TEMPERATURE AND PRESSURE.

SEE ALSO
❮ **96–97** Properties of materials
❮ **98–99** States of matter
Convection currents **189** ❯

Every substance has a standard state. This is its state (solid, liquid,
or gas) at 25°C (77°F)—just above room temperature. Increasing or
decreasing that temperature eventually leads to a change in state.

States and energy

Changes in state are the result of
energy being added to or removed
from a substance. Taking energy from
a gas results in it becoming a liquid
and then a solid. Adding energy has
the reverse effect. The energy within
a substance makes its basic units—
atoms or molecules—vibrate (wobble).
This vibration, called internal energy,
is measured when temperatures
are recorded.

SEAgel is a spongelike
solid made from
seaweed. It is so light
that it floats in thin air!

▷ **Melting and boiling point**
The temperatures at which a solid changes
into a liquid or gas are called, respectively, the
melting and boiling points. The temperatures
are specific to each substance. They are always
measured at standard atmospheric pressure.
Changes in pressure affect the temperatures
at which substances change state.

Sublimation
Sometimes a solid does not melt, but
turns straight into a gas in a process
called sublimation. Carbon dioxide
sublimates almost all the time, while
water ice changes straight into
vapour if the air is very dry.

Deposition
The opposite process to
sublimation, deposition,
occurs when a gas turns
into a solid without first
becoming a liquid. Ice
can be deposited from
the vapor in air in very
cold conditions.

solid

Freezing
In a liquid, the vibration or internal energy
is just enough to break a few bonds—in
fact, they are constantly breaking and
reforming. A liquid freezes into a solid
when its atoms or molecules no longer
have enough energy to keep breaking bonds.

REAL WORLD

Salting ice

Adding salt to ice lowers the melting point of water by a couple
of degrees. Salting roads in winter stops dangerous sheets of ice
from forming—although if the conditions are well below 0°C
(32°F) the water will still freeze. The salt dissolved in the water
gets in the way of the water molecules, making it harder for
them to form all the bonds they need to become ice.

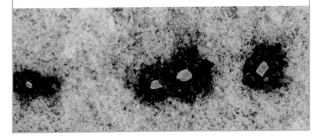

gas

Latent heat

Energy cannot be created from nothing, nor can it be made to disappear. So when a substance is condensing or freezing, rearranging its units into a lower-energy state, the unneeded energy is given out, warming the surroundings as latent heat. The same amount of heat moves the other way, from the surroundings to the substance, when it is boiling or melting and moving to a more energetic state.

▷ **Constant temperature**
This graph shows that the temperature stays constant at the melting and boiling points (when the change of state is taking place). The increase or decrease of energy at these points is the latent heat.

Temperature

Gas

boiling point —heat is used by atoms to break away

liquid

solid

melting point— heat is used to break bonds

A

Energy input

Condensation
The reverse of boiling, condensation, occurs when gas molecules are unable to escape and form bonds with other molecules that pass close by. Gradually the molecules gather together into larger droplets of liquid.

Boiling
A liquid boils into a gas when it has enough energy to break all of its bonds. Instead of vibrating around a fixed point, the molecules of gas are free to move in any direction.

liquid

Melting
The vibrations in solids are too weak to break the bonds connecting them. The solid melts only when its units have enough energy to break a few of the bonds and become a liquid. Substances with high melting points have strong bonds connecting their units, and so need a lot of heat energy to break them.

Changing states in mixtures

Mixtures contain ingredients that have different melting and boiling points. When a solid is dissolved in a liquid, such as the salt in seawater, the mixture looks and behaves like a liquid. However, when it is heated to boiling point, the mixture separates—the water evaporates, leaving behind the solid salt (which melts at a much higher temperature).

nonmelting chocolate chips

melting ice cream

▷ **Melting mixtures**
A chocolate chip ice cream is a mixture of ice, cream, and bits of chocolate. All of this is solid when the ice cream is served, but the ice and cream soon melt. The chocolate stays solid for longer, however.

Gas laws

THE GAS LAWS STATE HOW GASES RESPOND TO CHANGE.

SEE ALSO	
❰ 28–29 Respiration	
❰ 99 Gases	
Pressure	141 ❱
Pressure	184–185 ❱

The three laws relate the movements of molecules in a gas to its volume, pressure, and temperature, and state how each measure responds when the others change. Each gas law is named after its discoverer.

Boyle's law

This law is named after Robert Boyle, who lived in Britain and Ireland in the 17th century and was one of the world's first chemists. His law states that if the temperature of a gas stays the same, then its volume is inversely proportional to its pressure. In other words, forcing a gas into a smaller volume results in it exerting a higher pressure.

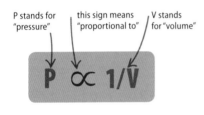

P stands for "pressure" · this sign means "proportional to" · V stands for "volume"

$$P \propto 1/V$$

△ **Equation for Boyle's law**
This equation shows the relationship between a gas's pressure and its volume. Increasing the pressure decreases the volume.

one weight produces pressure (P) in beaker

molecules spread evenly

△ **Diffusion**
The molecules in the gas spread out evenly to fill any container. This is called diffusion and means that molecules tend to move away from places where they are highly concentrated.

two weights produce double the pressure (2P) in beaker

high pressure squeezes molecules into half the original volume

△ **Pressure**
The force exerted on an area (its pressure) is caused by molecules in the gas hitting the inside of the container. Reducing the volume gives the molecules less room to move. They hit the sides more frequently, increasing the pressure.

REAL WORLD

Avogadro's law

There is a fourth gas law, which, although unrelated to the other three, was set out by the Italian Amedeo Avogadro (right) in 1811. It states that equal volumes of all gases at the same temperature and pressure contain the same number of molecules. Therefore a flask of hydrogen can contain the same number of molecules as an identical flask of oxygen, despite weighing a lot (16 times) less.

Robert Boyle was an alchemist and discovered his law while he was searching for a way to turn lead into gold.

Charles's law

This gas law, which is attributed to the French scientist Jacques Charles, states that the temperature of a gas is proportional to its volume. So if the gas is held in a container with an adjustable volume—a gas syringe, for example—increasing the temperature of the gas results in an increase in its volume.

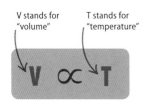

V stands for "volume"   T stands for "temperature"

$$V \propto T$$

△ **Equation for Charles's law**
This equation shows the relationship between a gas's volume and its temperature. Increasing the temperature increases the volume.

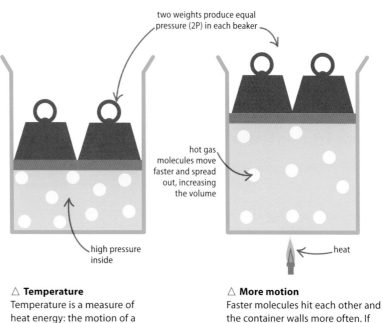

two weights produce equal pressure (2P) in each beaker

hot gas molecules move faster and spread out, increasing the volume

high pressure inside

heat

△ **Temperature**
Temperature is a measure of heat energy: the motion of a gas's molecules. Increasing the temperature of the gas increases the rate at which its molecules move.

△ **More motion**
Faster molecules hit each other and the container walls more often. If one wall is moveable, these impacts will push it outward, increasing the volume of the container.

Gay-Lussac's law

Named after French scientist Joseph Louis Gay-Lussac in 1808, this was the last of the three main gas laws to be formulated. It states that for a fixed volume of gas, the pressure is proportional to its temperature. In other words, when the temperature of a gas is increased, it also exerts a higher pressure. Similarly, squeezing a gas into a smaller volume increases its pressure (as per Boyle's law) and also raises the gas's temperature.

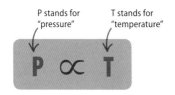

P stands for "pressure"   T stands for "temperature"

$$P \propto T$$

△ **Equation for Gay-Lussac's law**
This equation shows the relationship between the pressure of a gas and its temperature. Increasing the temperature increases the pressure.

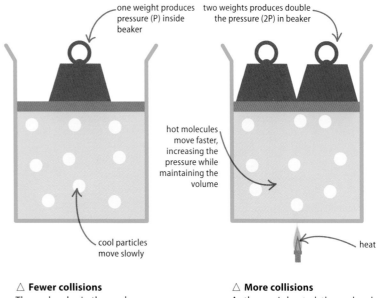

one weight produces pressure (P) inside beaker

two weights produces double the pressure (2P) in beaker

hot molecules move faster, increasing the pressure while maintaining the volume

cool particles move slowly

heat

△ **Fewer collisions**
The molecules in the cool gas move slowly and they hit the sides of the container infrequently. These few, weak collisions combine to create a low gas pressure, overall.

△ **More collisions**
As the gas is heated, the molecules move around faster and hit the sides of the container more often and with greater force. Thus the pressure goes up.

Mixtures

A MIXTURE IS A COMBINATION OF SUBSTANCES THAT
CAN BE SEPARATED BY PHYSICAL MEANS.

SEE ALSO	
Separating mixtures	**106–107 ❯**
Compounds and molecules	**110–111 ❯**
Water	**142–143 ❯**

Mixtures are classified as solutions, suspensions, or colloids based on
particle size. The substances in a mixture are not chemically linked.

Uneven and even

Every mixture has at least two
ingredients. The first ingredient is
known as the continuous medium.
Into this, the second ingredient,
known as the dispersed phase, is
mixed. In an even or homogeneous
mixture, the particles of the dispersed
phase are evenly distributed among
the molecules of the continuous
medium, so the concentrations of
each ingredient are constant. In an
uneven or heterogeneous mixture,
the dispersed phase is concentrated
in some places and not in others.
Some substances, normally liquids,
cannot be mixed together because
their molecules repel each other—
they are described as immiscible.

REAL WORLD

Lava lamp

A lava lamp makes
use of two immiscible
liquids. The clear liquid
is a mineral oil, while
the colored "lava" is a
wax. When the lamp is
turned on, light heats
the wax, reducing its
density so it begins
to rise up into the oil.
The wax does not mix,
however, and the
colored bubbles
rise and fall.

▽ **Seawater**
In seawater, the water is the continuous
medium, while salts—chiefly sodium
chloride—form the dispersed phase.
The salt mixes so thoroughly that it
dissolves and disappears from view.

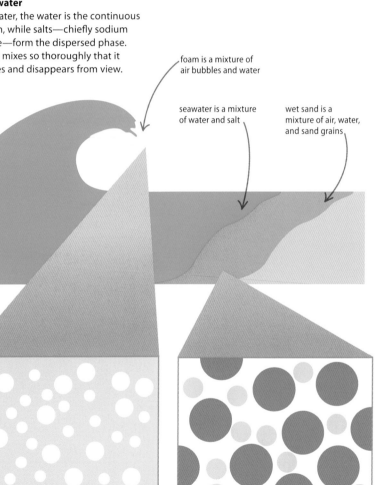

foam is a mixture of
air bubbles and water

seawater is a mixture
of water and salt

wet sand is a
mixture of air, water,
and sand grains

△ **Foam**
The foam of a breaking wave is a
heterogeneous mixture of air bubbles
and water. The white appearance of
the mixture is different from those
of both the constituents.

△ **Wet sand**
The sand grains are much larger than
the water molecules around them
so they remain distinct and are visible
when viewed close up. As the mixture
dries, the water is replaced by air.

Solutions

A homogeneous mixture is often referred to as a solution. The continuous medium is the solvent, and the dispersed phase is the solute. The solute disappears after it dissolves, although the color of the solvent may change.

SOLUTIONS			
Solvent	**Solute**	**Solution**	**Description**
helium	oxygen	deep-sea breathing gas	helium replaces other gases in the air
air	water	humid air	occurs on warm but damp days
air	smoke	smog	air pollution
water	carbon dioxide	soda water	fizzy water used in sodas
water	ethanoic acid	vinegar	sharp-tasting cooking ingredient
water	salt	seawater	salty water
palladium	hydrogen	palladium hydride	high-tech alloy used in industry
silver	mercury	amalgam	soft alloy used in dental fillings
iron	carbon	steel	high-strength alloy used in construction

Key

■ Solid ■ Liquid ■ Gas

Suspensions

A common type of heterogeneous mixture is a suspension. In contrast to solutions, where the solute breaks up into tiny particles, the particles of the dispersed phase are considerably larger than those of the continuous medium—at least one micrometer across. Everyday examples include the dust carried in wind, tiny droplets in the gas of an aerosol spray, or silt in river water.

▽ **Hanging around**
The particles of the dispersed phase are suspended—they are too small to sink quickly. There are three ways that the mixture can separate.

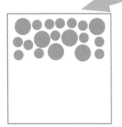

△ **Creaming**
If the suspended particles are less dense than the continuous phase, they will float. The particles will sit at the surface like cream floating on top of a cup of coffee.

△ **Sedimentation**
If the particles are denser than the continuous phase, they will sink. The particles will form a sediment, or layer, at the bottom of the mixture.

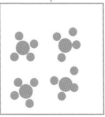

△ **Flocculation**
Sometimes the particles will clump to form larger particles, or floccs. Flocculation happens when the conditions change, or another substance is added to the mixture.

Colloids

A colloid is a mixture that is halfway between a solution and a suspension. The dispersed phase appears to be evenly distributed to the naked eye, but at a microscopic level the two constituents remain heterogeneously mixed. Ice cream, fog, and milk are examples of colloids.

▷ **Cloud**
A cloud is a colloid of liquid water droplets mixed into air. If the droplets grow beyond a certain size, they fall as rain.

fat and water are immiscible (they won't mix), so the fat forms tiny blobs

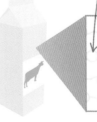

△ **Milk**
Milk is a colloid of fat in water. Colloids are often white, because the larger size of the dispersed phase causes light to scatter when it passes through the mixture.

Separating mixtures

MIXTURES ARE MADE UP OF SEPARATE SUBSTANCES.

The constituents of mixtures are not chemically joined. Since they remain distinct substances, they can be separated using only physical means. The precise method depends on the type of mixture.

Liquid mixtures

Dissolved solids can be separated by evaporating away the liquid solvent, leaving crystals of solid behind. This is how salt is separated from seawater. Collecting a pure sample of the solvent is more complicated. The vapor passes through a condenser, where it is cooled back into a liquid. A condenser is also used in distillation, which separates a mixture of two or more liquids.

Filtration

Silt in river water is an example of an uneven mixture—large, heavy solids are mixed into a continuous medium of much smaller particles. This kind of mixture can be separated using filters. A filter is a material that allows the smaller particles through but blocks the progress of the larger ones. Most laboratory filters are made from paper, but wire meshes can be used too.

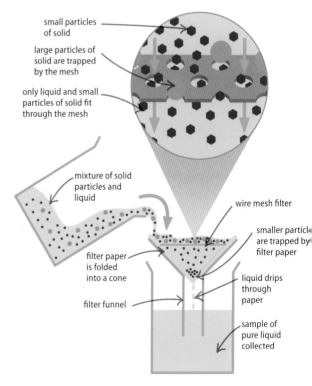

small particles of solid

large particles of solid are trapped by the mesh

only liquid and small particles of solid fit through the mesh

mixture of solid particles and liquid

wire mesh filter

smaller particles are trapped by filter paper

filter paper is folded into a cone

liquid drips through paper

filter funnel

sample of pure liquid collected

△ **Double filter**
The experiment above uses two different filters to separate out two different-sized consituents. Water and small particles pass through the first filter, made of wire mesh, but larger particles are trapped. The smaller particles are trapped in the second filter, made of paper, leaving just the pure liquid to drip into the beaker.

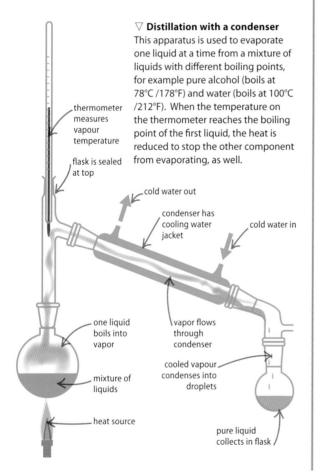

▽ **Distillation with a condenser**
This apparatus is used to evaporate one liquid at a time from a mixture of liquids with different boiling points, for example pure alcohol (boils at 78°C /178°F) and water (boils at 100°C /212°F). When the temperature on the thermometer reaches the boiling point of the first liquid, the heat is reduced to stop the other component from evaporating, as well.

thermometer measures vapour temperature

flask is sealed at top

cold water out

condenser has cooling water jacket

cold water in

one liquid boils into vapor

vapor flows through condenser

cooled vapour condenses into droplets

mixture of liquids

heat source

pure liquid collects in flask

Centrifugal force

Another way to separate an uneven mixture is by using a centrifuge. In a suspension, the solid particles are often still too small to sink to the bottom under the pull of gravity alone. So the mixture is spun around at high speed, creating a centrifugal force that pushes the solid material down to the bottom of the test tube.

solvent

centrifugal force

medium-density grain

spin of centrifuge

clear solvent

most dense grain

solids sorted by density

least dense grain

REAL WORLD

Butter churn

Butter is made by separating the solid fats from the liquid component in milk. This is done by mechanical disruption. The mixture is churned (spun) and this makes the blobs of fat stick together. The fat blobs get bigger and bigger until they separate from the water. The products from the churning are butter and buttermilk— a thinner, lower-fat liquid.

△ **Sorting mixtures**
The centrifugal force has the strongest effect on the densest particles in the mixture, so these move to the bottom fastest. This phenomenon can be used to sort suspended particles or grains—the densest ones form the lower layer, with successively less dense particles layered on top.

Chromatography

When the components of a mixture have the same particle sizes or similar boiling points, they are separated by chromatography. The mixture is dissolved in a solvent. The solvent, known as the mobile phase, is then drawn through a substance known as the stationary phase, often filter paper. The mobile phase moves forward, but each component in the mixture moves through the stationary phase at a different speed. As a result, each component becomes fixed in the stationary medium at a different point, forming separated samples of each substance.

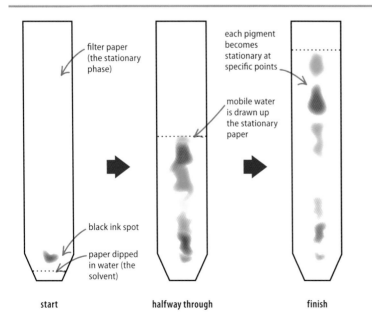

filter paper (the stationary phase)

each pigment becomes stationary at specific points

mobile water is drawn up the stationary paper

black ink spot

paper dipped in water (the solvent)

start   **halfway through**   **finish**

◁ **Separating black ink**
Black ink is a mixture of colored pigments in water. These can be separated using chromatography. The word means "writing with color," and the drop of ink forms bands of its individual color components.

Elements and atoms

EVERYTHING IS BUILT FROM ELEMENTS AND ATOMS.

In ancient times, people believed that our world was made from just a few elemental substances: earth, air, fire, and water. Chemists now know that it is made from 90 naturally occurring elements.

What is an element?

An element is a substance that cannot be broken down into simpler constituents. Therefore a pure sample of an element is made entirely from one type of atom. The structure of that atom defines the element's physical and chemical properties.

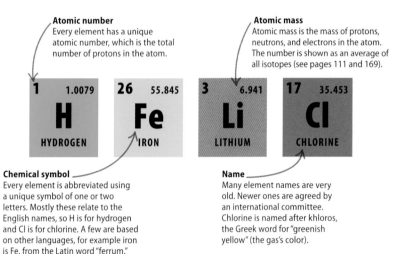

Atomic number
Every element has a unique atomic number, which is the total number of protons in the atom.

Atomic mass
Atomic mass is the mass of protons, neutrons, and electrons in the atom. The number is shown as an average of all isotopes (see pages 111 and 169).

1 1.0079	26 55.845	3 6.941	17 35.453
H	**Fe**	**Li**	**Cl**
HYDROGEN	IRON	LITHIUM	CHLORINE

Chemical symbol
Every element is abbreviated using a unique symbol of one or two letters. Mostly these relate to the English names, so H is for hydrogen and Cl is for chlorine. A few are based on other languages, for example iron is Fe, from the Latin word "ferrum."

Name
Many element names are very old. Newer ones are agreed by an international committee. Chlorine is named after khloros, the Greek word for "greenish yellow" (the gas's color).

Hennig Brand

The German Hennig Brand is the first historical figure known to have discovered a new element. In 1669, he found phosphorus after investigating the substances in his urine. The phosphorus glowed in the dark, making Brand think he had found a magic material.

Atomic structure

An atom is made up of positively charged protons and negatively charged electrons. The atoms of every element have a specific number of protons. Protons have a positive charge, but atoms are always neutral because the protons are balanced by an equal number of negatively charged electrons.

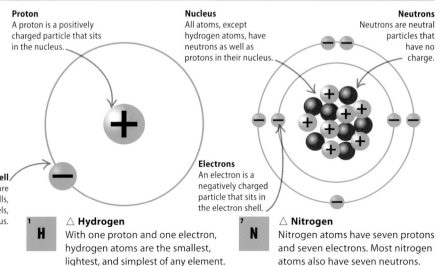

Proton
A proton is a positively charged particle that sits in the nucleus.

Nucleus
All atoms, except hydrogen atoms, have neutrons as well as protons in their nucleus.

Neutrons
Neutrons are neutral particles that have no charge.

Electron shell
The electrons are arranged in shells, or energy levels, around the nucleus.

Electrons
An electron is a negatively charged particle that sits in the electron shell.

1 H △ **Hydrogen**
With one proton and one electron, hydrogen atoms are the smallest, lightest, and simplest of any element.

7 N △ **Nitrogen**
Nitrogen atoms have seven protons and seven electrons. Most nitrogen atoms also have seven neutrons.

Electron configurations

As the atomic number increases, atoms get heavier and larger, because the electrons are arranged in shells positioned farther and farther out from the nucleus. The first shell can hold two electrons and the second can hold eight. Once the third shell has eight, the fourth shell starts to fill up, although in some cases these shells can hold many more than eight electrons (see page 124).

Shell shape
In reality the electron shells are not round. Scientists draw them like this so they are easy to see and compare.

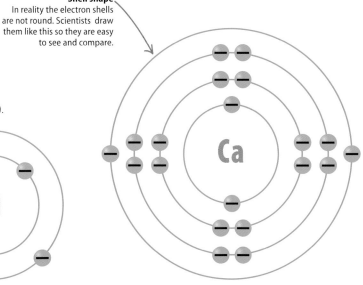

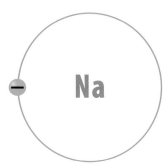

△ **Helium**
With an atomic number of 2, helium has two protons and therefore two electrons in a single shell.

△ **Lithium**
Lithium has an atomic number of 3. The first electron shell is full, so the third electron sits in another shell.

△ **Calcium**
With an atomic number of 20, the electrons in a calcium atom are arranged over four shells.

Outer shell

The electrons in an atom's outer shell are the ones that form bonds with other atoms and become involved in chemical reactions. So the number of outer electrons in an atom is a strong indicator of an element's physical and chemical properties. Atoms react with each other to achieve a full outer shell and therefore become more stable. The diagrams below show only the outer shell of each atom.

▽ **Octet rule**
Atoms need to have eight electrons in their outer shell to become stable. This is called the octet rule. They must either gain electrons to reach eight, or lose electrons so that the next shell down—which will be full—becomes their outer shell.

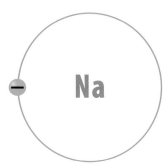

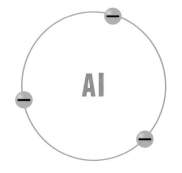

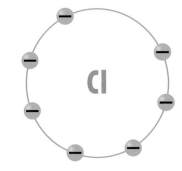

△ **Sodium**
A sodium atom has just one outer electron. To get a full outer shell, it must give that electron away.

△ **Aluminum**
An aluminum atom has three outer electrons. It has to lose all of these to become stable.

△ **Chlorine**
A chlorine atom has seven outer electrons. It has space for one more, which would fill its outer shell.

Compounds and molecules

ATOMS JOIN TOGETHER TO FORM COMPOUNDS AND MOLECULES.

Few elements exist naturally in their pure form. Gold is one example. Most other elements form compounds, when their atoms bond with those of other elements.

SEE ALSO

❮ **104–105** Mixtures

❮ **108–109** Elements and atoms

Ionic bonding   **112–113** ❯

Covalent bonding   **114–115** ❯

What is a compound?

Almost all everyday items are made up of chemical compounds, from the water coming out of the tap to the minerals in bricks and stones to the substances in the human body. A compound is a single substance made up of the atoms of two or more elements, which are chemically connected or bonded. This differentiates a compound from a mixture, which is made up of two or more separate substances.

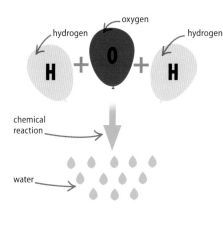

◁ **Fixed ratio**
A compound's constituent elements have a fixed ratio. Water (H_2O) has two parts of hydrogen (H) for every part of oxygen (O).

◁ **Chemical reactions**
A compound can only be formed when elements, or other compounds, react with each other.

◁ **Different properties**
A compound's properties are different to those of its constituent elements. Water is a liquid, for example, that is made up of two gases.

Molecules

A molecule is the smallest unit of a compound. Breaking the molecule down into simpler constituents would result in the compound ceasing to exist. The atoms in a molecule are connected by chemical bonds. The arrangement and strength of the bonds gives the molecule a certain shape.

At **very high pressures,** oxygen molecules transform into an eight-atom version that is bright red.

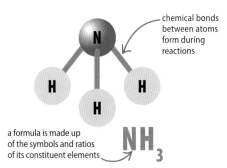

△ **Ammonia**
The molecules of this compound have a single nitrogen atom (N) bonded to three hydrogen atoms (H). Together they form a tetrahedron.

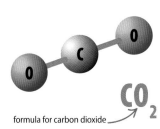

△ **Carbon dioxide**
As its name suggests, this compound has one carbon atom (C) bonded to two oxygen atoms (O). The three atoms form a straight molecule.

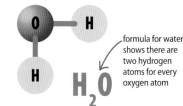

△ **Water**
Common compounds such as water have nonscientific names. Others, such as carbon dioxide, are named according to the elements they contain.

Molecular elements

Atoms get involved in reactions and form molecules to become more stable. So even when they are pure, most elements do not exist as single unbonded atoms. However, the molecules they form consist of only one type of atom.

▷ **Increased stability**
The oxygen in the air exists as a molecule of two oxygen atoms (O_2). When bonded together, the oxygen atoms are in a more stable state.

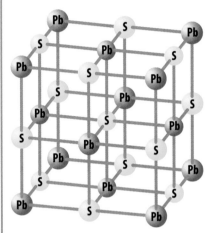

oxygen molecule

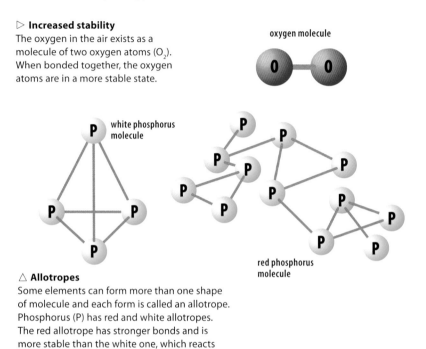

white phosphorus molecule

red phosphorus molecule

△ **Allotropes**
Some elements can form more than one shape of molecule and each form is called an allotrope. Phosphorus (P) has red and white allotropes. The red allotrope has stronger bonds and is more stable than the white one, which reacts very easily—even burning on contact with air.

Crystals

A crystal forms when the large numbers of molecules of an element or a compound are all joined together in a repeating pattern. For example, a diamond is a form of carbon made up of repeating tetrahedral units.

△ **Galena**
Galena is a compound of lead (Pb) and sulfur (S). Its formula is simply PbS and the lead and sulfur atoms form a cube of atoms. A galena crystal is made up of these cubic units.

Metallic bonds

Metal atoms lose their outer electrons easily, forming positively charged ions (see page 112) surrounded by a sea of shared electrons. The attraction between the negatively charged electrons and positively charged ions creates metallic bonds that "glue" the structure together.

▽ **Strong material**
The free electrons are shared by all neighboring ions and can slide past each other. This means that metal objects can be deformed, or bent, without breaking.

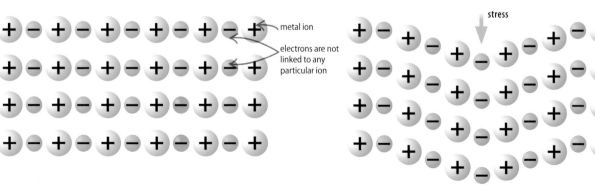

metal ion

electrons are not linked to any particular ion

stress

Ionic bonding

IONIC BONDING IS WHEN DIFFERENT ATOMS FORM BONDS
BY GAINING OR LOSING ELECTRONS.

Atoms bond with each other so that they can fill the spaces
in their outer electron shells. This makes them more stable.

What is an ion?

An atom has an equal number of protons and electrons so
it has no overall charge. If the atom loses or gains an electron, it
becomes a charged particle called an ion. Losing one electron
produces a positive ion with a charge of 1+; losing two results in
a charge of 2+. Ions formed by gaining electrons have negative
charges, so gaining one electron results in a charge of 1⁻.

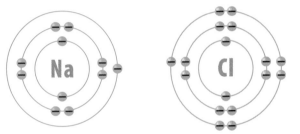

△ **A sodium atom** has one outer electron, while chlorine has
seven electrons in the outer shell, with room for one more.

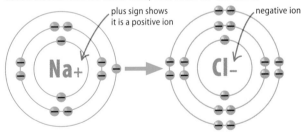

plus sign shows
it is a positive ion

negative ion

△ **Sodium loses its outer electron**, passing it to chlorine.
Both ions now have full outer shells.

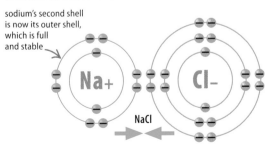

sodium's second shell
is now its outer shell,
which is full
and stable

NaCl

△ **The positive charge of the sodium** ion attracts the equal but
negative charge of the chloride ion to form an ionic bond between
the two. The resulting compound is sodium chloride (NaCl).

Octet rule

Atoms with low atomic numbers become full and stable
when their outer shells contain eight electrons—the so-called
octet rule (see page 109). Cations, or positive ions, are formed
when atoms lose electrons, while anions, or negative ions,
are formed when atoms gain electrons.

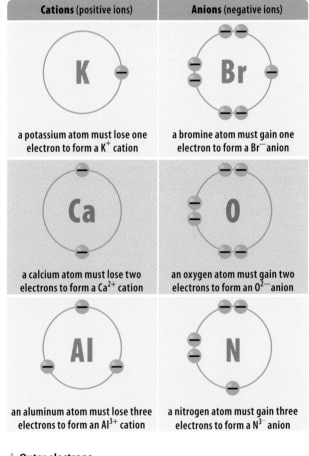

Cations (positive ions)	Anions (negative ions)
K	Br
a potassium atom must lose one electron to form a K⁺ cation	a bromine atom must gain one electron to form a Br⁻ anion
Ca	O
a calcium atom must lose two electrons to form a Ca²⁺ cation	an oxygen atom must gain two electrons to form an O²⁻ anion
Al	N
an aluminum atom must lose three electrons to form an Al³⁺ cation	a nitrogen atom must gain three electrons to form a N³⁻ anion

△ **Outer electrons**
Atoms with between one and three outer electrons will lose them,
whereas atoms with five to seven electrons will gain more. An atom
with a complete outer shell is stable, so does not lose or gain electrons.

Balancing charges

For a bond to form between ions, the positive and negative charges need to balance so that the overall molecule is neutral. As a result compounds are not always derived from one anion and one cation, but form with different proportions of ions.

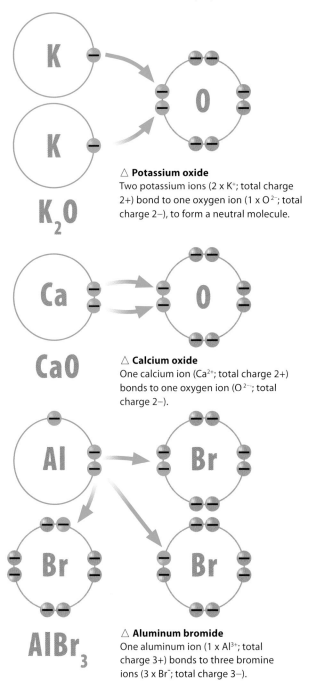

△ **Potassium oxide**
Two potassium ions (2 x K^+; total charge 2+) bond to one oxygen ion (1 x O^{2-}; total charge 2−), to form a neutral molecule.

△ **Calcium oxide**
One calcium ion (Ca^{2+}; total charge 2+) bonds to one oxygen ion (O^{2-}; total charge 2−).

△ **Aluminum bromide**
One aluminum ion (1 x Al^{3+}; total charge 3+) bonds to three bromine ions (3 x Br^-; total charge 3−).

Reactivity

Metal atoms give away electrons so they are electropositive. Nonmetals gain electrons so they are electronegative. Different atoms give away or gain electrons more easily than others.

▽ **Metal ions**
Magnesium (Mg) and sodium (Na) have three electron shells. However, magnesium has two outer electrons while sodium has one. It takes less energy to lose one electron than two, so sodium is more electropositive than magnesium. Potassium (K) also has one outer electron but it is in a fourth shell, farther away from the attractive pull of the nucleus. So potassium loses its outer electron more easily than sodium.

▽ **Nonmetal ions**
Oxygen (O) needs two electrons to complete the octet but nitrogen (N) needs three. It takes less energy to gain two electrons than three, so oxygen is more electronegative than nitrogen. Phosphorus (P) also needs three electrons but it has one more shell. The pull from the nucleus in this third shell is weaker than in a second shell, so it is harder for phosphorus to gain electrons than nitrogen.

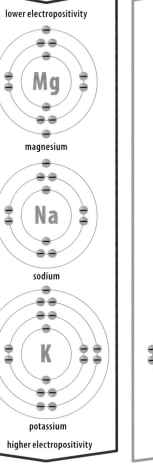

lower electropositivity

magnesium

sodium

potassium

higher electropositivity

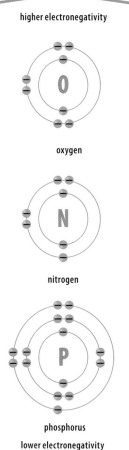

higher electronegativity

oxygen

nitrogen

phosphorus

lower electronegativity

Covalent bonding

COVALENT BONDING IS WHEN ATOMS FORM BONDS BY
SHARING ELECTRONS.

Rather than giving away or accepting electrons, some atoms
share their outer electrons to achieve full outer shells.

Sharing electrons

Covalent bonds are formed of pairs of electrons, one from each
atom. The pair is included in the outer electron shell of both
atoms at once. This allows the atoms to have a full set of eight
electrons in their outer shell and become stable. No electrons
leave their original atoms—so the atoms always remain neutral.

▽ **Double bond**
Oxygen gas is made up of O_2
molecules. They form when
oxygen atoms share not one
but two pairs of electrons in
what is known as a double bond.

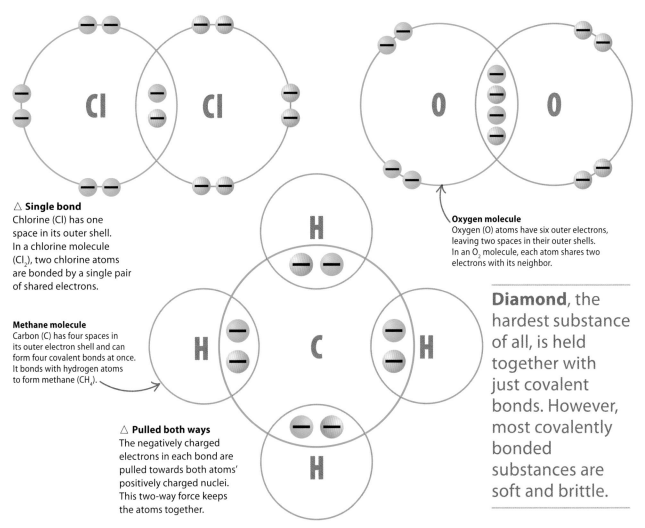

△ **Single bond**
Chlorine (Cl) has one
space in its outer shell.
In a chlorine molecule
(Cl_2), two chlorine atoms
are bonded by a single pair
of shared electrons.

Oxygen molecule
Oxygen (O) atoms have six outer electrons,
leaving two spaces in their outer shells.
In an O_2 molecule, each atom shares two
electrons with its neighbor.

Methane molecule
Carbon (C) has four spaces in
its outer electron shell and can
form four covalent bonds at once.
It bonds with hydrogen atoms
to form methane (CH_4).

△ **Pulled both ways**
The negatively charged
electrons in each bond are
pulled towards both atoms'
positively charged nuclei.
This two-way force keeps
the atoms together.

Diamond, the
hardest substance
of all, is held
together with
just covalent
bonds. However,
most covalently
bonded
substances are
soft and brittle.

Shapes and bonds

In a methane molecule (see page 114), every electron in the carbon atom's outer shell is shared with a hydrogen atom. However, in other molecules not all the electrons are involved in a bond. The ones that are not are called lone pairs. These lone pairs create a zone of electric charge that repels the bonded pairs and pushes them closer together. So the arrangement of bonded and lone pairs gives a molecule its shape.

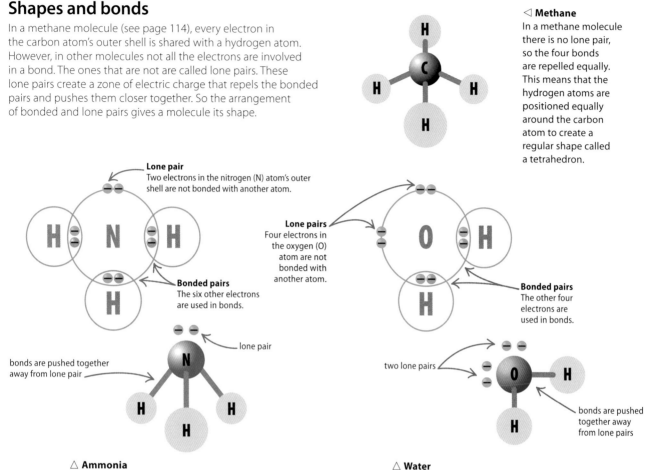

◁ **Methane**
In a methane molecule there is no lone pair, so the four bonds are repelled equally. This means that the hydrogen atoms are positioned equally around the carbon atom to create a regular shape called a tetrahedron.

Lone pair
Two electrons in the nitrogen (N) atom's outer shell are not bonded with another atom.

Lone pairs
Four electrons in the oxygen (O) atom are not bonded with another atom.

Bonded pairs
The six other electrons are used in bonds.

Bonded pairs
The other four electrons are used in bonds.

lone pair

bonds are pushed together away from lone pair

two lone pairs

bonds are pushed together away from lone pairs

△ **Ammonia**
The lone pair on nitrogen repels the three bonds so they are pushed closer together.

△ **Water**
In a water molecule, two lone pairs of electrons push the molecule into a V shape.

Intermolecular forces

Most simple covalent compounds are gases, because their molecules stay separate from each other in normal conditions. However, in a liquid or solid, weak intermolecular forces act between the molecules to hold them together. A common type is the dipole-dipole interaction, which occurs bewteen dipoles (molecules that have one negatively charged side and one positively charged side). A negative end of a dipole on one molecule will then attract a positive end of dipole on a neighbouring molecule, holding the two molecules together.

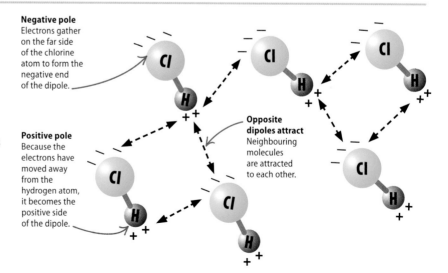

Negative pole
Electrons gather on the far side of the chlorine atom to form the negative end of the dipole.

Positive pole
Because the electrons have moved away from the hydrogen atom, it becomes the positive side of the dipole.

Opposite dipoles attract
Neighbouring molecules are attracted to each other.

Periodic table

CHEMISTS ORGANIZE THE ELEMENTS USING THE PERIODIC TABLE.

The elements are arranged according to their atomic structure.
Those with similar properties are grouped together.

Building the table

The periodic table we use today was formulated by Dmitri Mendeleev in 1869. The elements are arranged in rows in order of their atomic number. The atomic number is the number of protons each atom has in its nucleus (see page 108). By arranging the elements in this way, those with similar properties are grouped together. This means chemists can predict the likely characteristics of an element from its position in the table.

Periods
The table has seven horizontal rows called periods.

Precious metal

Gold was one of the first known elements. This was because gold is one of the few elements that occurs pure in nature, so it was easily discovered.

gold

atoms of the elements in Group 1 have one electron in their outer shells

▷ **Individual entry**
Every element is most easily identified by its symbol. The atomic number is the number of protons in the nucleus.

atomic number — relative atomic mass

1	1.0079
H	symbol
HYDROGEN	element name

Groups
The table has 18 columns called groups.

GROUP

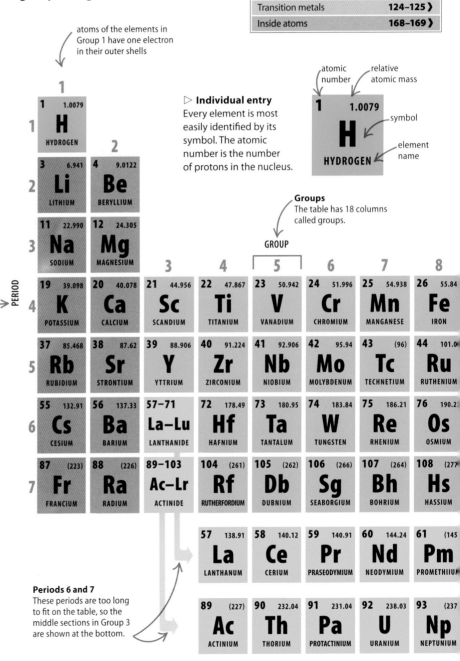

Periods 6 and 7
These periods are too long to fit on the table, so the middle sections in Group 3 are shown at the bottom.

PERIOD

1							
1 1.0079 **H** HYDROGEN	**2**						
3 6.941 **Li** LITHIUM	**4** 9.0122 **Be** BERYLLIUM						
11 22.990 **Na** SODIUM	**12** 24.305 **Mg** MAGNESIUM	**3**	**4**	**5**	**6**	**7**	**8**
19 39.098 **K** POTASSIUM	**20** 40.078 **Ca** CALCIUM	**21** 44.956 **Sc** SCANDIUM	**22** 47.867 **Ti** TITANIUM	**23** 50.942 **V** VANADIUM	**24** 51.996 **Cr** CHROMIUM	**25** 54.938 **Mn** MANGANESE	**26** 55.84 **Fe** IRON
37 85.468 **Rb** RUBIDIUM	**38** 87.62 **Sr** STRONTIUM	**39** 88.906 **Y** YTTRIUM	**40** 91.224 **Zr** ZIRCONIUM	**41** 92.906 **Nb** NIOBIUM	**42** 95.94 **Mo** MOLYBDENUM	**43** (96) **Tc** TECHNETIUM	**44** 101.0 **Ru** RUTHENIUM
55 132.91 **Cs** CESIUM	**56** 137.33 **Ba** BARIUM	**57–71** **La–Lu** LANTHANIDE	**72** 178.49 **Hf** HAFNIUM	**73** 180.95 **Ta** TANTALUM	**74** 183.84 **W** TUNGSTEN	**75** 186.21 **Re** RHENIUM	**76** 190.2 **Os** OSMIUM
87 (223) **Fr** FRANCIUM	**88** (226) **Ra** RADIUM	**89–103** **Ac–Lr** ACTINIDE	**104** (261) **Rf** RUTHERFORDIUM	**105** (262) **Db** DUBNIUM	**106** (266) **Sg** SEABORGIUM	**107** (264) **Bh** BOHRIUM	**108** (277) **Hs** HASSIUM

57 138.91 **La** LANTHANUM	**58** 140.12 **Ce** CERIUM	**59** 140.91 **Pr** PRASEODYMIUM	**60** 144.24 **Nd** NEODYMIUM	**61** (145 **Pm** PROMETHIUM
89 (227) **Ac** ACTINIUM	**90** 232.04 **Th** THORIUM	**91** 231.04 **Pa** PROTACTINIUM	**92** 238.03 **U** URANIUM	**93** (237 **Np** NEPTUNIUM

Building blocks

The table can be divided by column (group), by row (period), or by series.

3 — period runs left to right

△ **Period**
The elements in a period all have the same number of electron shells.

1
2
3
4

group runs top to bottom

△ **Group**
The elements in a group have the same number of outer electrons.

reactive metals

▷ **Series**
These are elements that react in a similar way.

transition elements

mainly nonmetals

rare earth metals

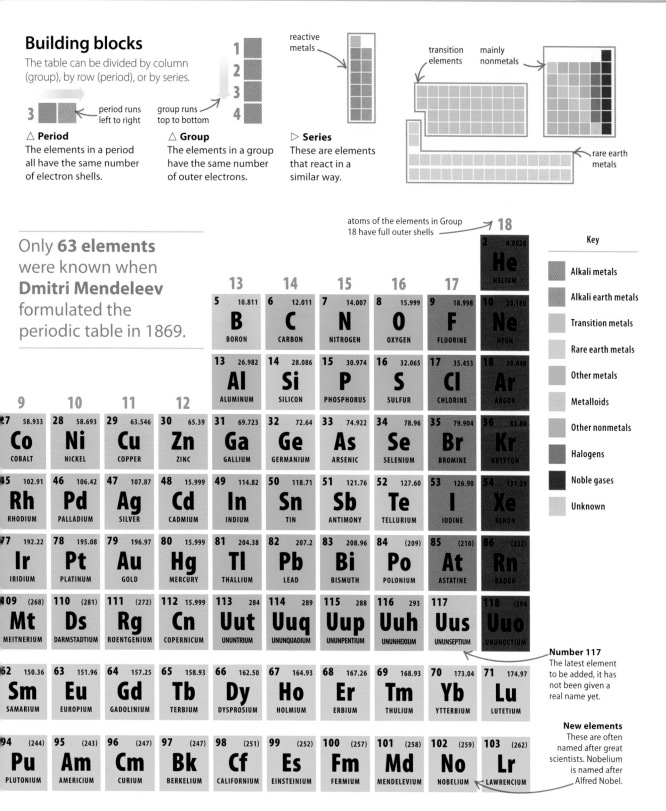

Only **63 elements** were known when **Dmitri Mendeleev** formulated the periodic table in 1869.

atoms of the elements in Group 18 have full outer shells

18

Key

Alkali metals
Alkali earth metals
Transition metals
Rare earth metals
Other metals
Metalloids
Other nonmetals
Halogens
Noble gases
Unknown

Number 117
The latest element to be added, it has not been given a real name yet.

New elements
These are often named after great scientists. Nobelium is named after Alfred Nobel.

2	4.0026
He	
HELIUM	

13	14	15	16	17
5 10.811 **B** BORON	6 12.011 **C** CARBON	7 14.007 **N** NITROGEN	8 15.999 **O** OXYGEN	9 18.998 **F** FLUORINE
13 26.982 **Al** ALUMINUM	14 28.086 **Si** SILICON	15 30.974 **P** PHOSPHORUS	16 32.065 **S** SULFUR	17 35.453 **Cl** CHLORINE

| 10   20.180 **Ne** NEON |
| 18   39.948 **Ar** ARGON |

9	10	11	12						
27 58.933 **Co** COBALT	28 58.693 **Ni** NICKEL	29 63.546 **Cu** COPPER	30 65.39 **Zn** ZINC	31 69.723 **Ga** GALLIUM	32 72.64 **Ge** GERMANIUM	33 74.922 **As** ARSENIC	34 78.96 **Se** SELENIUM	35 79.904 **Br** BROMINE	36 83.80 **Kr** KRYPTON
45 102.91 **Rh** RHODIUM	46 106.42 **Pd** PALLADIUM	47 107.87 **Ag** SILVER	48 15.999 **Cd** CADMIUM	49 114.82 **In** INDIUM	50 118.71 **Sn** TIN	51 121.76 **Sb** ANTIMONY	52 127.60 **Te** TELLURIUM	53 126.90 **I** IODINE	54 131.29 **Xe** XENON
77 192.22 **Ir** IRIDIUM	78 195.08 **Pt** PLATINUM	79 196.97 **Au** GOLD	80 15.999 **Hg** MERCURY	81 204.38 **Tl** THALLIUM	82 207.2 **Pb** LEAD	83 208.96 **Bi** BISMUTH	84 (209) **Po** POLONIUM	85 (210) **At** ASTATINE	86 (222) **Rn** RADON
109 (268) **Mt** MEITNERIUM	110 (281) **Ds** DARMSTADTIUM	111 (272) **Rg** ROENTGENIUM	112 15.999 **Cn** COPERNICUM	113 284 **Uut** UNUNTRIUM	114 289 **Uuq** UNUNQUADIUM	115 288 **Uup** UNUNPENTIUM	116 293 **Uuh** UNUNHEXIUM	117 **Uus** UNUNSEPTIUM	118 (294) **Uuo** UNUNOCTIUM

62 150.36 **Sm** SAMARIUM	63 151.96 **Eu** EUROPIUM	64 157.25 **Gd** GADOLINIUM	65 158.93 **Tb** TERBIUM	66 162.50 **Dy** DYSPROSIUM	67 164.93 **Ho** HOLMIUM	68 167.26 **Er** ERBIUM	69 168.93 **Tm** THULIUM	70 173.04 **Yb** YTTERBIUM	71 174.97 **Lu** LUTETIUM
94 (244) **Pu** PLUTONIUM	95 (243) **Am** AMERICIUM	96 (247) **Cm** CURIUM	97 (247) **Bk** BERKELIUM	98 (251) **Cf** CALIFORNIUM	99 (252) **Es** EINSTEINIUM	100 (257) **Fm** FERMIUM	101 (258) **Md** MENDELEVIUM	102 (259) **No** NOBELIUM	103 (262) **Lr** LAWRENCIUM

Understanding the periodic table

THERE ARE TRENDS IN THE PERIODIC TABLE.

SEE ALSO
❮ **113** Reactivity
❮ **116–117** Periodic table
Inside atoms **168–169** ❯

The periodic table arranges the elements according to the arrangment of their atoms' electrons. This means that similar elements are grouped together.

Size of atoms

Atoms get bigger as you move down the table because each period, or row, begins when a new shell is added to the atom. However, they get smaller from left to right as the number of outer electrons increases. This is because atoms with more outer electrons are held together with greater force, pulling them into smaller volumes.

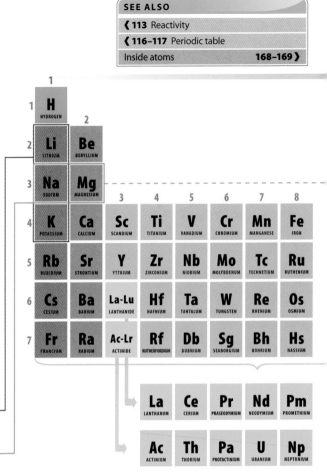

Two shells
A lithium atom has three electrons, with the third located in the second shell.

Three shells
The outer electron of the sodium atom starts a third shell, making the atom larger.

Four shells
Potassium, in Period 4, has a fourth electron shell located far out from the nucleus.

Metals and nonmetals

The left side of the periodic table is made up of metallic elements; the right side, nonmetallic. A metallic element has atoms that give up their outer electrons easily. The nonmetals hold firmly to their outer electrons and have very different properties from metals. Eight elements are semimetals, which have characteristics of both metals and nonmetals.

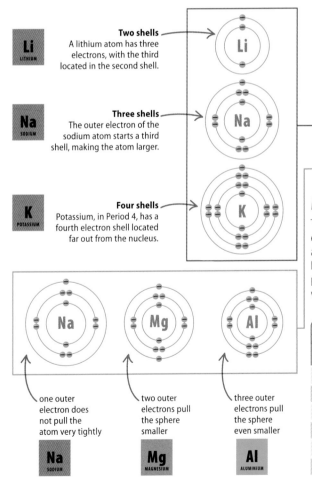

one outer electron does not pull the atom very tightly

two outer electrons pull the sphere smaller

three outer electrons pull the sphere even smaller

METALLIC AND NONMETALLIC	
Metallic	**Nonmetallic**
conducts heat	good insulator
conducts electricity	resists current
malleable and tough	brittle and crumbly
shiny and opaque	dull and translucent
high density	low density
low ionization energy	high ionization energy

◁ **Metallic vs nonmetallic**
Metal atoms have a certain set of characteristics due to their atomic structure. Nonmetals have an almost opposite set of characteristics.

Ionization energy

Ionization energy is the energy needed to take an electron out of an atom, making the atom a positively charged ion. The trend in the ionization energy of elements is the reverse of that of atomic size. In the periodic table, the required energy increases left to right and decreases top to bottom. Atoms with large numbers of outer electrons require more energy to ionize by losing an electron because their shells are held more tightly, closer to the nucleus. Large atoms, with outer electrons located far from the nucleus, lose them more easily.

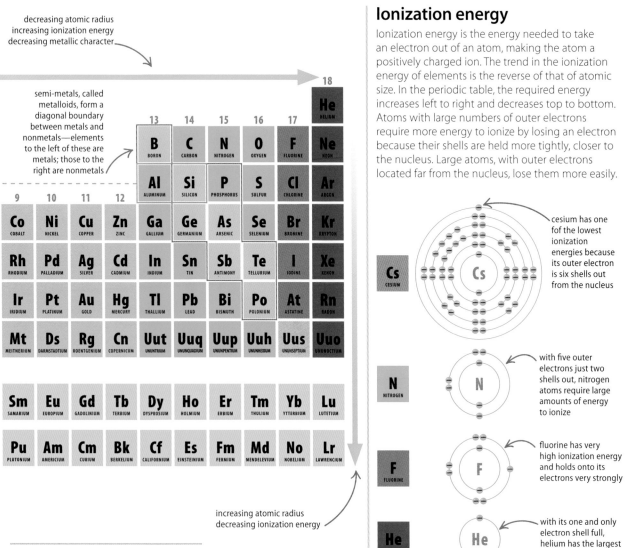

decreasing atomic radius
increasing ionization energy
decreasing metallic character

semi-metals, called metalloids, form a diagonal boundary between metals and nonmetals—elements to the left of these are metals; those to the right are nonmetals

increasing atomic radius
decreasing ionization energy

cesium has one fof the lowest ionization energies because its outer electron is six shells out from the nucleus

with five outer electrons just two shells out, nitrogen atoms require large amounts of energy to ionize

fluorine has very high ionization energy and holds onto its electrons very strongly

with its one and only electron shell full, helium has the largest ionization energy of any element

Trends in the table are not always followed: the atoms of **zirconium** and **hafnium** are almost identical in size, even though hafnium has 32 more electrons!

REAL WORLD

Ekasilicon

After developing the periodic table, Dmitri Mendeleev used it to predict the properties of elements that had yet to be discovered—left as gaps in the table. He described element 32 as ekasilicon, predicting its melting point, color, density, and chemical characteristics. In 1886, ekasilicon—eventually named germanium—was isolated, and matched Mendeleev's predictions.

Alkali metals and alkali earth metals

SIX ELEMENTS IN GROUP 1 OF THE PERIODIC TABLE ARE CALLED ALKALI METALS. THE SIX IN GROUP 2 ARE ALKALI EARTH METALS.

SEE ALSO	
❰ 112–113 Ionic bonding	
❰ 116–117 Periodic table	
❰ 118–119 Understanding the periodic table	
The halogens and noble gases	122–123 ❱
Transition metals	124–125 ❱
What is a base?	144 ❱

These elements get involved in chemical reactions with other elements easily because they have very few outer electrons.

Reactive metals

Elements in Group 1 have a single outer electron in their atoms, while those of Group 2 have two outer electrons. They form ions easily by losing these electrons, so they readily get involved in reactions, which makes them highly reactive. With just one electron to lose, a member of Group 1, such as potassium (K), will ionize more easily than a member of Group 2, which must lose two electrons.

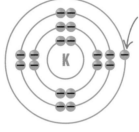

one outer electron

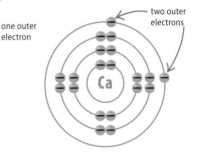

two outer electrons

△ **Potassium (Group 1)**
As the third alkali metal, potassium has one electron in its fourth and outermost shell.

△ **Calcium (Group 2)**
Calcium also has four electron shells, but there are two electrons in its outer shell.

Releasing hydrogen

These metals all react strongly with water, producing brightly colored flames. The metal ion swaps places with (displaces) a hydrogen ion in the water, forming a substance called a hydroxide. The displaced hydrogen is released as bubbles of gas. For example, when potassium is added to water, the products are potassium hydroxide (KOH) and hydrogen gas: $2K + 2H_2O \rightarrow 2KOH + H_2$.

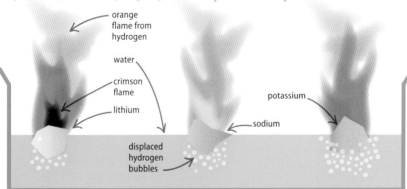

orange flame from hydrogen

water

crimson flame

lithium

potassium

sodium

displaced hydrogen bubbles

△ **Lithium**
When reacting with water, lithium burns with a crimson flame. The hydrogen released turns the flame orange.

△ **Sodium**
This metal produces an orange flame. It is the same color produced by sodium lamps used in street lights.

△ **Potassium**
Potassium burns with a lilac flame. It is more reactive than sodium and lithium and often explodes as it reacts.

REAL WORLD

In bodies

The alkali metals and alkali earth metals are common ingredients in living bodies. Sodium and potassium ions are used to create the electric pulses that fire through muscles and nerves, while calcium compounds are in bones, teeth, and the shells of snails.

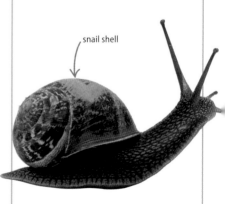

snail shell

Hydrogen and helium

Hydrogen is in Group 1 and has a single outer electron. However, it is not included in the alkali metals. This is because it has a distinct set of chemical properties compared to the other group members. Similarly, helium has two outer electrons, yet it is not included in Group 2. Instead it is in Group 18 with the noble gases, with which it shares most chemical properties.

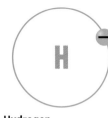

△ **Hydrogen**
Hydrogen has just one electron, which it loses less easily than other elements in its group.

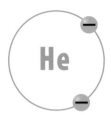

△ **Helium**
Helium has one electron shell, and with two electrons it is full. It does not form ions in chemical reactions.

Group trends

Members of Groups 1 and 2 become more reactive down the group as the atoms get bigger. This is because the atoms' negatively charged outer electrons are located farther away from the positively charged nucleus—and are held less strongly in the atom.

▽ **Alkali earth metals**
Earth metals are so called because they are found in compounds in the Earth's crust. Beryllium, for example, is found in gemstones such as emeralds. Although Group 2 metals react with water, they do so less strongly than those in Group 1.

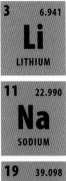

◁ **Lithium**
With the lowest density of any metal, lithium even floats in water.

◁ **Sodium**
The most abundant alkali metal, sodium compounds are found in many rocks.

◁ **Potassium**
Potassium is named after potash—potassium—containing compounds in the ash of burned wood.

◁ **Rubidium**
This metal would melt on a hot day and is so reactive that it catches fire in air.

◁ **Cesium**
Cesium melts at 28°C (82°F), which is only just above room temperature.

◁ **Francium**
This radioactive metal is extremely rare and little is known about it.

high reactivity

◁ **Beryllium**
This metal has a very low density so it is used to make high-speed aircraft and satellites.

◁ **Magnesium**
This metal is named after the region Magnesia in Greece, which has lots of magnesium compounds.

◁ **Calcium**
Calcium is common in Earth's rocks. Natural calcium compounds are often known as limes.

◁ **Strontium**
While most strontium is relatively stable, some forms are radioactive and dangerous.

◁ **Barium**
Barium compounds are added to fireworks to produce green explosions.

◁ **Radium**
This metal is highly radioactive and gives off a faint blue glow.

high reactivity

The halogens and noble gases

THESE ARE GROUPS 17 AND 18 OF THE PERIODIC TABLE.

SEE ALSO	
❮ **113** Reactivity	
❮ **118–119** Understanding the periodic table	
❮ **120–121** Alkali metals and alkali earth metals	
Radioactivity	**126–127** ❯
Types of reaction	**129** ❯

While the left side of the periodic table is dominated by metals, the right side—made up of Groups 17 and 18—are all nonmetals. The chemical characteristics of these two groups could not be more different.

The halogen group

There are five naturally occurring halogens. The reactivity decreases down the group as the atoms grow larger. The outer shell of a smaller atom is closer to the nucleus, so the electrons in it are held more strongly than in larger ones—and this includes the electron added during reactions to make an ion.

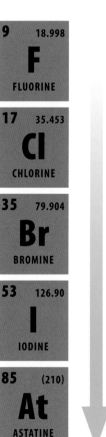

9	18.998
F	
FLUORINE	

◁ **Fluorine**
This pale yellow gas is the most reactive nonmetal element of all, and so forms compounds easily. Sodium fluoride is found in toothpaste.

17	35.453
Cl	
CHLORINE	

◁ **Chlorine**
This is a green gas that is used in many disinfectants and cleaning products, such as bleach. Chlorides are added to many swimming pools.

35	79.904
Br	
BROMINE	

◁ **Bromine**
This is the only nonmetal element found in liquid form in standard conditions. Its compounds are used in fireproofing.

53	126.90
I	
IODINE	

◁ **Iodine**
This purple-gray solid does not melt into a liquid at atmospheric pressure; it changes from a solid state straight into a purple gas.

85	(210)
At	
ASTATINE	

◁ **Astatine**
The heaviest halogen is highly radioactive and is very rare. Its atoms break up into other elements quickly.

low reactivity

The salt formers

The members of Group 17 are also known as the halogens, meaning "salt formers." The atoms of halogens have outer shells with seven electrons—out of a maximum of eight. As a result, all the halogens are very electronegative, meaning that they form negatively charged ions easily by attracting an electron each to fill their outer shells. They do this by reacting with metals (which form positively charged ions) to form stable ionic compounds. These substances are called salts.

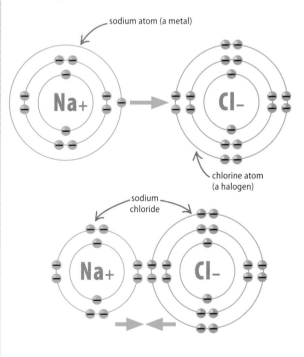

△ **Common salt**
It is perhaps no surprise that the most common salt is called common salt or table salt. This is a compound formed when the halogen chlorine (Cl) reacts with the metal sodium (Na), producing sodium chloride (NaCl).

Displacement

Halogens all react in the same way and form similar families of compounds. Therefore, a more highly reactive halogen will displace a less reactive one from its compounds. This can involve the two halogens swapping places in two compounds. When a pure halogen is used, the displaced element is also released in its pure form.

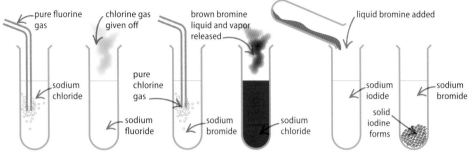

pure fluorine gas

chlorine gas given off

brown bromine liquid and vapor released

liquid bromine added

sodium chloride

pure chlorine gas

sodium fluoride

sodium bromide

sodium iodide

sodium bromide

solid iodine forms

sodium bromide

sodium chloride

△ **Fluorine displaces chlorine**
Fluorine displaces chlorine from sodium chloride (NaCl). It will also displace bromine and iodine.

△ **Chlorine displaces bromine**
Chlorine displaces bromine from sodium bromide (NaBr). It will also displace iodine but not fluorine.

△ **Bromine displaces iodine**
Bromine displaces iodine from sodium iodide (NaI). It will not displace chlorine or fluorine.

Inert gases

The noble gases form Group 18 of the periodic table. Apart from helium, they all have atoms with eight electrons in their outer shells—a full set. This makes them chemically inactive or inert. In other words, they are noble and do not mix with the other elements, and hardly ever take part in chemical reactions. Their atoms do not form molecules, even with themselves, and all Group 18 elements exist as gases made up of single atoms.

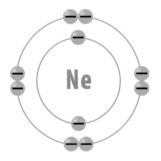

△ **Neon atom's shell**
Like all noble gases, neon does not bond ionically—it has no spaces to fill in its outer shell. Neon atoms do not share electrons in covalent bonds for the same reason.

2	4.0026
He	
HELIUM	

◁ **Helium**
Helium has just two outer electrons, filling a single shell around the nucleus.

10	20.180
Ne	
NEON	

◁ **Neon**
Discovered in 1898, this gas's name means "the new one."

18	39.948
Ar	
ARGON	

◁ **Argon**
The most common noble gas on Earth, it forms one percent of the atmosphere.

36	83.80
Kr	
KRYPTON	

◁ **Krypton**
Much rarer than neon, this gas's name means "the hidden one."

54	131.29
Xe	
XENON	

◁ **Xenon**
Xenon—"the strange one"—is a dense gas; a balloon of it falls straight down.

86	(222)
Rn	
RADON	

◁ **Radon**
All radon atoms are radioactive. The gas is formed naturally when uranium in rocks breaks down.

Neon lights

When heated, noble gases glow a specific color. Helium was discovered by its characteristic colors coming from the Sun—and it was named after the Greek word helios for Sun. Electrifying noble gases has the same effect, and these are used in gas-discharge lamps—or neon lighting.

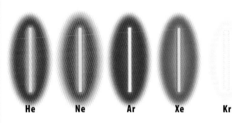

He   Ne   Ar   Xe   Kr

△ **Glowing gases**
In a neon light, electrical current runs through a tube of noble gas. As electrons are ripped off the atoms, they release a certain color of light.

Helium balloons

Helium is the second lightest gas in the Universe after hydrogen. Helium balloons float upward in the denser air gases. While hydrogen balloons and airships explode easily, helium balloons cannot burn, so they are safe to use in any size.

Transition metals

THE TRANSITION METALS ARE GROUPED IN A BLOCK
THAT FORMS THE CENTER OF THE PERIODIC TABLE.

The transition metals make up a block from Groups 3 to 12 in the
periodic table. They have distinct chemical properties because of
the unique way that their electrons are arranged inside the atoms.

SEE ALSO	
❰ **109** Electron configurations	
❰ **112–113** Ionic bonding	
❰ **118–119** Understanding the periodic table	
❰ **120–121** Alkali metals and alkali earth metals	
Redox reactions	**132–133** ❱

Inner and outer electrons

The transition elements only have one or two outer electrons. This is because they can
put more than eight electrons in the shell below the outermost shell. So as the atomic
number of the elements increases along each period in this block, the extra electrons
are not held in the atoms' outer shells, but are put in the next shell down, which can
hold up to 18 electrons. This is known as back-filling.

▽ **Adding electrons**
Calcium is not a transition element. It has
two outer electrons and eight in the next
shell down (the third). However, next along
in the periodic table is scandium—the first
transition element. It has one more electron
than calcium, but it sits in the third shell,
so a scandium atom still has two outer
electrons. Similarly, titanium has two outer
electrons but ten in the third shell.

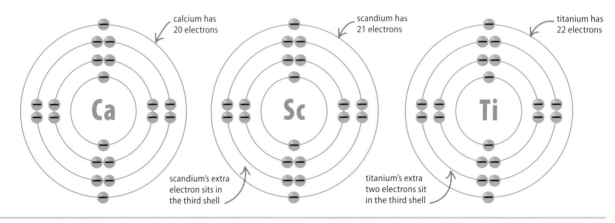

calcium has 20 electrons

scandium has 21 electrons

titanium has 22 electrons

scandium's extra electron sits in the third shell

titanium's extra two electrons sit in the third shell

Different charges

Like other metals, transition elements lose their outer electrons
easily to form positive ions. However, transition metals can then
continue to lose electrons from the next shell down and so can
form ions with a number of different charges, or oxidation states.
An ion's oxidation state indicates how many electrons have been
lost or gained: every electron lost increases the oxidation state by
one; +2 means two electrons have been lost (see page 132). For
example, manganese (Mn) forms ions with five common charges.

Oxidation state	Electrons lost by manganese
+2	two outer electrons
+3	two outer electrons and one inner electron
+4	two outer electrons and two inner elecrons
+6	two outer electrons and four inner electrons
+7	two outer electrons and five inner electrons

REAL WORLD
Most useful metals

Transition metals are less reactive than the alkali and alkali earth
metals in Groups 1 and 2. Because of this, they have been used
in technology for thousands of years. Iron is the most common
transition metal. It is a very strong construction material. Nickel,
another transition metal, is used in many coins.

Complex colors

A complex ion has a metal ion at its center with a number of other molecules or ions surrounding it. Transition elements form huge complex ions with molecules such as water, ammonia, and chlorine. The structures of these ions are very complicated and vary enormously. The wavelengths of light that these ions absorb and emit also varies enormously, so the compounds that they form come in a rainbow of different colors.

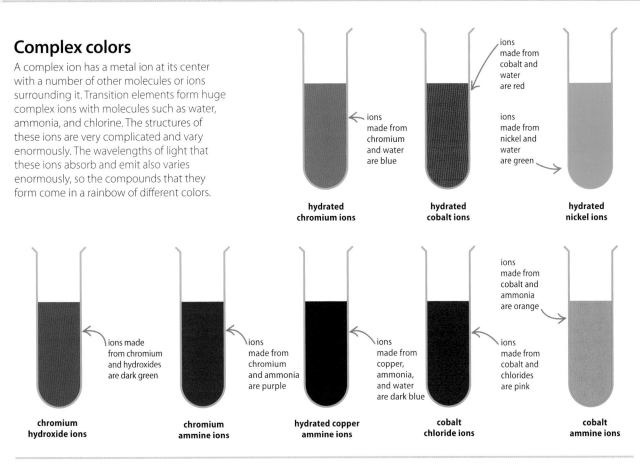

ions made from chromium and water are blue

hydrated chromium ions

ions made from cobalt and water are red

ions made from nickel and water are green

hydrated cobalt ions

hydrated nickel ions

ions made from chromium and hydroxides are dark green

chromium hydroxide ions

ions made from chromium and ammonia are purple

chromium ammine ions

ions made from copper, ammonia, and water are dark blue

hydrated copper ammine ions

ions made from cobalt and ammonia are orange

ions made from cobalt and chlorides are pink

cobalt chloride ions

cobalt ammine ions

Rare earth metals

The 30 rare earth metals are normally shown at the bottom of the periodic table, as there is no room to place them between Groups 2 and 3. They form in a similar way to the transition metals. Large atoms, from the sixth period on, grow by back-filling electrons, although this time the electrons are added two shells down, not one. The fourth and fifth atomic shells have room for 32 electrons.

▽ **Huge atoms**
Lanthanides are used to make high-tech alloys, while all of the actinides are radioactive. Uranium and thorium are used as nuclear fuels.

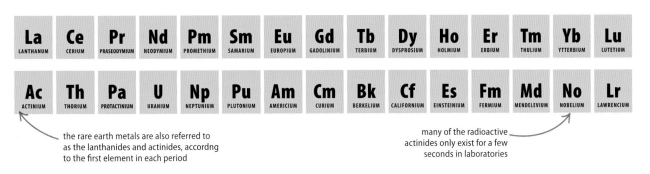

La	Ce	Pr	Nd	Pm	Sm	Eu	Gd	Tb	Dy	Ho	Er	Tm	Yb	Lu
LANTHANUM	CERIUM	PRASEODYMIUM	NEODYMIUM	PROMETHIUM	SAMARIUM	EUROPIUM	GADOLINIUM	TERBIUM	DYSPROSIUM	HOLMIUM	ERBIUM	THULIUM	YTTERBIUM	LUTETIUM

Ac	Th	Pa	U	Np	Pu	Am	Cm	Bk	Cf	Es	Fm	Md	No	Lr
ACTINIUM	THORIUM	PROTACTINIUM	URANIUM	NEPTUNIUM	PLUTONIUM	AMERICIUM	CURIUM	BERKELIUM	CALIFORNIUM	EINSTEINIUM	FERMIUM	MENDELEVIUM	NOBELIUM	LAWRENCIUM

the rare earth metals are also referred to as the lanthanides and actinides, accordng to the first element in each period

many of the radioactive actinides only exist for a few seconds in laboratories

Radioactivity

WHEN AN ATOM HAS AN UNSTABLE NUCLEUS, IT CAN BREAK
APART, EMITTING HIGH-ENERGY PARTICLES AND RADIATION.

SEE ALSO	
❰ 116–117 Periodic table	
Inside atoms	168–169 ❱
Electromagnetic waves	194–195 ❱
Energy from atoms	219 ❱
The Sun	232–233 ❱

A radioactive atom is generally very large, and its nucleus
has a different number of neutrons from a stable atom.
It is called a radioactive isotope of the element.

Radioactive decay

When an unstable nucleus breaks apart, or decays, it produces
radiation. Gamma rays are one type of radiation. They are
the highest energy waves in the electromagnetic spectrum.
Sometimes a nucleus will emit fast-moving particles. Losing
these alters the structure of the nucleus and produces a new
element. Alpha particles are made up of two protons and two
neutrons—the same as the nucleus of a helium atom. Beta
particles are generally single electrons.

▽ **Alpha decay**
An alpha particle is formed when the parent atom's nucleus loses
two protons and two neutrons. This decreases its atomic number
(the number of protons in an atom) by two. So radioactive uranium
(atomic number: 92) decays into thorium (90). The mass number
(the number of protons and neutrons) decreases by four.

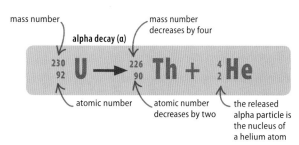

mass number

alpha decay (α)

mass number
decreases by four

$$^{230}_{92}\text{U} \rightarrow \, ^{226}_{90}\text{Th} + \, ^{4}_{2}\text{He}$$

atomic number

atomic number
decreases by two

the released
alpha particle is
the nucleus of
a helium atom

▽ **Beta decay**
A beta particle is formed when a neutron in the unstable
nucleus splits into a proton and electron. The proton stays
in the nucleus, raising the atomic number by 1, while the
electron is pushed out. Therefore radioactive carbon atoms
(atomic number: 6) form nitrogen (7). The nucleus has one less
neutron but one more proton, so the mass number stays the same.

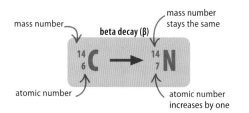

mass number

beta decay (β)

mass number
stays the same

$$^{14}_{6}\text{C} \rightarrow \, ^{14}_{7}\text{N}$$

atomic number

atomic number
increases by one

Dangerous radiation

Radioactive radiation is dangerous because it contains so much
energy that it can ionize—knock electrons off—the atoms in
living tissues. This damages the way cells work, causing them
to die in large numbers, and can trigger cancers. Large alpha
particles can only get into the body through food or drink,
but can then cause a lot of damage. Gamma rays shine right
through, but are less likely to hit a molecule and cause damage.

▽ **Penetrating power**
Alpha particles are blocked by the skin, although they can cause
radiation burns. Beta particles bounce off thin sheets of metal,
while it takes a thick layer of lead to shield against gamma rays.

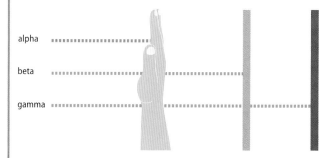

alpha

beta

gamma

REAL WORLD

Smoke detectors

Household smoke alarms
contain tiny—and safe—
amounts of americium, a
radioactive element made in
laboratories. The americium
ionizes the air inside.
A battery runs a current
through it. When smoke gets
in, the air is deionized and
the current is blocked,
triggering the alarm.

Half-life

Radioactive isotopes decay at a fixed rate that is measured as a half-life. This is the amount of time it takes for a sample to reduce its mass by half as it decays into other elements. Every radioactive isotope has a specific half-life. The more radioactive an isotope is, the shorter its half-life.

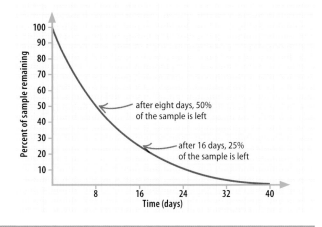

after eight days, 50% of the sample is left

after 16 days, 25% of the sample is left

▷ **Fixed decay rate**
The half-life of a radioactive substance is the same whether there is a lot of it or a little. Here, the half-life is eight days. After eight days, 50 percent of the original sample is left, after another eight days only 25 percent of the sample is left, and so on.

Decay series

Radioactive isotopes often decay into daughter atoms that are also radioactive. The decay continues through a series of radioactive isotopes until finally stable atoms are formed. Members of a decay series may exist for just a fraction of a second, while others are around for years, only gradually breaking down into the next element in line.

▷ **Uranium 238 series**
This is the most common isotope of uranium. It decays in a series of alpha and beta emissions producing a series of radioactive isotopes before reaching a stable form of lead (Pb, mass number: 206). Each isotope decays at a different rate so each has a different half-life.

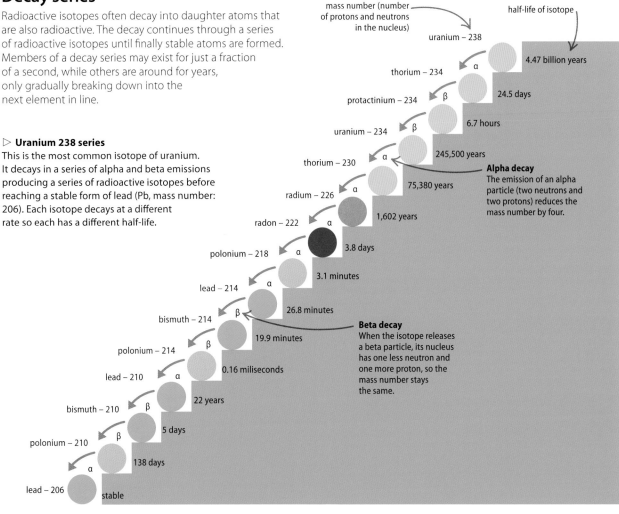

mass number (number of protons and neutrons in the nucleus)

half-life of isotope

uranium – 238

4.47 billion years

thorium – 234 α

protactinium – 234 β 24.5 days

uranium – 234 β 6.7 hours

thorium – 230 α 245,500 years

Alpha decay
The emission of an alpha particle (two neutrons and two protons) reduces the mass number by four.

radium – 226 α 75,380 years

radon – 222 α 1,602 years

polonium – 218 α 3.8 days

lead – 214 α 3.1 minutes

bismuth – 214 β 26.8 minutes

Beta decay
When the isotope releases a beta particle, its nucleus has one less neutron and one more proton, so the mass number stays the same.

polonium – 214 β 19.9 minutes

lead – 210 α 0.16 miliseconds

bismuth – 210 β 22 years

polonium – 210 β 5 days

lead – 206 α 138 days

stable

Chemical reactions

CHEMICAL REACTIONS ARE PROCESSES THAT CHANGE ONE
SET OF SUBSTANCES INTO ANOTHER.

New bonds are made and existing ones are broken in a chemical
reaction, which rearranges the atoms to form new substances.

Start and end points

At the starting point of a chemical reaction are substances called reactants.
Most reactions involve at least two reactants, although some involve only one
reactant. The reactants can be compounds or pure elements. When they come
into contact with each other, their ions and atoms are reorganized, resulting
in the formation of a new set of substances, known as the products.

Self-rising flour
produces gas bubbles
by a chemical reaction
to make cakes light.

▷ **Activating reaction**
During the reaction, the bonds
between atoms in the reactants
rearrange, which results in the
formation of the products. In
this case two reactants have
combined to form one product.

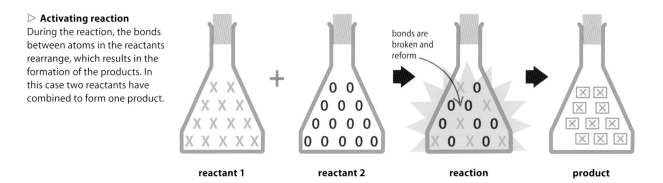

bonds are
broken and
reform

reactant 1   reactant 2   reaction   product

Conservation of matter

Atoms (or any other forms of matter) are neither created nor destroyed during chemical
reactions. Every atom that was part of the reactants is present in the products, even if
heat and flames are being released during the reaction. This principle is known as the
conservation of matter.

▷ **Reactants**
Sodium reacts with
water to produce the
compound sodium
hydroxide and bubbles
of hydrogen gas.

▷ **Products**
The weight of the
products—sodium
hydroxide and
hydrogen—is the
same as the weight
of the reactants.

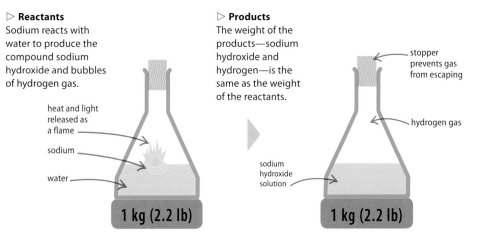

heat and light
released as
a flame

sodium

water

stopper
prevents gas
from escaping

hydrogen gas

sodium
hydroxide
solution

1 kg (2.2 lb)   **1 kg (2.2 lb)**

REAL WORLD

Sodas

The fizz of bubbles released
when a sparkling drink is
opened is produced by
a decomposition reaction.
Carbonic acid (H_2CO_3)
dissolves in the water and
breaks apart into carbon
dioxide gas—which makes
the refreshing bubbles—
and more water.

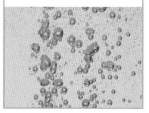

Equations

Chemists use equations to represent what is happening during chemical reactions. The formula of each reactant is written on the left-hand side, and those of the products are shown on the right. An arrow indicates the direction in which the reaction occurs.

▽ Chemical symbols

Instead of using the elements' names, their chemical symbols are shown.

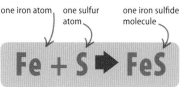

one iron atom — one sulfur atom — one iron sulfide molecule

$$Fe + S \rightarrow FeS$$

▽ Balanced equations

The number of atoms in the reactants is the same as the number of atoms in the products.

two hydrogen molecules have a total of four hydrogen atoms

$$2H_2 + O_2 \rightarrow 2H_2O$$

one oxygen molecule has two oxygen atoms — two water molecules still have four hydrogen atoms and two oxygen atoms

▽ Reaction conditions

The equation can also contain other information about the reaction, such as the state of the reactants and products.

two hydrogen chloride molecules — stands for "aqueous," or dissolved in water — two sodium atoms — stands for "solid" — two sodium chloride molecules — one hydrogen molecule has two hydrogen atoms — stands for "gas"

$$2HCl(aq) + 2Na(s) \rightarrow 2NaCl(aq) + H_2(g)$$

Types of reaction

There are three main types of chemical reactions. In a decomposition reaction, one complex product breaks apart into two (or perhaps more) simple products. In a synthesis reaction, two or more simple reactants join together to form a single, more complicated product. In displacement reactions, atoms or ions of one type swap places with those of another, forming new compounds.

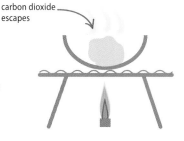

carbon dioxide escapes

◁ Decomposition reaction

Calcium carbonate ($CaCO_3$) decomposes into calcium oxide (CaO) and carbon dioxide (CO_2) when heated.

calcium carbonate — calcium oxide — carbon dioxide

$$CaCO_3 \rightarrow CaO + CO_2$$

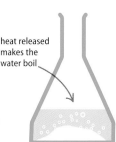

heat released makes the water boil

◁ Synthesis reaction

Calcium oxide (CaO) powder and water (H_2O) combine in a synthesis reaction to form calcium hydroxide $Ca(OH)_2$, which dissolves in the remaining water.

calcium oxide — water — calcium hydroxide

$$CaO + H_2O \rightarrow Ca(OH)_2$$

brackets show that there are two hydroxides (OH) joined to each calcium atom

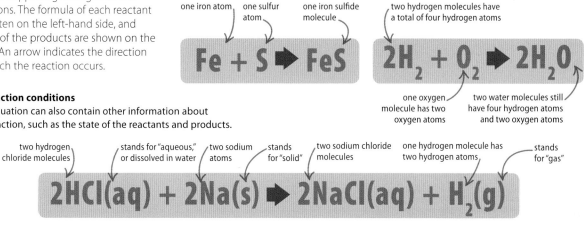

carbon dioxide is added by bubbling it through the water

◁ Displacement reaction

When heated gently, carbon dioxide (CO_2) displaces the hydroxide in calcium hydroxide ($Ca(OH)_2$) to make calcium carbonate ($CaCO_3$) and water (H_2O).

calcium hydroxide — carbon dioxide — calcium carbonate — water

$$Ca(OH)_2 + CO_2 \rightarrow CaCO_3 + H_2O$$

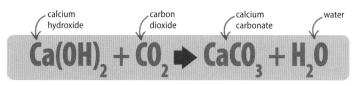
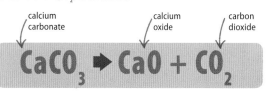
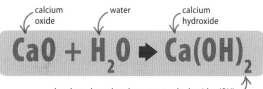

Combustion

COMBUSTION IS A REACTION THAT PRODUCES HEAT AND
LIGHT IN THE FORM OF FLAMES AND EXPLOSIONS.

Most combustion reactions involve oxygen, heat, and fuel. All of
these components are needed for the reaction to continue.

Heat and light

A flame is an area of hot glowing gases that have been released
by a combustion reaction. For example, a candle wick is soaked
with hot liquid wax that is a fuel and burns (undergoes a
combustion reaction with oxygen in the air). The products of
the reaction are carbon dioxide gas and water vapor. These
glow briefly as they are released, contributing to the flame.

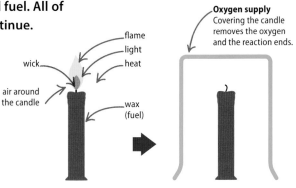

Oxygen supply
Covering the candle
removes the oxygen
and the reaction ends.

flame
light
heat
wick
air around
the candle
wax
(fuel)

Gas tests

The gases commonly produced
in chemistry experiments often
look exactly the same. It may
be dangerous to smell them—
even if they have a characteristic
odor. Chemists use combustion
tests to identify the three most
easily confused gases—hydrogen,
oxygen, and carbon—in a safe way.
A sample of each gas is exposed
to a burning splint—a strip of
dried wood used in a lab—and
the gas can be identified by the
characteristic way it combusts.

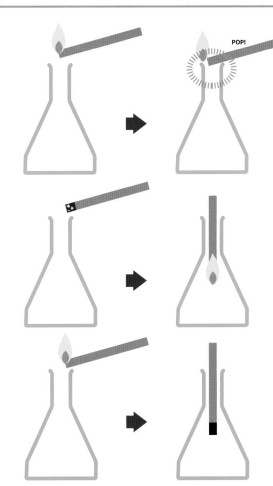

POP!

◁ **Hydrogen**
Hydrogen is very flammable and
burns very quickly. A burning
splint will pop before it even
enters the flask as the hydrogen
rushes out to the flame.

◁ **Oxygen**
Oxygen is the gas that fuels
combustion. If a smouldering
splint is exposed to a flask filled
with oxygen, the wood will
reignite and burst into flames.

◁ **Carbon dioxide**
Carbon dioxide is a common
product of combustion
reactions but it does not
burn itself. A burning splint
will go out when it is exposed
to carbon dioxide.

Combustion—and
its fire—was the first
chemical reaction
that humans learned
to control.

Fuels

A fuel is a substance that burns readily and releases useful energy in the form of heat. Most fuels are carbon compounds. All fuels need to be handled with care so they do not burn too fast. Uncontrollably fast combustions create explosions in which large amounts of energy are released in a very short time.

△ **Wood**
Probably the first fuel used by humans, wood is largely cellulose, made from carbon molecules. Most dried wood burns at about 300°C (572°F), although some types get a lot hotter than this. Other materials are released as smoke when wood burns.

△ **Coal**
Coal is a flammable rock made from the remains of ancient trees exposed to pressure and heat over time. Its main constituent is pure carbon, although there are many other impurities, including sulfur. Most coals can burn at about 700°C (1,292°F).

△ **Methane**
Methane, or natural gas, is a simple gas made from hydrocarbons (see page 158). It is found in underground gas fields. It is also produced by natural processes in marshy areas and in the stomachs of herbivores. Its abundance makes it a very popular fuel.

△ **Gasoline**
Gasoline is a flammable liquid made from hydrocarbons, chiefly octane (C_8H_{18}). It is refined from crude oil (see page 157). The liquid is easy to store in tanks and pump around. It also ignites more easily than other fuels, even from its fumes.

△ **Paraffin wax**
Paraffin wax is a solid and is also made from refined crude oil. The solid does not burn easily, but when melted the liquid wax will ignite. Once lit, the process is self-sustaining—the heat of the combustion melts more of the solid into flammable liquid.

Fire control

Firefighters tackle fires using an understanding of combustion reactions. The fire triangle is a simple way of expressing the three things needed for combustion to continue: oxygen, heat, and fuel. Taking one of these components away will make the reaction end—and the fire go out.

▷ **Oxygen**
This gas is one of the reactants in the combustion reaction. Firefighters smother a fire with foam, sand, or a blanket to cut off the oxygen supply.

◁ **Heat**
The heat released by the reaction is used to power the combustion of more fuel and oxygen. Adding water will cool the reaction and reduce its energy.

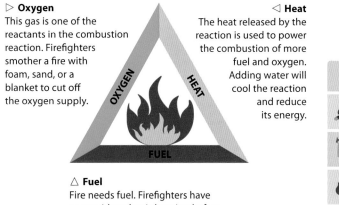

△ **Fuel**
Fire needs fuel. Firefighters have to consider what is burning before deciding how best to extinguish the fire safely and effectively.

▽ **Fire extinguishers**
Different fire extinguishers are designed to tackle fires fueled by different types of substances. For example, water is not used on burning liquid because the hot fuel bubbles up through it, making the water boil and spray the burning fuel into the air.

Type of fire	Fire extinguishers			
	Water	Foam	CO$_2$	Powder
paper, wood, textiles, and plastics	✓	✓		✓
flammable liquids		✓	✓	✓
flammable gases				✓
electrical equipment			✓	✓

Redox reactions

IN A REDUCTION-OXIDATION (REDOX) REACTION, ELECTRONS ARE TRANSFERRED FROM ONE ATOM TO ANOTHER.

SEE ALSO

❬ **112–113** Ionic bonding	
Electrochemistry	**148–149** ❭
Refining metals	**152–153** ❭
Electric currents	**203** ❭

A redox reaction is one in which the oxidation state of one reactant rises as the oxidation state of the other falls to balance it. The oxidation state is the number of electrons added to or taken from an atom.

Oxidation states

Chemical reactions occur because most atoms have an incomplete outer shell of electrons, which makes them electrically unstable. To fill their outer shell they form bonds with other atoms in which they accept or donate electrons. The number of electrons that an atom needs to lose or gain to make itself stable is called its oxidation state. Any uncombined element has an oxidation state of zero.

+3							
+2	Cu^{2+}	Zn^{2+}					
+1			Na^+				
0	Cu	Zn	Na	0	Cl		
-1					Cl^-		
-2				O^{2-}			

oxidation (loss of electrons) → reduction (gain of electrons)

△ **Positive or negative?**
The oxidation state, or number, shows the number of electrons that are gained or lost when an atom changes to an ion (see pages 112–113). The oxidation state of all uncombined elements is zero, as is the sum of the oxidation numbers in a neutral compound. For simple ions, the oxidation state is the same as the electrical charge of the ion.

Changing oxidation state

When atoms or ions undergo a reaction, their oxidation state changes. For example, the oxidation state of sodium changes from 0 to +1 because it has one electron in its outer shell to give away. It is easier for it to donate the electron than to try to fill up its shell with seven more electrons. On the other hand, chlorine has an oxidation state of -1 because it is lacking one electron to complete its outer shell.

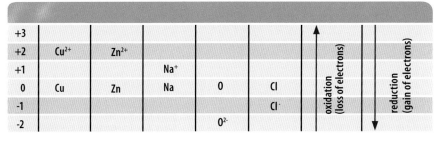

sodium has one electron in its outer shell

chlorine lacks one electron in its outer shell

$$Na^+ + Cl^- \rightarrow NaCl$$

△ **Sodium chloride**
Sodium and chlorine make an ideal pair to form a compound because sodium has an electron to donate to chlorine.

Oxidation and reduction

When an atom (or ion) loses electrons during a chemical reaction it is said to have been "oxidized." The term "oxidation" originally applied to reactions where oxygen had combined with another substance, but now it is used in any reaction where electrons are donated. The atom or ion that gains electrons is said to have been "reduced." All redox reactions happen in pairs—for every reduction reaction there is a corresponding oxidation reaction.

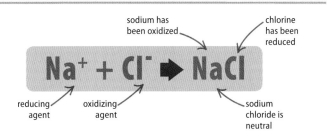

sodium has been oxidized

chlorine has been reduced

$$Na^+ + Cl^- \rightarrow NaCl$$

reducing agent

oxidizing agent

sodium chloride is neutral

△ **Oxidizers and reducers**
The atom or ion that accepts the electrons is called the oxidizing agent, oxidant, or oxidizer. The atom or ion that donates the electrons is called the reducing agent.

Electrochemistry

The exchange of electrons that occurs in redox reactions can be used to create an electric current in an apparatus called an electrochemical cell. The current forms when electrons are released from the oxidation reaction and made to travel to the reduction reaction, which will absorb the electrons. In this experiment, a piece of zinc metal is dipped in zinc sulfate and a piece of copper is put in copper sulfate. These two metals are connected by a conducting wire. The sulfate solutions have free ions, which can carry an electric current (see page 148).

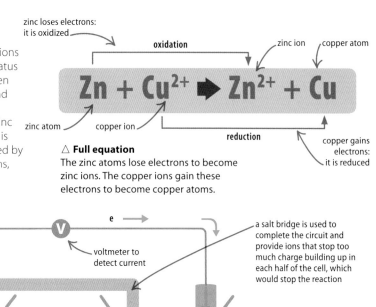

zinc loses electrons: it is oxidized

oxidation

zinc ion   copper atom

$$Zn + Cu^{2+} \rightarrow Zn^{2+} + Cu$$

zinc atom   copper ion

reduction

copper gains electrons: it is reduced

△ **Full equation**
The zinc atoms lose electrons to become zinc ions. The copper ions gain these electrons to become copper atoms.

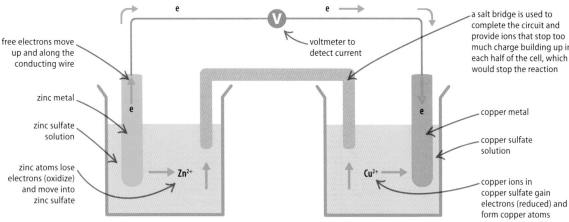

free electrons move up and along the conducting wire

voltmeter to detect current

a salt bridge is used to complete the circuit and provide ions that stop too much charge building up in each half of the cell, which would stop the reaction

zinc metal

zinc sulfate solution

zinc atoms lose electrons (oxidize) and move into zinc sulfate

Zn^{2+}

Cu^{2+}

copper metal

copper sulfate solution

copper ions in copper sulfate gain electrons (reduced) and form copper atoms

△ **Oxidation cell**
In the left half of the cell, oxidation occurs. The zinc metal atoms lose electrons to form zinc ions. The free electrons travel up and along the conducting wire to the reduction cell. The zinc ions (Zn^{2+}) move into the zinc sulfate solution.

△ **Reduction cell**
In the right half of the cell, reduction happens. Copper ions (Cu^{2+}) in the copper sulfate solution move to the piece of copper and accept two electronseach that have arrived from the oxidation cell. The copper ions thus become copper atoms.

Corrosion

A familiar phenomenon involving redox reactions is corrosion, in which metals and other materials are oxidized. Corrosion takes place in damp conditions, involving a reaction with oxygen or carbon dioxide and occasionally with pollutants such as hydrogen sulfide.

▷ **Types of corrosion**
The products of the reaction, such as rust, cause discolouration and weaken the original object.

Metal	Corrosion	Chemical name	Description
iron	rust	hydrated iron oxide	flaky rust expands and cracks the metal
copper	verdigris	copper carbonate	turns objects gray-green
aluminum	alumina	aluminum oxide	forms a dull layer on metal
silver	tarnish	silver sulfide	makes silver dark and dull
gold	no corrosion	none	gold always stays shiny

Energy and reactions

A LOOK AT THE WAY ENERGY IS INVOLVED
IN CHEMICAL REACTIONS.

All chemical reactions require energy. Energy is needed to begin breaking and reforming atomic bonds. Most reactants need energy added to them before they will react.

Activation energy

The activation energy is the amount of energy that a reaction needs to begin. It is like a hill that the reactants have to get over. A reaction between a strong acid and alkali has low activation energy. It occurs as soon as the reactants are mixed because the molecules have enough energy already. The combustion of coal has a higher activation energy, so coal must be heated (adding energy) before it will burst into flames.

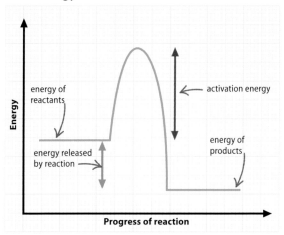

▷ **Energy graph**
When the energy involved in a reaction is shown as a graph, the activation energy forms a hump, over which the reactants must pass to form products.

Exothermic reaction

Chemical reactions need energy to begin, but they also release energy as the reactants reorganize into products. When the amount of energy released is greater than the activation energy, the reaction is exothermic. Exothermic reactions, such as the combustion of fuels, heat the surroundings with this release of energy.

▽ **Energy released**
In an exothermic reaction the energy in the products is lower than that in the reactants. This is because energy is lost as heat during the reaction.

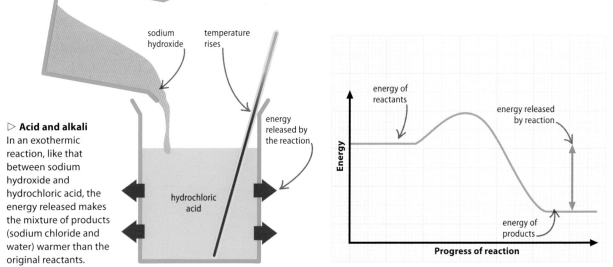

▷ **Acid and alkali**
In an exothermic reaction, like that between sodium hydroxide and hydrochloric acid, the energy released makes the mixture of products (sodium chloride and water) warmer than the original reactants.

Endothermic reaction

When the amount of energy released during a reaction is less than the activation energy, the process is decribed as endothermic. Because more energy is going into the reaction than is coming out, the reaction mixture and its surroundings become colder as their energy is taken in by the reaction.

▽ **Energy taken in**
In an endothermic reaction the products have more energy than the reactants. This is because energy is taken in from the surroundings during the reaction.

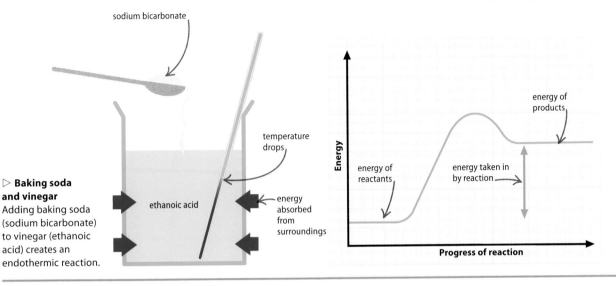

sodium bicarbonate

temperature drops

ethanoic acid

▷ **Baking soda and vinegar**
Adding baking soda (sodium bicarbonate) to vinegar (ethanoic acid) creates an endothermic reaction.

energy absorbed from surroundings

energy of reactants

energy taken in by reaction

energy of products

Energy

Progress of reaction

Calorimeter

All the energy used during a chemical reaction can be measured using a calorimeter. A reaction takes place in a central chamber, which is surrounded by water. The calorimeter is completely cut off from the outside, so any changes (rises and falls) in the water temperature can only be a result of the reaction taking place.

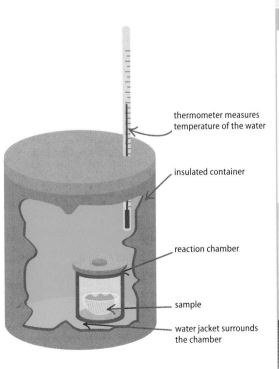

thermometer measures temperature of the water

insulated container

reaction chamber

sample

water jacket surrounds the chamber

▷ **Bomb calorimeter**
This device is used to measure the energy in substances, including different foods. The sample is burned in pure oxygen and the amount of energy released is proportional to the rise in water temperature.

Hand warmers

Exothermic reactions are a convenient way of producing heat. Hand-warming packets and self-cooking cans contain two reactants in separate containers. When the hand warmer is bent in half, the containers rupture, mixing the reactants. Their reaction produces harmless products and enough heat to keep hands warm on cold days.

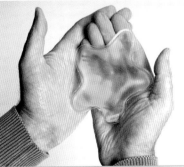

Rates of reaction

REACTANTS TURN INTO PRODUCTS AT DIFFERENT RATES.

Reaction rates depend on the substances involved. Dynamite burns so rapidly that it explodes in a fraction of a second, while an iron nail takes years to turn to rust.

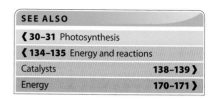
Measuring rates

To understand what controls the speed of a reaction, a chemist first needs to be able to measure its rate—how quickly the reactants are converting into products. Only one product needs to be measured, since any others are produced at the same rate.

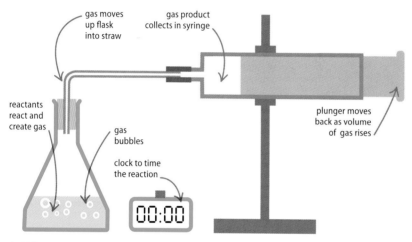

gas moves up flask into straw

gas product collects in syringe

plunger moves back as volume of gas rises

reactants react and create gas

gas bubbles

clock to time the reaction

00:00

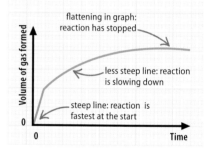

Volume of gas formed

flattening in graph: reaction has stopped

less steep line: reaction is slowing down

steep line: reaction is fastest at the start

0

0   Time

△ **Using a syringe to measure gas**
Measuring the increase in the volume of a gas product is relatively simple using a syringe. The measurements can be taken at regular intervals, timed by the clock. The volume will increase at a rate that is proportional to the reaction.

△ **Rate of reaction graph**
The volume measurements can be plotted against time on a graph. The steep line at the beginning shows that the rate of reaction starts very fast but then tails off.

Reactivity and temperature

Every reaction has an activation energy, which is the amount of energy the reactants need in order to break and reform atomic bonds. When a reaction has low activation energy, more of the reactants have the amount of energy needed and so the reaction occurs more quickly than a reaction with a higher activation energy. Heating the reactants—and increasing the pressure—adds energy and increases the rate of reaction.

▷ **Magnesium in water**
In cold water, magnesium reacts very slowly, forming magnesium oxide and bubbles of hydrogen. Heating the water to near its boiling point makes the same reaction run more quickly, making the water fizz with hydrogen bubbles.

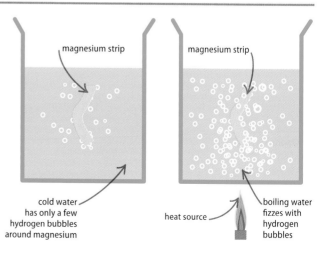

magnesium strip

magnesium strip

cold water has only a few hydrogen bubbles around magnesium

heat source

boiling water fizzes with hydrogen bubbles

Concentration

Concentration is a measure of how much of a substance is present in a certain volume of a mixture. The rate of reaction is proportional to the concentrations of the reactants. Even if one reactant is present in large amounts, the reaction will only speed up as more of the other is added.

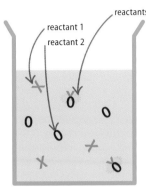

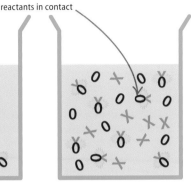

reactants in contact

reactant 1
reactant 2

△ **Low concentration**
Reactants must make contact with each other to react. When reactants are mixed in low concentrations, they are widely dispersed and come into contact with each other infrequently.

△ **High concentration**
When reactants are mixed in higher concentrations, their molecules are less spread out and come into contact with each other more frequently. As a result, the rate of reaction is higher.

Particle size

When a solid reactant is added to a liquid or dissolved reactant, the reaction will proceed faster if the solid is crushed into powder rather than added as a single lump. The liquid reactant reacts with the surface of the solid, and the powdered reactant has a larger combined surface area than the single mass.

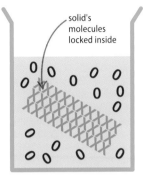

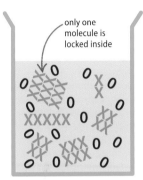

solid's molecules locked inside

only one molecule is locked inside

△ **Large solid, small area**
When the solid reactant is added as a big lump, the liquid reactant has fewer opportunities to react it. This is because many of the solid's molecules are locked away inside, out of reach of the liquid reactant.

△ **Small solids, large area**
The same amount of solid reacts much more quickly when broken up into smaller sizes. This is because more of its molecules are made available to take part in the reaction.

Light

Some reactions speed up when exposed to bright light or other higher energy forms of radiation, such as ultraviolet (UV) light. The reactants absorb the energy from certain wavelengths and this is enough to give them the activation energy to begin reacting. These reactions are called photochemical reactions. Photosynthesis, used in plants to turn carbon dioxide and water into glucose, needs light. Without it, the rate of reaction is negligible.

▷ **Ozone layer**
The reaction that creates ozone, a form of oxygen with three atoms per molecule happens mostly where high-energy light hits the high atmosphere. The ozone forms a layer in the high atmosphere and helps to absorb the dangerous UV rays in sunlight.

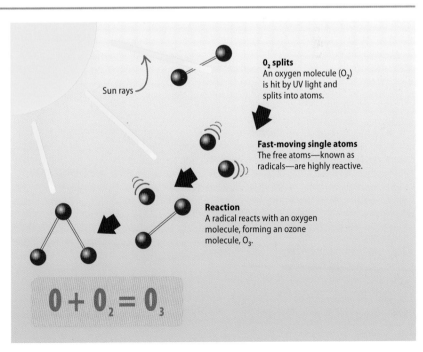

Sun rays

O_2 splits
An oxygen molecule (O_2) is hit by UV light and splits into atoms.

Fast-moving single atoms
The free atoms—known as radicals—are highly reactive.

Reaction
A radical reacts with an oxygen molecule, forming an ozone molecule, O_3.

$$O + O_2 = O_3$$

Catalysts

CATALYSTS SPEED UP REACTIONS BY LOWERING
THE ACTIVATION ENERGY REQUIRED.

SEE ALSO

❮ **67** Digestive chemicals

❮ **134** Activation energy

❮ **136** Reactivity and temperature

Energy   **170–171** ❯

Various catalysts used in laboratories and industry make chemical reactions
run faster and allow unreactive materials to get involved in reactions.
The enzymes that control reactions in cells are also catalysts.

Less energy needed

Many reactions have activation energies
that are so high that the reactions
never happen on their own—or
happen so slowly that they are hardly
noticeable. Catalysts make such
reactions possible by reducing the
activation energy needed.

▷ **Energy graph**
Catalysts reduce the energy barrier
between reactants and products.
In industry, a catalyst can make
reactions more economical.

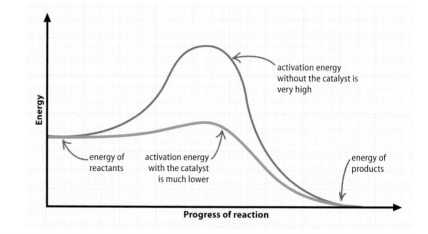

activation energy
without the catalyst is
very high

energy of
reactants

activation energy
with the catalyst
is much lower

energy of
products

Progress of reaction

Energy

How catalysts work

Catalysts are a highly varied group
of materials. Many are porous
substances with tiny spaces inside
where the reactants are brought
together in such a way that they
react without the need for a lot of
energy. The precise mechanisms
vary but usually a catalyst facilitates a
reaction by providing an intermediary
phase between the reactants and the
products. The catalyst is involved in
the reaction but not consumed by it.

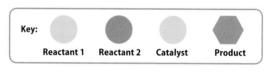

Key: Reactant 1   Reactant 2   Catalyst   Product

▷ **Reactant 1 bonds with catalyst**
One reactant bonds temporarily
with the catalyst, forming a
complex molecule.

▷ **Reactant 2 joins in**
The molecule bonds with the
other reactant, bringing the
reactants close together.

▷ **Product forms**
While held in this way, the reactants
need much less energy to react.
The product can form easily.

▷ **Catalyst breaks away**
The product breaks from the catalyst,
which is unchanged by the reaction
and available to repeat its role.

The word **"catalyst"**
comes from the
Greek word meaning
"to untie."

Enzymes

Most of the chemical reactions that take place inside living bodies would not happen without the catalytic effect of enzymes. Enzymes are protein molecules that are highly folded into shapes specific to their roles. These shapes create an area known as the enzyme's active site. The reactants—known as substrates in biochemistry—are also molecules with complex shapes. They fit onto the enzyme's active site, where the reaction takes place. Enzymes are used in the digestive system to break down large molecules of food into smaller ones.

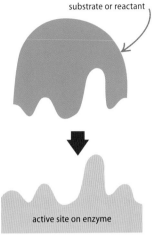

substrate or reactant

bonds in substrate are weakened

products leave the active site

active site on enzyme

active site on enzyme

active site on enzyme

△ **Active site**
Only a specific substrate can bond to a specific enzyme's active site, like a key fitting into a lock.

△ **Catalyzed reaction**
When joined to the enzyme, the chemical reaction can take place. Bonds within the substrate are weakened.

△ **Products produced**
The substrate divides into two products. These break off the active site, leaving it free to collect a new substrate.

Catalytic converter

Every new car is fitted with a catalytic converter, or "cat." The engine exhaust passes through this device before it enters the air. Inside is a honeycomb-shaped ceramic coated with a thin layer of a platinum and rhodium alloy, which is the catalyst. The catalyst changes dangerous gases, such as carbon monoxide (CO), nitrogen oxides (NO), and unburned hydrocarbons, into comparatively harmless ones—carbon dioxide (CO_2), nitrogen (N_2), and water (H_2O).

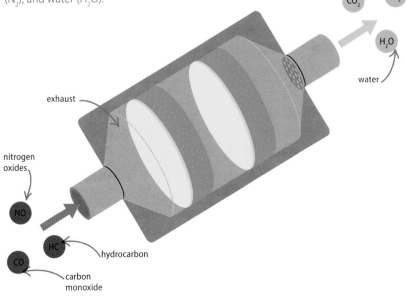

carbon dioxide

nitrogen gas

CO_2

N_2

H_2O

water

exhaust

nitrogen oxides

NO

HC

hydrocarbon

CO

carbon monoxide

Margarine

Margarine is made using a catalyst. The starting materials are vegetable oils: long chain molecules made from carbon and hydrogen. The oils are unsaturated—their molecules have room for more hydrogen atoms. Hydrogen is bubbled through the oil over a nickel catalyst, which saturates the oil molecules, turning them into a butterlike solid.

Reversible reactions

SOME REACTIONS CAN BE REVERSED.

In general, chemical reactions run in just one direction. The energy that is required to turn the products back into reactants is just too great for it to happen. However, some reactions are easily reversible.

SEE ALSO	
❮ **100–101** Changing states	
❮ **102–103** Gas laws	
❮ **128–129** Chemical reactions	
❮ **134–135** Energy and reactions	
Pressure	**184–185** ❯

Two-way reactions

A reversible reaction is one that can go backward as well as forward—products that form can easily be turned back into the original reactants. The amounts of energy needed to make the reaction run in either direction are rarely equal, but there is not a large difference between the two. A common reversible reaction is to use heat to drive water from a solid. This is reversed by simply adding water.

▷ **Hydrating crystals**
Copper sulfate crystals are blue due to water molecules locked inside them. Heating the crystals drives out water and they turn into the white anhydrous (without water) form. However, adding water easily reverses the process.

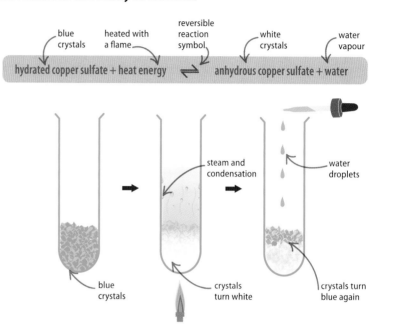

blue crystals — heated with a flame — reversible reaction symbol — white crystals — water vapour

hydrated copper sulfate + heat energy ⇌ anhydrous copper sulfate + water

steam and condensation — water droplets

blue crystals — crystals turn white — crystals turn blue again

Dynamic equilibrium

Reversible reactions do not normally run one way and then the other. They run in both directions simultaneously. However, it is the rate of reaction in each direction that dictates the yield (the proportions of reactants and products). When the rate of reaction in both directions is the same, the reaction is in equilibrium.

Key   Reactants   Products

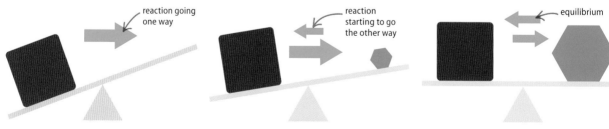

reaction going one way

reaction starting to go the other way

equilibrium

△ **One way**
At the start of the reaction, few products have formed so the reaction runs in one direction. The high concentration of reactants makes the rate of reaction high.

△ **Mostly one direction**
Although there is still a high concentration of reactants, the reaction is starting to go backward. However, the concentration of products is increasing.

△ **Dynamic equilibrium**
Products are being made at the same speed as they turn back into reactants but the reactions continue. This stage in the reaction is called dynamic equilibrium.

Temperature

If a change such as temperature is made to a reaction in equilibrium, the speed of the forward or backward reaction will adjust to counter the effects of that change. Every reversible reaction has an exothermic direction (giving out energy) and an endothermic one (taking in energy). If heated, the reaction that takes in heat (the endothermic direction) will speed up to balance the effect of the heating.

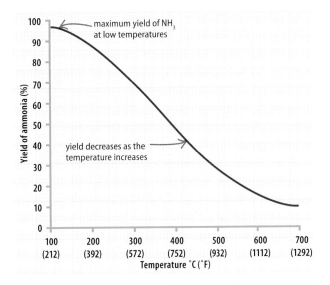

maximum yield of NH_3 at low temperatures

yield decreases as the temperature increases

Yield of ammonia (%)

Temperature °C (°F)

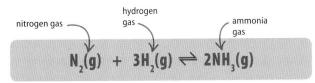

nitrogen gas — hydrogen gas — ammonia gas

$$N_2(g) \; + \; 3H_2(g) \rightleftharpoons 2NH_3(g)$$

▷ **Making ammonia**
Nitrogen and hydrogen give out heat when they react to form ammonia. Adding heat reduces the yield of ammonia because more of the compound decomposes back into hydrogen and nitrogen.

Pressure

Pressure also affects the equilibrium in reactions involving gases. Pressure is caused by gas molecules hitting the sides of the container. The more molecules there are, the higher the pressure in the container. Increasing the pressure during a reversible reaction shifts the equilibrium toward the side with fewer molecules. In the equation for making ammonia there are four molecules of reactants (one molecule of nitrogen and three molecules of hydrogen) and two molecules of product (ammonia). An increase in pressure favors the forward direction of the reaction, which produces ammonia, rather than the reverse.

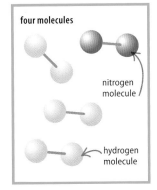

four molecules

nitrogen molecule

hydrogen molecule

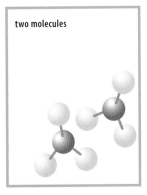

two molecules

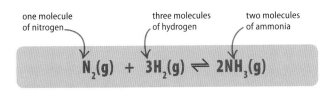

one molecule of nitrogen — three molecules of hydrogen — two molecules of ammonia

$$N_2(g) \; + \; 3H_2(g) \rightleftharpoons 2NH_3(g)$$

△ **Reactants**
Increasing the pressure pushes the hydrogen and nitrogen molecules together and drives the reaction to produce ammonia.

△ **Products**
Only two ammonia molecules are produced, which take up less space than the reactants. This makes the pressure fall.

Photosynthesis is a reversible reaction. If there is too much sugar and oxygen in a plant cell, **photorespiration** occurs, turning them back into carbon dioxide and water.

REAL WORLD

Quicklime

Quicklime is a chemical made by heating calcium carbonate. The heat makes the carbonate decompose into quicklime and carbon dioxide. However, these two products can then react back into calcium carbonate. To stop this, the kiln draws carbon dioxide away from the quicklime.

Water

ONE OXYGEN ATOM AND TWO HYDROGEN ATOMS
BOND TO FORM THE COMPOUND WATER.

Water is one of the very few natural substances that
are liquid in everyday conditions. Its unusual properties
stem from the oxygen atom in its molecules.

Hydrogen bonds

In addition to the covalent bonds that join
the hydrogen and oxygen atoms in a water
molecule, there are bonds between the
molecules themselves. The electrons in the
covalent bond are pulled closer to the oxygen
atom than to the hydrogen atoms. This makes
the oxygen atom negatively charged and
the hydrogen atoms positive. These opposite
charges on different molecules attract each
other in what is known as a hydrogen bond.

▷ **Dipole** The areas of charge
in the water molecule are called
poles. A hydrogen bond is
formed between the negative
end of one molecule and
the positive end of another.

the oxygen atom forms
the negative end of
the dipole

positive and
negative ends
attract each other
and form a
hydrogen bond
between the
molecules

each hydrogen atom
forms the positive end
of the dipole

States of water

On Earth, water is mainly liquid—
it covers 70 percent of the planet's
surface. However, the other states
of water are just as familiar—polar
regions are covered in ice, while the
atmosphere is filled with water vapor.
Like all gases, water vapor has a lower
density than water. However, almost
uniquely among natural substances,
when water freezes into ice (solid)
it expands and has a lower density.
As a result, ice floats on water. With
other substances, their solid states
have higher densities and sink
under their liquid states.

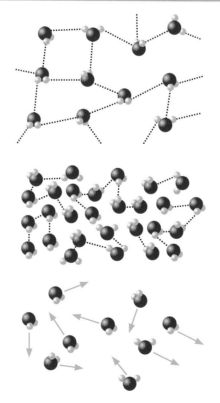

◁ **Ice**
An ice crystal is held together with
hydrogen bonds. As the crystal forms,
the molecules spread out so each molecule
can bond with three others. This makes
the molecules take up a larger volume.

◁ **Liquid water**
In liquid water there are fewer hydrogen
bonds. The molecules can get closer to each
other, taking up less volume. The bonds
break and reform, allowing the molecules
to move around.

Water is **densest at 4°C
(39°F)**, which is the
temperature at the
bottom of the deepest
ocean floors.

◁ **Water vapor**
In the gaseous form, the water molecules
of water are independent of each other
and can move around freely. Water vapor
forms below the boiling point of water,
while steam is technically vapor above
100°C (212°F).

Universal solvent

Water is sometimes known as the universal solvent because so many substances dissolve in it. The property is another result of water molecules' polarity. Ionic substances are made up of charged particles bonded together. When they are added to water, the ions split from their opposite partners and form bonds with the positive and negative ends of the water molecules.

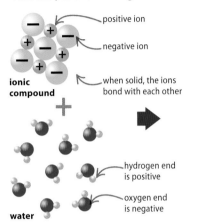

positive ion

negative ion

ionic compound

when solid, the ions bond with each other

hydrogen end is positive

oxygen end is negative

water

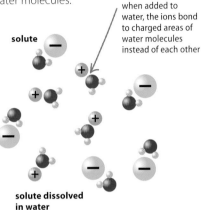

solute

when added to water, the ions bond to charged areas of water molecules instead of each other

solute dissolved in water

The Dead Sea

When things dissolve in water they make it more dense. Seawater is more dense than freshwater because it has salt dissolved in it. The water in the Dead Sea is so salty that it is much denser than the human body. That is why bathers can float so easily.

Water hardness

Hardness is the term used to describe how many minerals are dissolved in water. Temporary hardness is largely due to dissolved calcium hydrogen carbonate. When the water is heated it decomposes into carbon dioxide, water, and solid calcium carbonate. This solid is called limescale and builds up on heating equipment like kettles. Other calcium (and magnesium) compounds form permanent hardness. They can affect the taste of drinking water and the action of soaps. A water softener replaces minerals that cause hardness with sodium ions, which do not cause as many problems.

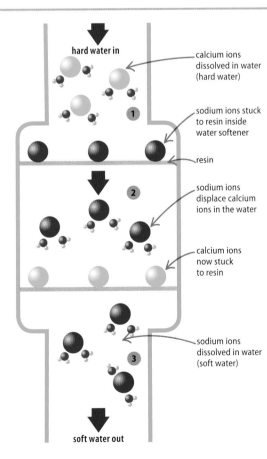

hard water in

calcium ions dissolved in water (hard water)

sodium ions stuck to resin inside water softener

resin

sodium ions displace calcium ions in the water

calcium ions now stuck to resin

sodium ions dissolved in water (soft water)

soft water out

Key   Calcium ion   Sodium ion

1. Hard water in
Dissolved calcium ions make the water hard. This hard water is fed into the softener before it reaches the tap.

2. Inside the softener
The softener contains a porous resin filled with sodium ions. As the hard water flows though, the sodium ions displace (swap places with) the less reactive calcium ions in the water.

3. Soft water out
The water flowing out of the softener contains sodium ions, while the calcium ions are left behind in the resin. The resin needs to be replaced regularly or washed with a sodium solution to restock the sodium ions.

Acids and bases

ACIDS AND BASES ARE CHEMICAL OPPOSITES, BUT THESE
TWO TYPES OF COMPOUNDS ARE CLOSELY RELATED.

SEE ALSO

❰ **89** Acid rain

❰ **112–113** Ionic bonding

❰ **114–115** Covalent bonding

❰ **120** Why alkali?

❰ **136–137** Rates of reaction

The chemistry of acids and bases is driven by hydrogen ions. Acids
are substances that produce hydrogen ions; bases are substances
that react with acids by accepting these hydrogen ions.

What is an acid?

Acids are compounds that release positively charged particles of hydrogen,
called hydrogen ions (H^+), when dissolved in water. These ions are highly reactive
and can bond to other substances and have a corrosive effect on them. The
strength of an acid depends on the number of hydrogen ions that it can release.

DNA, the chemical that
carries genetic code, is
a type of acid.

▽ **Strong acids**
The most powerful acids are ionically bonded compounds (see
page 112). They split into hydrogen and other ions completely when
dissolved in water, thus releasing large quantities of free hydrogen ions.

Name	Formula	Where it is found
hydrochloric acid	HCl	the stomach
sulfuric acid	H_2SO_4	car batteries
nitric acid	HNO_3	process to make fertilizers

▽ **Weak acids**
Acids which have a covalent structure (see page 114) do not break up
into ions so easily. They have complex molecules, but sometimes the
bond holding a certain hydrogen ion weakens, allowing it to break off.

Name	Formula	Where it is found
citric acid	$C_6H_8O_7$	lemon juice
ethanoic acid	CH_3COOH	vinegar
formic acid	HCOOH	ant stings

What is a base?

A base is a compound that reacts with an acid by accepting its
hydrogen ions. The most reactive bases are soluble compounds
called alkalis. As it dissolves, an alkali releases hydroxide ions
(OH^-). The hydrogen and hydroxide ions combine very readily
to form water, so the reaction between an acid and an alkali is
often vigorous.

▽ **Common alkalis**
Any compound with a hydroxide ion is known as an alkali.
Alkalis are used in industry to make soaps and are added to
waste to help it decay more rapidly.

Name	Formula	Where it is found
sodium hydroxide	NaOH	oven cleaner
magnesium hydroxide	$Mg(OH)_2$	indigestion tablets
potassium hydroxide	KOH	soap

Neutralization

The reaction between an acid and an alkali (or other base,
such as an oxide) is called neutralization, because it results in
products that are neither acid nor alkali. One of the products is
always water. The other, known as the salt, is a compound
formed from the left-over ions.

▽ **General equation**
An acid and base always react to produce a salt and water. The salt
produced when hydrochloric acid (HCl) reacts with sodium hydroxide
(NaOH) is sodium chloride (NaCl)— better known as common salt.

$$\text{acid + base} \longrightarrow \text{salt + water}$$
$$\text{HCl(aq) + NaOH(aq)} \longrightarrow \text{NaCl(aq)} + H_2O(l)$$

Measuring acidity

Acidity is measured in pH, which stands for "power of hydrogen." Neutral substances such as water have a pH of 7. A substance with a pH lower than this is acidic; one with a pH higher than this (up to 14) is alkaline. The pH measures the concentration of hydrogen ions. Each whole pH number on the scale is ten times more acidic or basic than the previous number. For example, a substance with a pH of 6 has ten times more hydrogen ions in it than water, which has a pH of 7.

▷ **Indicators**

Chemicals used to test whether a substance is acid or alkaline are called indicators. Litmus was the first indicator used to show pH. It produces a red color for acid and blue for alkali. However its range of colors is limited, making it hard to judge the precise pH. A dye called universal indicator is more practical and produces a wide range of colors indicating where the solution fits into the pH scale.

REAL WORLD

Indigestion tablets

The discomfort of indigestion is caused by digestive acids leaking out of the stomach into the esophagus. This causes a burning sensation as it attacks the soft lining of the throat. Antacid tablets contain alkalis—often magnesium hydroxide—that neutralize these acids into harmless salts.

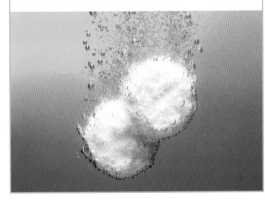

Rainwater is slightly acidic because carbon dioxide dissolves in it, making carbonic acid.

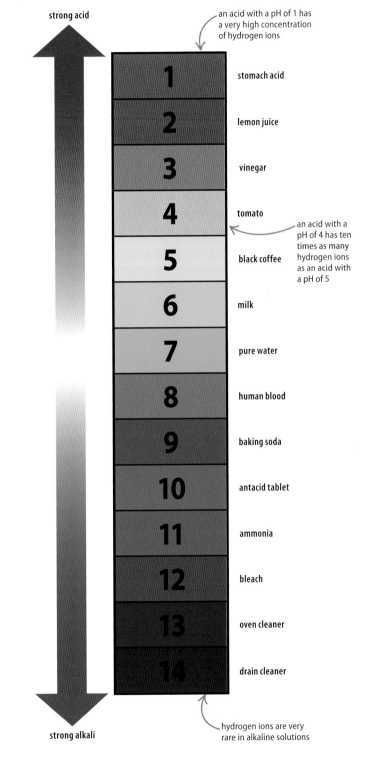

strong acid

an acid with a pH of 1 has a very high concentration of hydrogen ions

1	stomach acid
2	lemon juice
3	vinegar
4	tomato
5	black coffee
6	milk
7	pure water
8	human blood
9	baking soda
10	antacid tablet
11	ammonia
12	bleach
13	oven cleaner
14	drain cleaner

an acid with a pH of 4 has ten times as many hydrogen ions as an acid with a pH of 5

strong alkali

hydrogen ions are very rare in alkaline solutions

Acid reactions

ACIDS REACT WITH A RANGE OF OTHER SUBSTANCES
IN PREDICTABLE WAYS.

Although acids come in many forms, they all react in the
same way. When any acid is added to metals, oxides, or other
compounds, the reaction generates the same set of products.

Acids and metals

If a metal is more reactive than the hydrogen in an acid, they
will react to form a salt and hydrogen gas. The most reactive
metals, such as potassium, even do this with water—which
contains hydrogen but is, by definition, neutral. Metals such
as copper and gold are less reactive than hydrogen, so they
do not react with most acids.

▽ **General equation**
Iron displaces the hydrogen in the sulfuric acid
(H_2SO_4), forming a salt, iron sulfate ($FeSO_4$), which
is an ionically bonded compound. The hydrogen has
nothing else to react with, so it is released as a gas.

$$acid + metal = salt + hydrogen$$
$$H_2SO_4(aq) + Fe\ (s) \blacktriangleright FeSO_4(aq) + H_2(g)$$

▽ **Iron plus sulfuric acid**
When solid iron is added to the acid, the mixture begins to
fizz with hydrogen bubbles. The iron sulfate salt dissolves
forming a green solution.

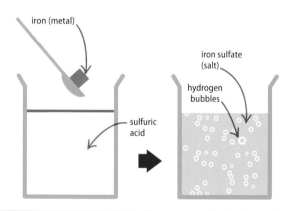

iron (metal)

iron sulfate
(salt)

hydrogen
bubbles

sulfuric
acid

REACTION OF METALS			
Name	**Reacts with water**	**Reacts with most acids**	**Level of reaction**
potassium	yes	yes	high
sodium	yes	yes	
lithium	yes	yes	
calcium	yes	yes	
magnesium	no	yes	
aluminum	no	yes	
zinc	no	yes	
iron	no	yes	
tin	no	yes	
lead	no	yes	
copper	no	no	
mercury	no	no	
silver	no	no	
gold	no	no	no reaction

REAL WORLD

Acid rain

Rainwater is naturally slightly acidic
because carbon dioxide gas in the
air dissolves in it, making weak
carbonic acid. When this acidic
rain falls on certain stones it reacts
with the chemicals in the stones,
gradually eating away at them
in a process called weathering.

Acids and oxides

When an acid reacts with an oxide (a compound with oxygen), it forms a salt and water. The hydrogen ions from the acid and the oxide ions form the water molecules. The cation (the positively charged portion of the oxide)—generally a metal ion—then forms a salt with the anion (the negative part of the acid).

▽ **General equation**
The acid–oxide reaction has the same products as an acid-base reaction (see page 144).

acid + oxide = salt + water
$$H_2SO_4(aq) + CuO(s) \rightarrow CuSO_4(aq) + H_2O(l)$$

▽ **Sulfuric acid (H_2SO_4) plus copper oxide (CuO)**
The black copper oxide powder added to colorless sulfuric acid produces copper sulfate salt ($CuSO_4$) dissolved in the water (H_2O), a blue solution.

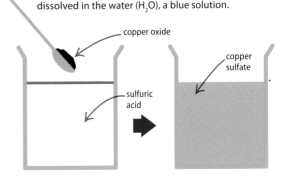

copper oxide

copper sulfate

sulfuric acid

Acids and carbonates

When an acid reacts with a carbonate, the products are a salt, water, and carbon dioxide. A carbonate is an ionic compound in which the anion is CO_3^{2-}. The carbonate ion is displaced in the reaction by the anion from the acid. The carbonate ion then reacts with the free hydrogen ion to form water and carbon dioxide.

▽ **General equation**
The acid-carbonate reaction produces a salt, water, and also carbon dioxide gas.

acid + carbonate = salt + carbon dioxide + water
$$H_2SO_4(aq) + CaCO_3(s) \rightarrow CaSO_4(aq) + CO_2(g) + H_2O(l)$$

▽ **Sulfuric acid (H_2SO_4) + calcium carbonate ($CaCO_3$)**
White calcium carbonate powder added to sulfuric acid produces carbon dioxide (CO_2) and calcium sulfate ($CaSO_4$), which is insoluble and forms a sediment.

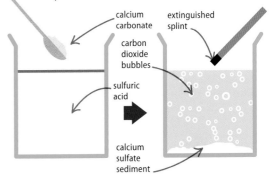

calcium carbonate

extinguished splint

carbon dioxide bubbles

sulfuric acid

calcium sulfate sediment

Acids and sulfites

A sulfite is a compound made up of at least one cation, often a metal, and a sulfite SO_3^{2-} ion. When a sulfite reacts with an acid, the products are a salt, water, and sulfur dioxide. The sulfite ion is displaced in the reaction by the anion from the acid. The sulfite ion then reacts with the free hydrogen to form water and sulfur dioxide gas.

▽ **General equation**
This reaction is very similar to that of the acid-carbonate reaction, except sulfur dioxide gas is formed instead of carbon dioxide.

acid + sulfite = salt + sulfur dioxide + water
$$2HCl(aq) + CuSO_3(s) \rightarrow CuCl_2(aq) + SO_2(g) + H_2O(g)$$

▽ **Hydrochloric acid (HCl) plus copper sulfite ($CuSO_3$)**
Blue copper sulfite crystals added to clear hydrochloric acid produces the green salt copper chloride ($CuCl_2$), dissolved in water, and smelly sulfur dioxide (SO_2).

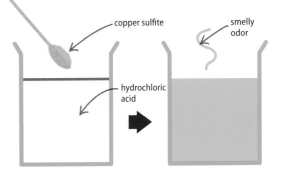

copper sulfite

smelly odor

hydrochloric acid

Electrochemistry

ELECTRICITY IS USED IN CHEMISTRY TO ALTER
COMPOUNDS OR TRANSFER MATERIALS.

The energy carried by electric currents can be used in chemistry.
Currents are frequently used to force compounds apart, by
converting ions back into atoms to produce pure elements.

Electrolytes

An electrolyte is a liquid that conducts
electricity. It is an ionic compound and
has to be liquid (molten or in a solution)
so that its component ions are free to
move. A power source is connected
to two electrodes that are placed in
the electrolyte. This creates a positive
charge at one electrode (the anode)
and a negative charge at the other
(the cathode). The positive and negative
ions in the electrolyte then move toward
the electrode with the opposite charge,
where they receive or donate electrons.
This flow of ions carries electricity
through the liquid.

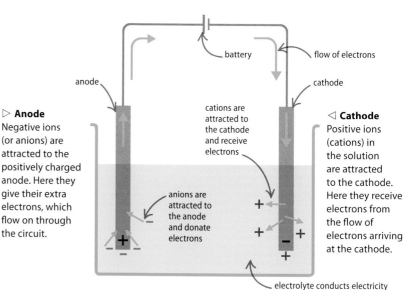

▷ **Anode**
Negative ions
(or anions) are
attracted to the
positively charged
anode. Here they
give their extra
electrons, which
flow on through
the circuit.

battery   flow of electrons

anode   cathode

cations are
attracted to
the cathode
and receive
electrons

anions are
attracted to
the anode
and donate
electrons

◁ **Cathode**
Positive ions
(cations) in
the solution
are attracted
to the cathode.
Here they receive
electrons from
the flow of
electrons arriving
at the cathode.

electrolyte conducts electricity

Electrolysis

Passing an electric current through
an ionic compound will split it into
its component elements. This is
called electrolysis and was the
process used to isolate many new
elements for the first time. When
the power source is turned on, the
positive and negative ions in the
compound are attracted to their
oppositely charged electrodes.
At the cathode, positive ions receive
electrons and, at the anode, negative
ions lose electrons, so the ions
become neutral atoms again. The
pure elements build up at each
electrode and can be collected.
Water is an ionic compound (H_2O)
that can be split into hydrogen (H)
and oxygen (O) in this way.

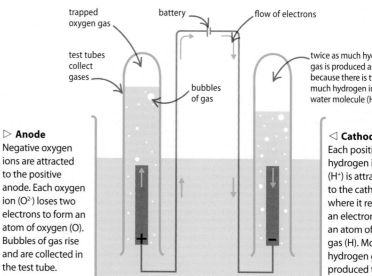

trapped
oxygen gas   battery   flow of electrons

test tubes
collect
gases

bubbles
of gas

twice as much hydrogen
gas is produced as oxygen
because there is twice as
much hydrogen in each
water molecule (H_2O)

▷ **Anode**
Negative oxygen
ions are attracted
to the positive
anode. Each oxygen
ion (O^{2-}) loses two
electrons to form an
atom of oxygen (O).
Bubbles of gas rise
and are collected in
the test tube.

◁ **Cathode**
Each positive
hydrogen ion
(H^+) is attracted
to the cathode,
where it receives
an electron to create
an atom of hydrogen
gas (H). More
hydrogen gas is
produced than
oxygen gas.

Purifying metals

Electrochemistry can be used to remove impurities from a metal and make an extremely pure sample. The piece of impure metal is used as the anode. A pure sliver of the same metal is the cathode. When the current is switched on, the metal ions in the impure metal leave the anode and dissolve in the electrolyte, a copper sulfate solution. Here they are free to move to the cathode, where they receive electrons and turn back into metal atoms.

Electrolysis hair removal uses electricity to turn salts in the hair into alkalis that kill the roots.

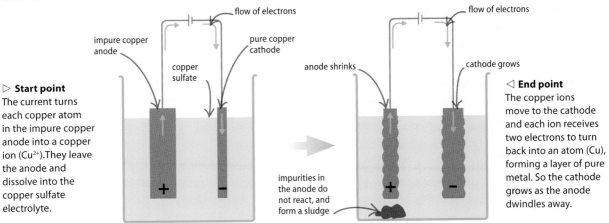

flow of electrons

impure copper anode

pure copper cathode

copper sulfate

anode shrinks

cathode grows

flow of electrons

▷ **Start point**
The current turns each copper atom in the impure copper anode into a copper ion (Cu²⁺).They leave the anode and dissolve into the copper sulfate electrolyte.

impurities in the anode do not react, and form a sludge

◁ **End point**
The copper ions move to the cathode and each ion receives two electrons to turn back into an atom (Cu), forming a layer of pure metal. So the cathode grows as the anode dwindles away.

Electroplating

A thin layer of precious metal can be added to a less expensive metal object using a process called electroplating. A piece of precious metal, such as gold or silver, is used as the anode. The item to be plated forms the cathode. The electrolyte also contains ions of the precious metal. The current makes the anode gradually dissolve and the precious metal ions transfer to the cathode, where they coat the object.

REAL WORLD
Galvanization

Electroplating can be used to coat and protect steel with a layer of zinc to make galvanized steel. This zinc-plated steel is more resistant to corrosion than iron (the main constituent of steel).

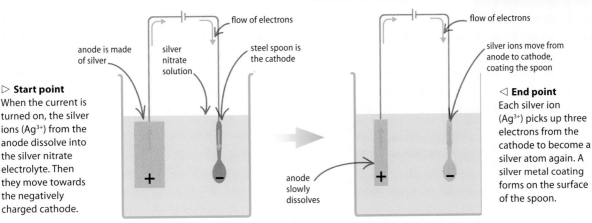

anode is made of silver

silver nitrate solution

steel spoon is the cathode

flow of electrons

silver ions move from anode to cathode, coating the spoon

▷ **Start point**
When the current is turned on, the silver ions (Ag³⁺) from the anode dissolve into the silver nitrate electrolyte. Then they move towards the negatively charged cathode.

anode slowly dissolves

◁ **End point**
Each silver ion (Ag³⁺) picks up three electrons from the cathode to become a silver atom again. A silver metal coating forms on the surface of the spoon.

Lab equipment and techniques

A GUIDE TO THE BASIC APPARATUS IN A CHEMISTRY LAB AND HOW IT IS USED.

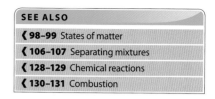
Every chemistry lab has some basic apparatus that can be used for heating substances, observing reactions, and finding out more about materials.

Bunsen burner

This simple gas burner was designed in the late 19th century by German chemist Robert Bunsen. It is used as the primary source of heat in chemistry experiments. The burner has two main settings that can be adjusted by opening and closing an air hole on its base. When the air hole is closed, the burner has a luminous flame. When the hole is opened, it lets in more air, which creates a very hot and roaring blue flame.

▷ **Different flames**
The roaring blue flame is used to heat reactants and boil liquids during experiments. The luminous flame, which is taller and not so hot, is used to ignite splints and burn powders in flame tests (see page 130).

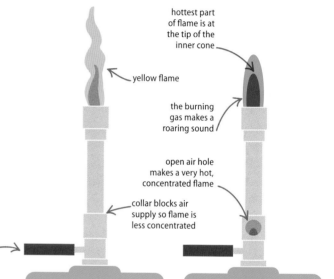

hottest part of flame is at the tip of the inner cone

yellow flame

the burning gas makes a roaring sound

open air hole makes a very hot, concentrated flame

collar blocks air supply so flame is less concentrated

gas supply

Measuring liquids

Chemists must be careful when measuring the volume of liquids. The surfaces of liquids are not flat, but have a curved edge called the meniscus. Water, like most liquids, has a concave meniscus. Other liquids, such as mercury, have a convex surface.

▷ **Measure at eye level**
To measure a volume, the eye should be level with the meniscus.

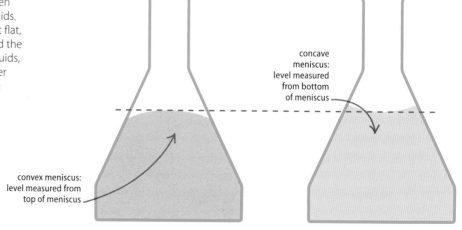

concave meniscus: level measured from bottom of meniscus

convex meniscus: level measured from top of meniscus

mercury

water

Moles

Chemists measure quantities of reactants and products in moles. A mole is a standard unit of atoms. It is defined as the number of atoms in 12 g (0.5 oz) of carbon. This mass is known as the relative atomic mass (RAM) of carbon. A mole of anything else contains this same number of atoms (roughly 6.02×10^{23}), but because atoms have different masses, a mole of one element will have a different mass from a mole of another. Compounds have relative formula masses (RFM), which are calculated by adding up the RAM of their constituents.

Element	RAM
hydrogen	1
carbon	12
oxygen	16
sodium	23
sulfur	32
iron	56
gold	197

△ **Relative atomic mass**
Atoms of different elements have different masses, so moles of different substances have different masses too. For example, one mole of carbon weighs twelve times more than one mole of hydrogen.

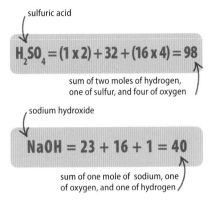

sulfuric acid

$$H_2SO_4 = (1 \times 2) + 32 + (16 \times 4) = 98$$

sum of two moles of hydrogen, one of sulfur, and four of oxygen

sodium hydroxide

$$NaOH = 23 + 16 + 1 = 40$$

sum of one mole of sodium, one of oxygen, and one of hydrogen

△ **Relative formula mass**
The RFM of a compound is the sum of the RAM of each of its constituent atoms.

Apparatus diagram

Chemists often draw the apparatus they use in experiments so that other scientists can replicate the equipment. The apparatus is drawn in a simple, two-dimensional style that makes each item easy to recognize. Test tubes, beakers, conical flasks, and other items of glassware are used to hold reactants during heating. They are frequently positioned on a steel tripod that holds them above the Bunsen burner.

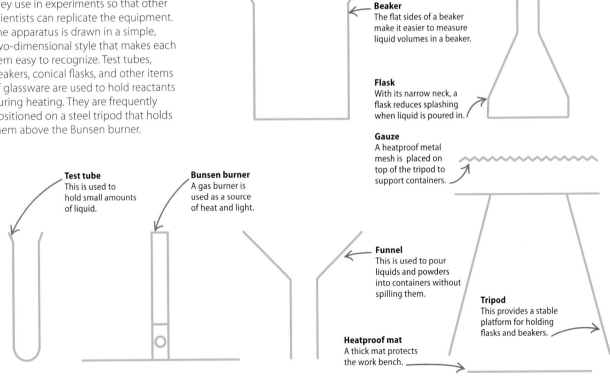

Beaker
The flat sides of a beaker make it easier to measure liquid volumes in a beaker.

Flask
With its narrow neck, a flask reduces splashing when liquid is poured in.

Gauze
A heatproof metal mesh is placed on top of the tripod to support containers.

Test tube
This is used to hold small amounts of liquid.

Bunsen burner
A gas burner is used as a source of heat and light.

Funnel
This is used to pour liquids and powders into containers without spilling them.

Tripod
This provides a stable platform for holding flasks and beakers.

Heatproof mat
A thick mat protects the work bench.

Refining metals

THE CHEMICAL PROCESSES THAT EXTRACT PURE METALS FROM ORES.

SEE ALSO
❮ 124–125 Transition metals
❮ 129 Types of reaction
❮ 132–133 Redox reactions
❮ 148–149 Electrochemistry

Few metals are found pure in nature. Most exist in ores, compounds rich in metals that have to be chemically altered to remove the pure metal.

Iron smelting

The most common iron ores are oxides (in which iron is bonded to oxygen), such as hematite (Fe_2O_3). The ores have their oxygen removed in a process called smelting, which takes place in a blast furnace. The reducing agent (the substance that removes the oxygen) is carbon monoxide, a gas that is sformed by burning coke, a form of coal. The heat from the combustion of coke also powers the various reactions taking place. Impurities such as silicon dioxide are also removed in the process.

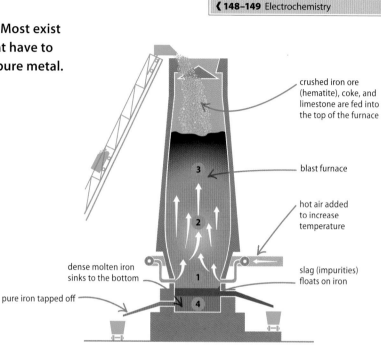

crushed iron ore (hematite), coke, and limestone are fed into the top of the furnace

blast furnace

hot air added to increase temperature

dense molten iron sinks to the bottom

slag (impurities) floats on iron

pure iron tapped off

1. $2C + O_2 \Rightarrow CO_2 + C \Rightarrow 2CO$
The coke is more or less pure carbon. It burns near the bottom of the furnace with oxygen, forming carbon dioxide. The carbon dioxide then reacts with more carbon to form carbon monoxide (CO).

2. $3CO + Fe_2O_3 \Rightarrow 3CO_2 + 2Fe$
The carbon monoxide rises and reacts with the hot ore in the middle of the furnace. Because the carbon in the gas is more reactive than iron, it takes the oxygen ions from the ore, forming pure iron and carbon dioxide gas.

3. $CaCO_3 \Rightarrow CaO + CO2$
Calcium carbonate (limestone) is also added to the furnace. The heat from the combustion at the bottom makes the calcium carbonate decompose into calcium oxide (quicklime) and carbon dioxide.

4. $CaO + SiO_2 \Rightarrow CaSiO_3$
The quicklime moves to the bottom, where molten iron is being formed. Quicklime is very reactive and reacts with impurities in the iron, such as silicon dioxide, removing them to form a waste product called slag.

Thermite process

Another way to extract pure iron from its ores is to burn it with pure aluminum, a more reactive metal. This very rapid reaction is called the thermite process and it is exothermic (see page 134). Aluminum is more reactive than iron so it snatches the oxygen from the iron ore, leaving free elemental iron and aluminum oxide.

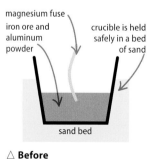

magnesium fuse
iron ore and aluminum powder
crucible is held safely in a bed of sand
sand bed

large flames and sparks
sand bed

molten iron
gray ash
sand bed

△ **Before**
Powdered iron ore and aluminum are mixed in a heat-resistant crucible. The reaction is ignited with a strip of magnesium that burns white hot.

△ **Reaction**
The aluminum snatches the oxygen from the iron, forming aluminum oxide. The reaction releases a large amount of heat with sparks and flames.

△ **After**
The heat melts the iron and it sinks to the bottom of the crucible. The molten iron is surrounded by the gray crystals of aluminum oxide.

Aluminum production

Aluminum cannot be reduced easily like iron. It is too reactive so there are no suitable reducing agents. Therefore this extremely useful metal is extracted from its ore—generally bauxite (Al_2O_3)—by electrolysis (see page 148). The ore is dissolved in molten cryolite (a mineral compound of sodium, aluminum, and fluorine). This electrolyte (liquid that can conduct electricity) is more than 1,000°C (1,832°F) and is held in a tank or cell, lined with graphite carbon. This lining acts as the negatively charged cathode, while more graphite blocks are used as the positively charged anodes.

Before the invention of electrolysis in the 1880s, pure **aluminum was more expensive than gold.**

▷ **The Hall-Héroult process**
This process is named after Martin Hall and Paul Héroult, who invented it independently of each other in the late 1880s. When an electric current is passed through the elecrolyte, it is broken down into positively charged aluminum ions and negatively charged oxygen ions, which are free to move.

1. Al^{3+} (l) + 3e⁻ ➡ Al (l)
Positive aluminum cations are attracted to the negative cathode. Here, each ion (Al3+) receives three electrons from the cathode lining, changing it into an atom of aluminum. The liquid aluminum pools on the cathode at the bottom of the cell and is drained off regularly.

2. $2O^{2-}$ (l) + C (s) ➡ CO_2 (g) + 4e⁻
Negative oxygen ions are attracted to the positive anodes, where each ion loses two electrons to form an oxygen atom. The oxygen reacts with the carbon in the anode to produce carbon dioxide gas, which bubbles out of the liquid. As the carbon is used up, the anodes gradually corrode and need to be replaced.

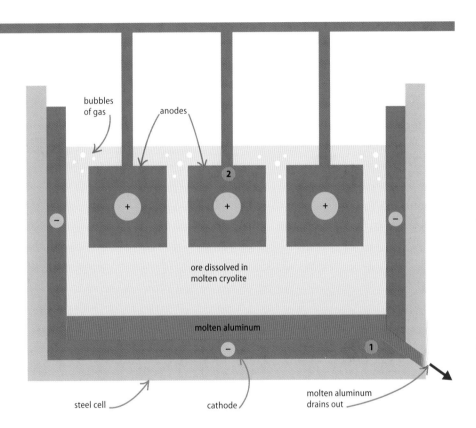

bubbles of gas

anodes

2

ore dissolved in molten cryolite

molten aluminum

steel cell

cathode

molten aluminum drains out

Alloys

Two or more metals—and sometimes other elements—are mixed together to form an alloy. Alloys exhibit some of the properties of all their individual constituents, so they can be adapted to suit many applications. The first metal implements created by humans were made of bronze, a mixture of copper and tin, two metals that were easy to extract from ores.

COMMON ALLOYS				
Name	**Main metal**	**Other metal**	**Properties**	**Uses**
carbon steel	iron	carbon	high strength	construction
stainless steel	iron	chromium	resistant to corrosion	eating utensils
bronze	copper	tin	easily worked	bells
brass	copper	zinc	does not corrode	zippers, keys
solder	tin	lead	low melting point	soldering
invar	iron	nickel	does not expand when hot	precision
amalgam	mercury	silver	starts soft, then hardens	dental fillings

Chemical industry

SOME CHEMICAL REACTIONS ARE PERFORMED ON
A HUGE SCALE TO PRODUCE VALUABLE SUBSTANCES.

SEE ALSO

❬ **78–79** Cycles in nature
❬ **128–129** Chemical reactions
❬ **144–145** Acids and bases
❬ **148–149** Electrochemistry

Many of the raw materials that humans need exist in nature. They are
refined from ores or separated from mixtures such as seawater. Some
compounds, however, are made in factories using chemical reactions.

The Haber process

Also known by its full name, the Haber-Bosch, this process turns
nitrogen and hydrogen gas into ammonia (NH_3). Ammonia is used
to make crop fertilizers and explosives, such as TNT and dynamite.
Nitrogen is the most common gas on Earth—it makes up 78
percent of the air—but it is very unreactive. The Haber process
uses a catalyst (see page 138) to make the reaction occur.

1. Gases mixed
A mixture of hydrogen (H_2) and
nitrogen (N_2) gases is pumped
into the reactor. Three times as
much hydrogen is added as
nitrogen to create the correct
ratio for ammonia (3:1).

2. In the reactor The gases are
passed over an iron catalyst,
which brings them together so
they react to form ammonia. The
reaction takes place at 450°C
(842°F) and at 200 times the
atmospheric pressure.

3. Product separated
The ammonia gas leaving the
reactor moves into the
condenser, where it is cooled
so that it turns into liquid
ammonia that can then be
tapped off.

4. Reactants recycled Not all
of the reactants turn into
ammonia. The unused nitrogen
and hydrogen gases rise out of
the condenser and are recycled
back into the reactor.

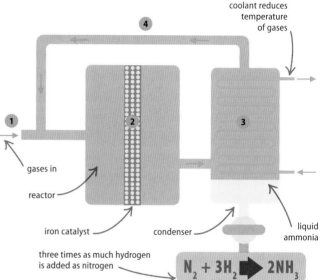

coolant reduces
temperature
of gases

gases in

reactor

iron catalyst

condenser

liquid
ammonia

three times as much hydrogen
is added as nitrogen

$$N_2 + 3H_2 \rightarrow 2NH_3$$

Nitric acid production

One of the chemicals that is made from ammonia is nitric acid (HNO_3). This acid
reacts with bases to form nitrate salts, which plants need to make proteins.
Nitric acid is mainly used to make fertilizers, but it is also used as a rocket fuel
and is one of the few solvents that can dissolve gold.

1. Converter
In the converter, ammonia (NH_3) and
oxygen (O_2) react at 800°C (1,472°F)
using a platinum catalyst to make
nitric oxide (NO) and water.

2. Oxidation chamber
The gases from the converter are cooled
to 100°C (212°F). More oxygen is added,
some of which will react with the nitric
oxide to make nitrogen dioxide (NO_2).

3. Absorption tower
Water trickles down through the quartz
crystals while the gases rise. The
nitrogen dioxide and left-over oxygen
react with the water to form nitric acid.

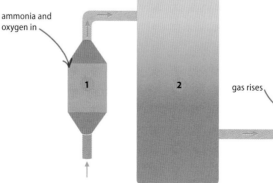
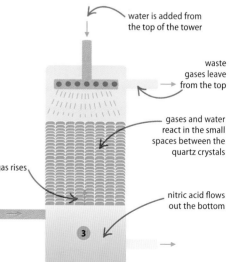

ammonia and
oxygen in

gas rises

water is added from
the top of the tower

waste
gases leave
from the top

gases and water
react in the small
spaces between the
quartz crystals

nitric acid flows
out the bottom

Contact process

This is the industrial process for making sulfuric acid. Sulfur dioxide gas reacts with water using a catalyst to produce the acid in a multistage process. Sulfuric acid is one of the most powerful acids. It is used in car batteries and its salts (the sulfates) are used in fertilizers. It is also used in papermaking.

1. Furnace
In the furnace, sulfur (S) is burned with oxygen (O) from the air to form sulfur dioxide (SO_2).

2. Cleaning the gas
In the next three chambers, the gas is filtered, washed, and dried to remove any impurities that could interfere with the catalyst.

3. Reactor
The vanadium oxide catalyst makes the sulfur dioxide react with more oxygen to form sulfur trioxide (SO_3).

4. Absorption tower
The sulfur trioxide is dissolved in a little sulfuric acid. This makes it safe to dilute with water to make a lot more sulfuric acid—the end product.

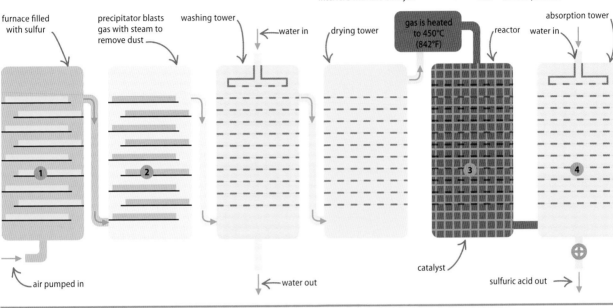

furnace filled with sulfur

precipitator blasts gas with steam to remove dust

washing tower

←water in

drying tower

gas is heated to 450°C (842°F)

reactor

water in

absorption tower

catalyst

air pumped in

←water out

sulfuric acid out →

Downs cells

Pure chlorine gas (Cl_2) and sodium metal (Na) are made by the electrolysis of sodium chloride (common salt: NaCl). This takes place on an industrial scale in a large tank called a Downs cell. The tank contains liquid salt—it is heated to more than 600°C (1,112°F) so that the salt melts. When an electric current is run through the liquid, the molten salt breaks up into sodium and chloride ions, which move to the electrodes and turn into atoms. The pure elements can then be collected.

1. At the iron cathode ($2Na^+ + 2e^- \rightarrow 2Na$)
Positive sodium ions (Na^+) move to the cathode, where they gain an electron each to form sodium atoms (Na). This metal is less dense than the sodium chloride electrolyte so it floats to the surface where it can be collected.

2. At the carbon anode ($2Cl^- \rightarrow Cl_2 + 2e^-$)
Negative chlorine ions (Cl^-) are attracted to the positively charged anode, where they lose an electron each to form chlorine atoms (Cl). The element bubbles out of the liquid electrolyte as chlorine gas (Cl_2).

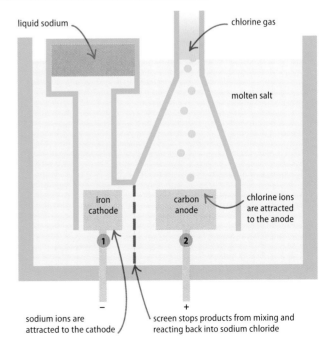

liquid sodium

chlorine gas

molten salt

iron cathode

carbon anode

chlorine ions are attracted to the anode

−   +

sodium ions are attracted to the cathode

screen stops products from mixing and reacting back into sodium chloride

Carbon and fossil fuels

CARBON AND ITS COMPOUNDS FORM THE BASIS FOR
ALL FOSSIL FUELS.

After hydrogen, carbon is the most common element in living things. When
organisms die, their remains are preserved underground. Over millions of
years are transformed into useful, carbon-rich compounds called fossil fuels.

Forms of carbon

Pure carbon exists in different forms, or allotropes (see page 111). The carbon atoms
in each allotrope link up in different ways, which gives the allotropes very different
properties. Diamond is an extremely hard and sparkling gem. The arrangement of the
atoms in graphite, however, make it a slippery gray solid, often used as pencil lead.

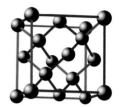

△ **Diamond**
The carbon atoms are arranged in a very
rigid crystal network based on repeating
tetrahedra of four atoms.

△ **Fullerene**
The atoms link together to form a
ball-shaped cage. Fullerenes may
contain 100, 80, or 60 carbon atoms.

△ **Graphite**
The atoms are arranged in sheets of
hexagons. The sheets are only loosely
bonded, so they slip over each other.

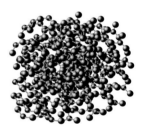

△ **Soot**
The atoms in this allotrope are arranged
randomly. Soot forms from the uncombusted
carbon released when fossil fuels burn.

Coal

Coal is a carbon-rich sedimentary rock made from the remains
of trees. Most of the coal mined today formed from forests that
grew around 300 million years ago. The plant material was
buried in the absence of oxygen, so huge amounts were
preserved as sediments, gradually forming coal.

▷ **Coal formation**
The process begins when plant remains sink in
waterlogged, boggy soil. The lack of oxygen prevents
the wood from decaying. These remains form a dense
soil called peat, which can itself be used as a fuel
when dried. Over time the peat is buried, and the
increased pressure drives out the water to form
lignite (soft, brown rock). Deeper down, heat
hardens the lignite into coal.

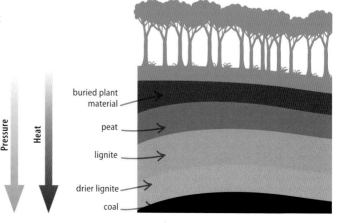

Pressure

Heat

buried plant
material

peat

lignite

drier lignite

coal

Petroleum

Petroleum—meaning "rock oil"—is a mixture of natural compounds known as hydrocarbons, which are made from carbon and hydrogen. Petroleum forms from a thick ooze of dead microorganisms that covered the beds of ancient seas. After being buried by other sediments, the biological material broke down into hydrocarbons over millions of years.

▷ **Oil and gas fields**
Petroleum oil or gas is a natural product that percolates up through porous rocks to the surface. When the petroleum's passage is blocked by nonpermeable rock, it accumulates as an oil (or gas) field.

oil or gas field

impervious
cap rock

porous
reservoir
rock

organic, rich source
rock exposed to
heat and pressure

Crude oil distillation

The mixture of hydrocarbons collected from underground reservoirs of petroleum is known as crude oil. It contains thousands of mostly liquid compounds—the gas given off is known as natural gas. Crude oil is separated into useful fractions: groups of compounds that have similar boiling points, indicating that their molecules have a similar size.

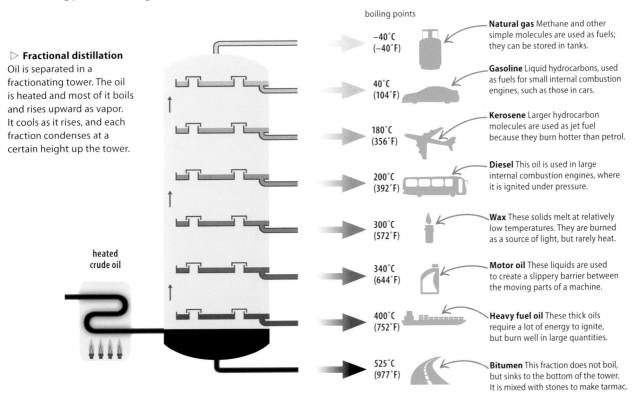

▷ **Fractional distillation**
Oil is separated in a fractionating tower. The oil is heated and most of it boils and rises upward as vapor. It cools as it rises, and each fraction condenses at a certain height up the tower.

heated
crude oil

boiling points

−40°C
(−40°F)

40°C
(104°F)

180°C
(356°F)

200°C
(392°F)

300°C
(572°F)

340°C
(644°F)

400°C
(752°F)

525°C
(977°F)

Natural gas Methane and other simple molecules are used as fuels; they can be stored in tanks.

Gasoline Liquid hydrocarbons, used as fuels for small internal combustion engines, such as those in cars.

Kerosene Larger hydrocarbon molecules are used as jet fuel because they burn hotter than petrol.

Diesel This oil is used in large internal combustion engines, where it is ignited under pressure.

Wax These solids melt at relatively low temperatures. They are burned as a source of light, but rarely heat.

Motor oil These liquids are used to create a slippery barrier between the moving parts of a machine.

Heavy fuel oil These thick oils require a lot of energy to ignite, but burn well in large quantities.

Bitumen This fraction does not boil, but sinks to the bottom of the tower. It is mixed with stones to make tarmac.

Hydrocarbons

THE DIFFERENT FAMILIES OF COMPOUNDS MADE
PURELY FROM CARBON AND HYDROGEN.

SEE ALSO	
❰ 110–111	Compounds and molecules
❰ 114–115	Covalent bonding
❰ 156–157	Carbon and fossil fuels

Hydrocarbons are the simplest compounds in
living things. The study of chemicals found
in living things is called organic chemistry.

Hydrocarbon chains

Carbon atoms can form up to four
covalent bonds. This allows carbon to
form intricate hydrocarbon molecules.
The carbon atoms are chained together,
with hydrogen atoms bonded to the
spare electrons. When hydrogen atoms
are not available, two carbon atoms may
form double, and even triple, bonds.

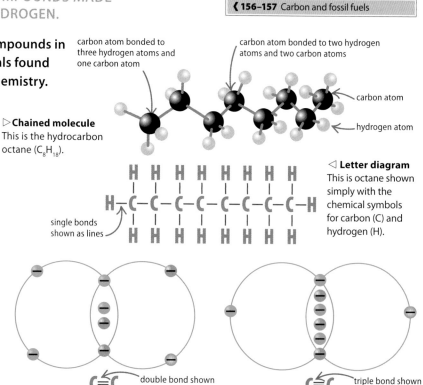

carbon atom bonded to
three hydrogen atoms and
one carbon atom

carbon atom bonded to two hydrogen
atoms and two carbon atoms

◁ carbon atom

◁ hydrogen atom

▷ **Chained molecule**
This is the hydrocarbon
octane (C_8H_{18}).

◁ **Letter diagram**
This is octane shown
simply with the
chemical symbols
for carbon (C) and
hydrogen (H).

single bonds
shown as lines

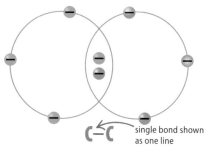

single bond shown
as one line

△ **Single bond**
The normal carbon-to-carbon
covalent bond involves sharing
a single pair of electrons.

double bond shown
as two lines

△ **Double bond**
This bond has two pairs of electrons
shared between the atoms. It is
less stable than a single bond.

triple bond shown
as three lines

△ **Triple bond**
This very unstable bond contains
three shared pairs of electrons
to form a triple bond.

Naming system

Hydrocarbons with chained and
branched molecules are known
as aliphatics. They are named with
a prefix that is specific to the
number of carbon atoms in their
longest chain. Side branches on
the main chain are also named
using the same prefixes. These
branches are known as alkyl
groups, and so the prefix is
followed by the suffix "–yl" to
show they relate to a branch
other than the main chain. For
example a methyl chain is a side
branch with one carbon atom.

Prefix	Number of carbon atoms
meth	1
eth	2
prop	3
but	4
pent	5
hex	6

△ **Prefixes**
The first four prefixes are
specific to hydrocarbons,
while from five onward the
prefixes are based on Latin
and Greek numbers.

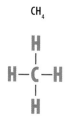

△ **Methane**
The simplest
hydrocarbon is also
known as natural
gas and is burned
as a fuel.

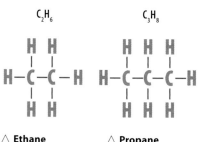

△ **Ethane**
Ethane has two
carbon atoms, and
is the compound
used to make
polythene plastic.

△ **Propane**
With three carbon
atoms, propane gas
is the fuel supplied
in the tanks used in
camping stoves.

Alkanes, alkenes, and alkynes

Chained hydrocarbons form families according to how their carbon atoms are bonded. Hydrocarbons with single bonds are called alkanes. Chains with at least one double bond are alkenes. A triple-bonded compound is an alkyne.

Suffix	Contains
ane	carbon–carbon single bonds
ene	carbon–carbon double bonds
yne	carbon–carbon triple bonds

◁ **Suffix**
Compounds are given a suffix to show which family they belong to.

C_2H_6

◁ **Ethane**
The single bond in ethane makes this hydrocarbon relatively stable and unreactive.

C_2H_4

◁ **Ethene**
The double bond makes ethene more flammable than ethane.

C_2H_2

◁ **Ethyne**
The triple bond is very unstable. Ethyne and all alkynes are highly flammable and reactive.

reactivity increasing

Isomers

Aliphatic compounds can have the same formula—the number of carbon and hydrogen atoms—but be arranged in different ways. These similar compounds are known as isomers. Side branches change the way isomers behave, making them react differently and have different melting and boiling points.

two methyl groups attached to second and third carbon atoms

C_6H_{14}

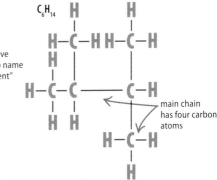

main chain has six carbon atoms, so name has the prefix "hex"

carbon atoms have single bonds, so name ends with "ane"

C_6H_{14}

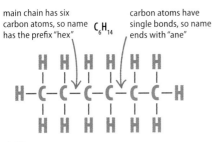

methyl group

main chain has five carbon atoms, so name has the prefix "pent"

C_6H_{14}

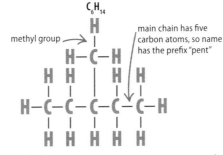

main chain has four carbon atoms

△ **Hexane**
This liquid alkane has six carbon atoms in a single chain. It has a total of four isomers and its main use is in petrol.

△ **Methylpentane**
The longest chain in this compound is a pentane. A methyl group (side chain with one carbon) adds the sixth carbon atom.

△ **2,3-Dimethylbutane**
The longest chain in this isomer is a butane. Two methyl groups are attached to second and third carbon atoms in the butane.

Aromatics

Hydrocarbons can also form ringed molecules called aromatics. The simplest of these is benzene (C_6H_6), which has six carbon atoms linked with alternating single and double bonds. The electron pairs forming the three double bonds are free and shared between all six carbon atoms, forming a ring-shaped "delocalized" bond.

◁ **Benzene ring**
The shared electrons form a circular bond and give the molecule its shape.

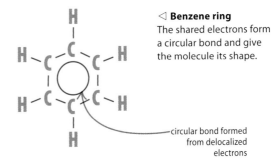

circular bond formed from delocalized electrons

Functional groups

HYDROCARBONS CAN REACT WITH OTHER ELEMENTS.

SEE ALSO
❰ **28–29** Respiration
❰ **78–79** Cycles in nature
❰ **144–145** Acids and bases

These "functional groups" of additional elements dominate the compound's chemical behavior.

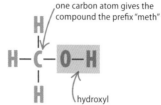

one carbon atom gives the compound the prefix "meth"

hydroxyl

two carbon atoms give the compound the prefix "eth"

Alcohols

These are organic molecules in which an oxygen and hydrogen (–OH) is added to the carbon chain, in the place of a hydrogen atom. Ethanol —the alcohol with two carbon atoms—is the compound in alcoholic drinks. It is produced by natural fermentation processes and can be metabolized by the body. However, all other alcohols are much more poisonous.

△ **Methanol**
This simplest alcohol is used as an antifreeze and solvent.

△ **Ethanol**
This alcohol is found in beer and wine and is purified into liquors.

three carbon atoms give the compound the prefix "prop"

hydroxyl on second carbon atom

R indicates rest of compound

◁ **Hydroxyl**
The functional group with a hydrogen atom and an oxygen atom is called a hydroxyl.

△ **Propan-2-ol**
This compound is so named because the functional group is on the second carbon atom.

hydrogen in hydroxyl

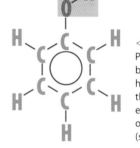

◁ **Phenol**
Phenol is acidic because the hydrogen in the hydroxyl easily breaks off and reacts (see page 144).

Carboxylic acids

Organic acids have a carboxyl group (COOH). The hydrogen breaks off and reacts with alkali compounds and metals. The rest of the molecule forms a carboxylate ion with a charge of –1. The salts produced when the acid reacts are called carboxylates. Most carboxylic acids are weak and have a maximum pH of around 3 or 4.

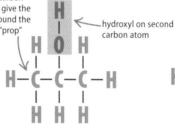

◁ **Carboxyl**
The carboxyl group is formed from the carbon at the end of a chain joined to one oxygen atom with a double bond and to a hydroxyl group with a single bond.

CFCs or chlorofluorocarbons, the chemicals that damage the ozone layer, are organic halide compounds.

one carbon atom gives the compound the prefix "meth"

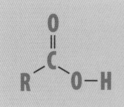

two carbon atoms give the compound the prefix "eth"

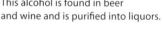

△ **Methanoic acid**
Also known as formic acid, this is the simplest carboxylic acid. It is used to soften animal hides into leather.

△ **Ethanoic acid**
Also known as acetic acid, this is the sour-tasting ingredient in vinegar. It forms naturally from ethanol due to the action of bacteria.

Esters

When a carboxylic acid reacts with an alcohol, they form an ester. The functional group of the ester links the two original molecules together. The fats and oils in living things—including the lipids that form cell membranes—are esters. Soaps, oils, and fats are also all types of ester.

△ **Functional group**
The oxygen from alcohol bonds to the carboxyl (carbon and oxygen group) from the acid.

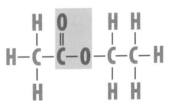

△ **Ethyl ethanoate**
This ester has a strong pear drop smell and is used as nail varnish remover.

△ **Thiol smells**
The odor in garlic as well as the noxious smell sprayed by skunks is due to a thiol. Its functional group is called a sulfydryl.

△ **Amine smells**
The smell of fish is due to the presence of a compound called trimethylamine. Its functional group includes nitrogen.

Thiols and amines

Thiols are similar to alcohols, except the functional group has a sulfur instead of an oxygen atom. The word "thiol" is a mixture of the Latin words for sulfur and alcohol. These compounds have strong smells. Amines are another smelly group of organic compounds. They have a functional group with one nitrogen and two hydrogen atoms. When an amine group attaches to a carboxylic acid, it forms an amino acid, one of the building blocks of proteins.

Ant stings

Some insect venoms, especially the stings of fire ants, have methanoic acid as their active ingredient. A fire ant squirts the acid onto attackers, causing small but painful burns. The original name, formic acid, is derived from the Latin word for "ant."

Organic halides

Members of the halogen group (see page 122) form only one bond in chemical reactions, just like hydrogen, but they are much more reactive. Halogens replace hydrogen atoms in hydrocarbon molecules, forming organic halides.

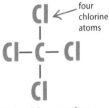

◁ **Chloromethane**
With just a single chlorine atom, this is the most reactive of this family of compounds. One of its uses is to make silicone rubbers.

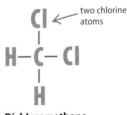

△ **Dichloromethane**
This sweet-smelling liquid is used in paint strippers, aerosol sprays, and to decaffeinate coffee.

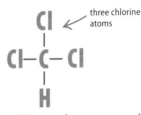

△ **Trichloromethane**
Better known as chloroform, this compound was one of the first anesthetics.

△ **Tetrachloromethane**
Also known as carbon tetrachloride, this toxic liquid is banned in some countries.

Polymers and plastics

COMPOUNDS FORMED FROM LONG CHAINS OF SMALLER
MOLECULES ARE CALLED POLYMERS.

SEE ALSO	
❰ **84–85** Genetics	
❰ **96–97** Properties of materials	
❰ **158–159** Hydrocarbons	
Stretching and deforming	**174–175** ❱

Plastics and other artificial fibers, such as nylon, are familiar types of
polymers. However, these long-chained molecules are also widespread
in nature. Many of the chemicals in food are polymers too.

Monomers

The repeating units in a polymer are called monomers. A polymer may contain a
single type of monomer or have two or more types of repeating units—known as
copolymers. Monomers are held together by covalent bonds (see page 114). Many
artificial polymers are derived from alkene monomers, which have double bonds
that can be broken and reformed to make the chains.

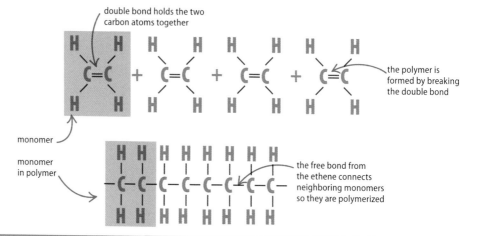

double bond holds the two
carbon atoms together

▷ **Ethene monomer**
One of the most common
plastics is made from chains of
ethene monomers. Ethene is the
simplest alkene molecule. Its
polymer is called polythene.

the polymer is
formed by breaking
the double bond

monomer

▷ **Polythene polymer**
While ethene is a gas, polythene
(also known as polyethylene)
is a transparent solid. It can
be formed from an unlimited
number of ethene monomers.

monomer
in polymer

the free bond from
the ethene connects
neighboring monomers
so they are polymerized

Natural polymers

The natural world contains many polymers. Living things frequently digest these
large compounds, breaking them into their monomers, which are absorbed and
then rebuilt into different polymers.

△ **Protein**
Muscles and many other
features in a living body
are made from proteins,
which are polymers of
amino acid monomers.

△ **DNA**
DNA is a complex copolymer.
The sides are formed from
chains of sugar, while the
crosslinks are pairs of four
monomers called nucleic acids.

△ **Cellulose**
The wall around a plant cell is
made from a polymer of glucose
called cellulose. Cellulose
forms tough fiber and is a major
component of wood and paper.

△ **Starch**
Found in potatoes and bread,
starch is also made from glucose
monomers. However, they are
chained together differently to
form globules rather than fibers.

Plastics

Many artificial polymers are plastics. A plastic is an incredibly useful material that can be molded into any shape while hot, becoming solid when cool. It can also be pulled into thin films and used as a protective coating. Plastic is made from monomers derived from crude oil.

▽ **Common plastics**
Several plastics have become very familiar over the last few decades, because they have a huge range of applications.

Polymer	Monomer	Properties of polymer
polythene (polyethylene)	ethene	makes flexible plastics; is used in packaging and to insulate electrical wires
polystyrene	styrene	used to make Styrofoam; is also added to other polymers to make them waterproof
PVC (polyvinyl chloride)	chloroethene (vinyl chloride)	makes very tough plastics; is not damaged by strong chemicals; is a good insulator
teflon (polytetrafluoroethylene)	tetrafluoroethylene	a very slippery substance that is used on nonstick pans

Properties of plastics

It is easier to shape plastic polymers while they are warm or melted into a liquid. There are two main types of plastic. Thermoplastics can be molded, melted, and reshaped repeatedly. Thermosets can only be molded once; after they have set, they will burn without melting if reheated.

▷ **Polymer properties**
The properties of a polymer result from the shape of the monomers. Thermosets form crosslinks when solid, which make the polymer into a rigid lattice.

Before stretching	During stretching	After stretching
unbranched and straight chains	polymers slip a little	polymers stay stretched
branched chains	polymers slip easily	polymers stay stretched
coiled chains	polymers lengthen and slip past each other	polymers shorten again, but the shape is stretched
crosslinked coils	coils lengthen but do not slip past each other	polymers spring back to original shape
crosslinked straight chains	polymers move only very slightly	polymers spring back to original shape

REAL WORLD

Rubber

The bark of rubber trees produces an oily liquid called latex that contains the compound isoprene. Adding an acid makes the isoprene in the liquid polymerize into solid rubber, which can be made into sheets or molded before it dries out.

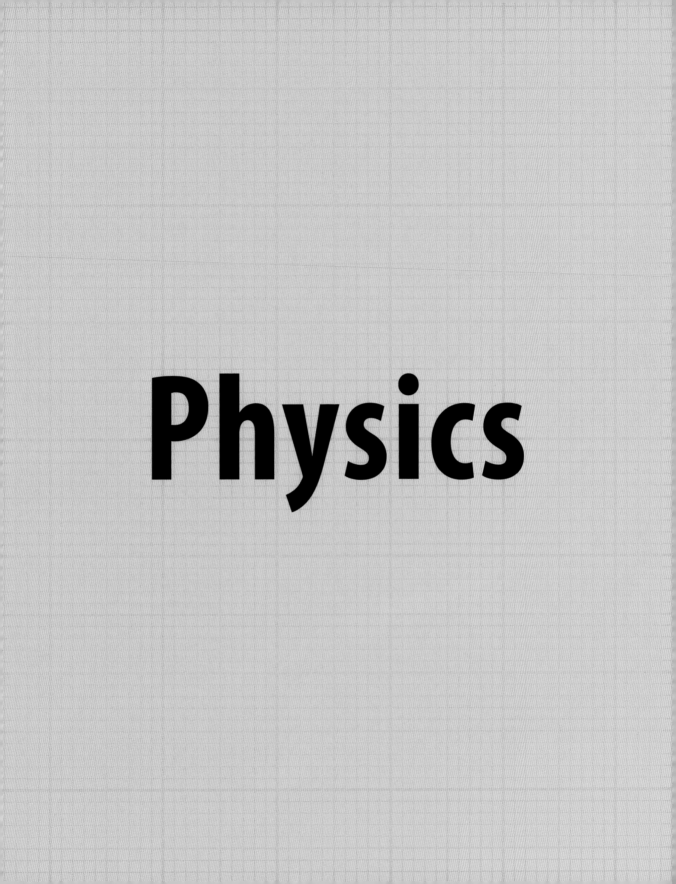

Physics

What is physics?

THIS FIELD OF SCIENCE SEEKS TO REVEAL THE WORKINGS
OF THE UNIVERSE ON THE LARGEST AND SMALLEST SCALES

The word "physics" means "nature" in ancient Greek, and physicists
tackle the most fundamental subjects in the Universe, such as the
nature of energy, space, and even time.

Foundation of knowledge

Physics is the foundation of all scientific knowledge. Chemistry,
biology, and other sciences are built on an understanding of
physics. For example, physicists have revealed the structure of
the atom, which chemists use to understand how chemicals
react with each other. Meanwhile, physics has also explained
how energy behaves, which is crucial knowledge for biologists
figuring out how organisms stay alive. A few physicists, such
as Albert Einstein and Isaac Newton, have become famous
because their discoveries have had such a far-reaching effect.

an apple falls due to gravity
(a force of attraction)

▷ **Falling objects**
Physics explains many
everyday phenomena. For
example, Newton's theory of
gravitation (see pages 178–179)
explains why an apple—or any
object—falls to the ground.

Energy, mass, space, and time

Physics can express everything in the Universe
in terms of mass, energy, and force, from the
workings of a giant star to a raindrop falling from
a cloud. A mass is an object that is affected by forces.
What a force does is transfer energy from one mass
to another, which changes the way the masses
move or are shaped. For example, throwing a ball
or stretching a rubber band requires force—even
light shining on an object exerts a tiny force on it!

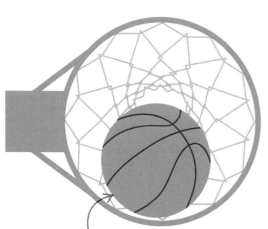

by throwing the ball, a basketball player
applies a force that propels the ball at speed
in a certain direction (hopefully into a basket)

◁ **In motion**
They may not know it, but basketball
players use physics. They push the ball
in just the right direction and with just
the right force to score a basket.

Machines

Physics allows us to build machines that harness forces and the energy they transfer to do useful work. A machine is a device that carries out a task by changing forces in some way. Machines need not be complex; in fact, a piece of high-tech machinery, such as a robot or an engine, is really a series of much simpler machines working together. Simple machines include levers, wheels, screws, ramps, and pulleys. Machines make work easier by converting small forces into big ones.

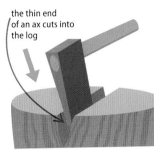

the thin end of an ax cuts into the log

△ **Focusing force**
Even the blade of an ax is a machine. The force pressing on the wide end of the wedge-shaped blade is focused into the sharp edge so it slices through solid objects.

Radiation

People are often confused by the term "radiation," thinking it refers to the dangerous particles blasted out by nuclear reactions. However, in physics the word "radiation" normally refers to waves of light, heat, and other invisible rays that travel across the Universe. Together they form the electromagnetic spectrum, which is made up of mostly familiar types of radiation. As well as light, the spectrum includes radio waves, gamma rays, ultraviolet light, infrared (or heat), and X-rays. These are all examples of electromagnetic waves.

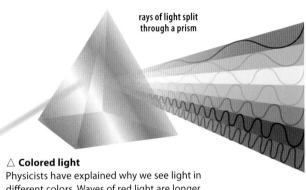

rays of light split through a prism

△ **Colored light**
Physicists have explained why we see light in different colors. Waves of red light are longer (and have less energy) than waves of violet light. All the other colors are in between.

Electricity

Thanks to physicists researching sparks and magnetic forces, most machines are powered by electric currents. This process began in ancient times, when early scientists examined magnetic, iron-rich stones that stuck to each other. Over the centuries, it was discovered that magnetism and electricity are linked—an area of physics called electromagnetism. This field also involves atomic structure and where radiation comes from.

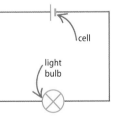

cell

light bulb

◁ **Electric circuits**
Electricity can be put to work using a circuit of different components. For example, a light bulb turns electric current into light when a cell is connected.

Astronomy

In many ways, astronomy was the first science of all, because ancient people saw patterns in the movement of the planets, Moon, and Sun. Modern astronomy still involves stargazing, but high-tech telescopes are used to gather light and other radiation from farther out in space than ever before. The laws of physics discovered on Earth work in just the same way on the other side of the Universe. Therefore, astronomers can use their knowledge to understand the many different objects they see out in space—and even figure out how the Universe came into existence.

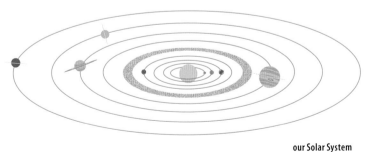

our Solar System

△ **Meet the neighbors**
Observations of the eight planets in our Solar System have taught us much about our own world. Astronomers are now searching for Earth-like planets around more distant stars.

Inside atoms

ATOMS ARE TOO SMALL TO SEE, EVEN WITH SOME OF THE MOST POWERFUL MICROSCOPES.

SEE ALSO

❮ **98–99** States of matter
❮ **108–109** Elements and atoms
❮ **116–117** The periodic table
❮ **126–127** Radioactivity
Forces and mass **172–173** ❯
Electricity **202–203** ❯

Everything we can see in the Universe, from the stars to our own bodies, is made up of atoms.

What is an atom?

Atoms are not all the same. There are 92 different types that occur naturally—and a few more that are made by scientists in laboratories. Each atom belongs to a specific element, a substance that cannot be purified further into simpler ingredients. Familiar elements include hydrogen, carbon, and lead.

▽ **Different atoms**
The atoms of every element have a unique size and mass. The mass varies with the number of protons and neutrons in the nucleus.

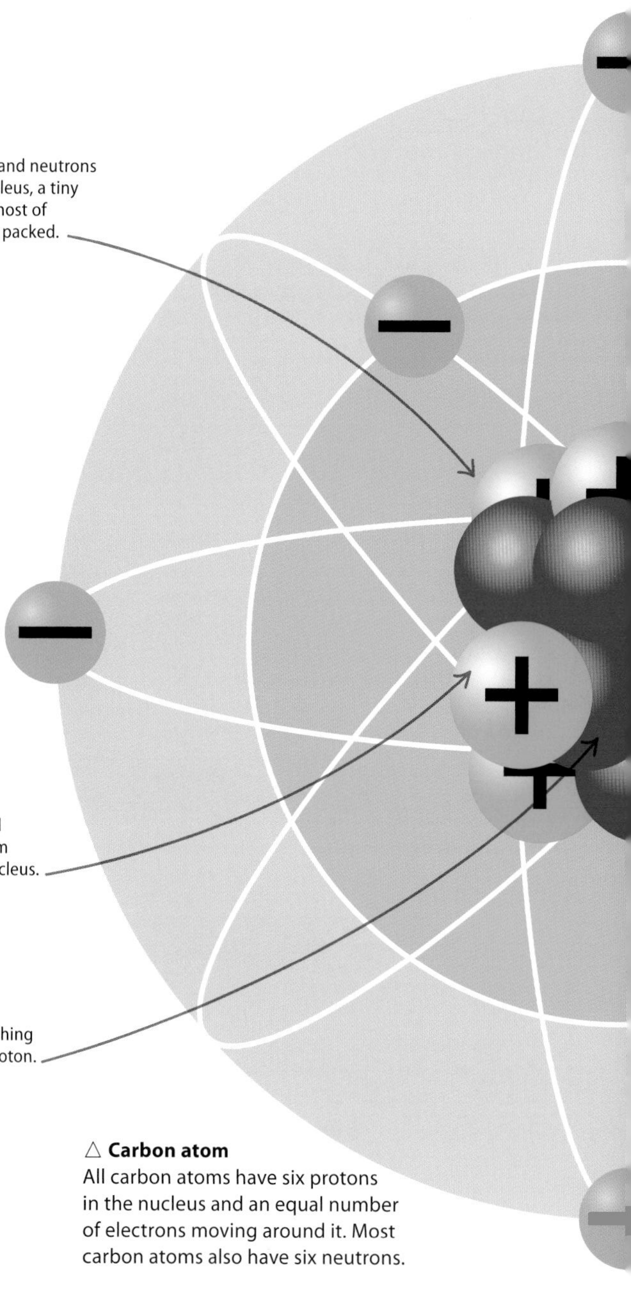

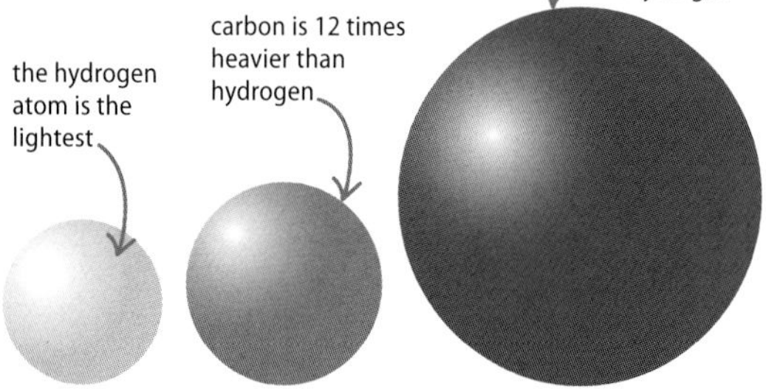

lead is 207 times heavier than hydrogen
carbon is 12 times heavier than hydrogen
the hydrogen atom is the lightest
hydrogen
carbon
lead

Subatomic structure

Atoms are made up of even smaller particles called protons, neutrons, and electrons. The atoms of a certain element have a unique number and arrangement of particles, which is what gives the element its distinct properties—making it a gas or metal, for example. An atom always has the same number of protons as electrons. Each proton has a positive charge, which is matched by the negative charge of an electron, making the whole atom neutral.

Nucleus
The protons and neutrons form the nucleus, a tiny core where most of the matter is packed.

Proton
Protons have positive charges that attract the negatively charged electrons, holding them in place around the nucleus.

Neutron
These particles have no charge. They make up the rest of the mass of the atom, each weighing slightly more than a proton.

△ **Carbon atom**
All carbon atoms have six protons in the nucleus and an equal number of electrons moving around it. Most carbon atoms also have six neutrons.

Electron shell
The electrons move around the nucleus, arranged in shells. Shells have a fixed number of spaces for electrons. In most cases, when one shell becomes full, another begins farther away from the nucleus.

Isotopes

Atoms occur in different forms. While an element's atomic nucleus always has a certain number of protons, many contain different numbers of neutrons. These alternative versions of the atom are called isotopes. Atoms of different isotopes have varying weights.

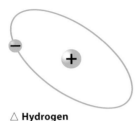
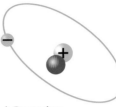
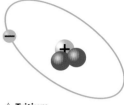

△ **Hydrogen**
The main isotope of hydrogen has no neutrons in its nucleus.

△ **Deuterium**
With one extra neutron, this atom weighs twice as much as the main hydrogen isotope.

△ **Tritium**
This hydrogen isotope is three times heavier than the main hydrogen isotope.

REAL WORLD

Radiocarbon dating

Scientists use the carbon-14 isotope to measure the age of ancient artifacts that are made from organic materials, such as wood or cotton. When new, the cotton wrapping of this mummy had a certain amount of carbon-14 in it. The isotope breaks down at a slow but fixed rate, and the amount left in the wrapping can tell scientists how old it is.

Atomic forces

There are three forces at work inside atoms. The first type is a strong force that pulls the particles in the nucleus together. The second types occurs when the electrons are bonded to the atom by an electromagnetic force, which also acts over a much larger distance outside of the atom. The third type is a weak force that is involved in radioactivity, pushing particles out of the nucleus.

Electron
The electron has a negative charge that is equal and opposite to that of the proton. However, the mass of an electron is just a tiny fraction of a proton's mass.

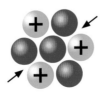
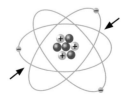

△ **Strong force**
This is the strongest force in nature, but it acts only over tiny distances.

△ **Electromagnetic force**
This force is involved in light and electricity, and holds atoms together.

△ **Weak force**
This force causes radioactive decay in atoms.

Energy

WE RELY ON ENERGY TO MAKE OUR WORLD FUNCTION.

Energy is what makes things happen. It is everywhere and in everything, giving objects the ability to move or glow with heat.

Measuring energy

Energy can be put to work. To a physicist, the word "work" means the amount of energy involved in moving an object. Work is calculated as the amount of force multiplied by the distance. Since force is measured in newtons (N) and distance in meters (m), such a calculation results in a unit of work called a newton meter (Nm).

▽ **One joule of energy**
One joule (J) is the amount of energy transferred to an object by a force of 1 N over a distance of 1 m (3¼ ft). This is roughly equivalent to lifting an apple up 1 m (3¼ ft).

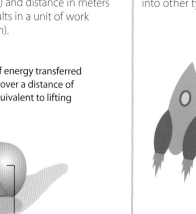

1 m (3¼ ft)

Types of energy

Energy can be seen working in many ways. Although they are given different names and appear in a wide range of contexts—from the energy released by an exploding star to the energy in a bouncing ball—all types of energy are closely related, and each one can change into other types (see the opposite page for examples).

◁ **Kinetic energy**
This is the energy of motion. As an object speeds up, it contains more kinetic energy.

◁ **Thermal energy**
The air blowing out from a hairdryer is hot because electrical energy is converted into thermal energy.

◁ **Electrical energy**
This type of energy is carried by an electric current that supplies all kinds of appliances.

◁ **Chemical energy**
This is the form of energy released when chemical reactions take place, such as burning fuel.

◁ **Radiant energy**
This is the form of energy carried by light and other types of electromagnetic radiation.

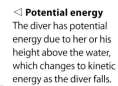

◁ **Nuclear energy**
This form of energy is released when atoms split apart (fission) or join together (fusion).

◁ **Sound energy**
This is a type of energy that objects produce when they vibrate in a medium, such as air.

◁ **Potential energy**
The diver has potential energy due to her or his height above the water, which changes to kinetic energy as the diver falls.

some of the rider's energy is converted to thermal energy, making him hot

climbing the hill, the rider's kinetic energy is converted into potential energy, which will be released (converted back into kinetic energy) when the rider freewheels down the other side

chemical energy in the rider's muscles makes the legs move

kinetic energy is transferred via the pedals and the chain to the back wheel

some of the wheels' kinetic energy becomes thermal energy, heating the bicycle's tires as they rub against the ground

Conservation of energy

The first law of thermodynamics (the study of how heat behaves) states that energy cannot be created or destroyed, but it can be transferred from one object to another and converted into different forms.

△ **Energetic bicycling**
The person is making the bicycle move by pushing on the pedals. As the bicycle goes faster, it gains kinetic energy. This is possible due to the chemical energy released in the rider's muscles that makes the legs move. At some point, the rider can use this same chemical energy in the muscles to stop the bike.

All machines will gradually lose energy, which, unfortunately, makes **perpetual motion impossible**.

REAL WORLD

Perpetual motion

For many years inventors have tried to develop a machine that could run forever. This machine shown right, designed by the German Ulrich von Cranach in 1664, was driven by cannonballs falling into the large wheel at the right. These would drop onto a curved track that fed them into an Archimedes screw. Powered by the wheel, the screw lifted the balls to the starting position. However, like all perpetual motion machines before and since, this clever design could not overcome the slowing effect of friction (see page 173).

Forces and mass

ALL MOTION IS CAUSED BY FORCES ACTING ON MASSES.

SEE ALSO
❰ 38–39 Movement
❰ 170–171 Energy
Gravity   178–179 ❱
Electricity   202–203 ❱

The effect of a force depends on the mass of the object. The greater the object's mass, the lower its resultant acceleration.

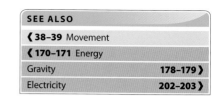

What is a force?

A force can affect an object in different ways. First, it may change the object's speed, so it moves faster or slower; second, a force can change the direction in which the object moves; third, the force may deform the shape of the object. Forces are measured in newtons (N). A force of 1 N results in a mass of 1 kilogram (kg) or 2.2 pounds (lb) reaching a speed of 1 meter (m) per second in one second.

ball reacts to the force of the golf club

golf club connects with the ball

◁ **Changing speed**
The force of the golf club increases the speed of the ball from zero to a high speed, sending it down the golf course.

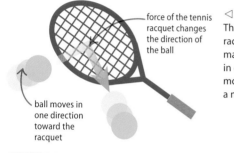

force of the tennis racquet changes the direction of the ball

ball moves in one direction toward the racquet

◁ **Changing direction**
The force of the tennis racquet on the ball makes it stop traveling in one direction, and moves the ball in a new direction.

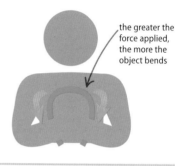

the greater the force applied, the more the object bends

◁ **Changing shape**
Depending on the toughness of the object, and strength of the person or machine used, the force exerted on an object may be able to change its shape.

What is mass?

Mass is a measure of how much an object resists a force. An object with a large mass contains more matter than one with a smaller mass. A force applied to a large mass results in a smaller acceleration than if it were applied to a small mass.

The precise kilogram unit is based on a single cylinder of the elements **platinum** and **iridium** kept in a safe in Paris, France.

tablecloth moves over smooth tabletop

mass of the china resists movement, and it does not move

hands jerk the tablecloth, and it moves as a result of this pulling force

◁ **Inertia**
One of the properties of mass is inertia. This is a tendency for objects to remain where they are at rest—or keep traveling at the same speed and in the same direction if on the move.

Friction and drag

Nothing in nature is perfectly smooth, so when objects slide past one another, their uneven surfaces push back against the direction of motion. This resistance force is known as friction. Drag is a similar phenomenon that occurs when an object pushes its way through air or water. The air or water pushes back, resisting the motion.

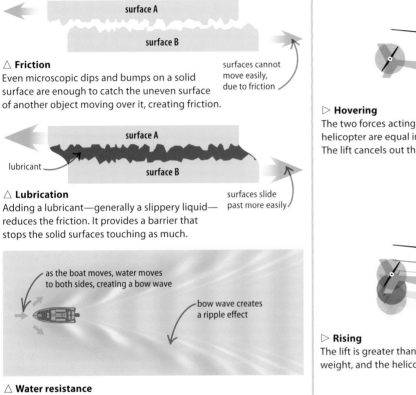

△ Friction
Even microscopic dips and bumps on a solid surface are enough to catch the uneven surface of another object moving over it, creating friction.

surfaces cannot move easily, due to friction

lubricant

surfaces slide past more easily

△ Lubrication
Adding a lubricant—generally a slippery liquid—reduces the friction. It provides a barrier that stops the solid surfaces touching as much.

as the boat moves, water moves to both sides, creating a bow wave

bow wave creates a ripple effect

△ Water resistance
The boat is moving through the water, and must push the water in front out of the way. The water resists and rises up as a bow wave.

REAL WORLD

Tire tread

Travel would be impossible without friction. For example, a tire has a rough tread to increase the friction force between the wheel and the road, preventing the car from sliding. Running shoes have rough soles for the same reason.

Resultant forces

Several different forces may act on one object at the same time, but sometimes the object cannot respond to each one individually. In these cases, the forces are combined to produce a single effect, so it appears that the object is being moved by a single force in one direction—this is the resultant force.

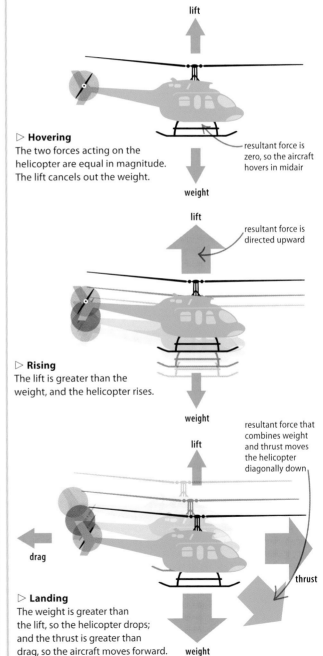

lift

weight

▷ Hovering
The two forces acting on the helicopter are equal in magnitude. The lift cancels out the weight.

resultant force is zero, so the aircraft hovers in midair

lift

resultant force is directed upward

weight

▷ Rising
The lift is greater than the weight, and the helicopter rises.

resultant force that combines weight and thrust moves the helicopter diagonally down

lift

drag

thrust

▷ Landing
The weight is greater than the lift, so the helicopter drops; and the thrust is greater than drag, so the aircraft moves forward.

weight

Stretching and deforming

AS WELL AS MOVING OBJECTS FROM PLACE TO PLACE,
FORCES CAN ALSO MAKE OBJECTS CHANGE SHAPE.

SEE ALSO

❰ 96–97 Properties of materials
❰ 98–99 States of matter
❰ 163 Plastics
❰ 172–173 Forces and mass

When a force acts on an object that cannot move, or when a number of different
forces act in different directions, they make the object's molecules (or other small
parts of it) move closer together or further apart, so the whole object changes shape.

Types of distortion

The type of distortion an object undergoes depends on the
number, directions, and strengths of forces acting on it, and
also on its structure and composition. Many objects simply
snap or shatter when strong forces act on them. Those that
do not are referred to as deformable, such as modeling clay.

Graphene is one of the **strongest**
and **most elastic** materials. It is made
of sheets of carbon atoms that are
connected together in hexagons.

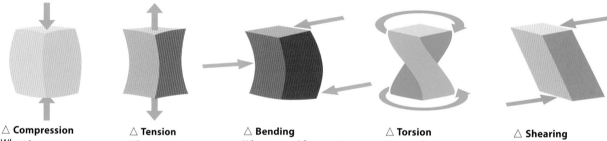

△ **Compression**
When two or more
forces act in opposite
directions and meet
at the same point inside
an object, the object
will compress and bulge
out on all sides.

△ **Tension**
When two or more
forces act in opposite
directions and pull
away from a object,
they apply tension,
and elastic objects will
stretch in response.

△ **Bending**
When several forces act
on an object in different
places, the object will
either snap (if it is brittle)
or bend (if it is malleable).
Many materials, like wood,
bend a little, and then snap.

△ **Torsion**
Turning forces, or
torques, that act in
opposite directions,
but affect different
parts of an object,
result in the object
being twisted.

△ **Shearing**
When forces act in
opposite directions
at the ends of an
object that is not
free to spin, its ends
will move in two
different directions.

Deformation

Forces that change the shape of an object
are known as stresses. The change in shape
of the object in response is called a strain.
When an object is stressed, three things
may happen: it may break; it may change
shape permanently (in which case it
is said to be "plastic"); or it may change
shape until the stress is removed—and
then return to its original shape (elastic).

▷ **Stress–strain curve**
Many materials behave elastically for small
distortions, and then will start to behave
plastically. Finally, they will break. The forces
required vary a lot, depending on the material.

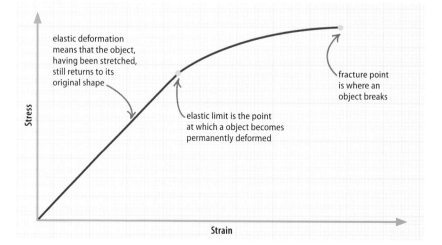

elastic deformation
means that the object,
having been stretched,
still returns to its
original shape

fracture point
is where an
object breaks

elastic limit is the point
at which a object becomes
permanently deformed

Stress

Strain

Hooke's Law

The English scientist Robert Hooke (1635–1703) discovered the law of elasticity. Hooke's Law states that the amount of stretch of a spring, or other stretchy object, is directly proportional to the force acting on it. The law is only true if the elastic limit of the spring has not been reached. If the elastic limit has been reached, the spring will not return to its original shape and may eventually break.

REAL WORLD

Bungee jump

If you fell a long distance while attached to a rope, you would stop suddenly, with a dangerous jerk. Elastic cords, such as those used by bungee jumpers, slow you down more gradually because the energy of the fall is slowly transferred to the cord as it stretches.

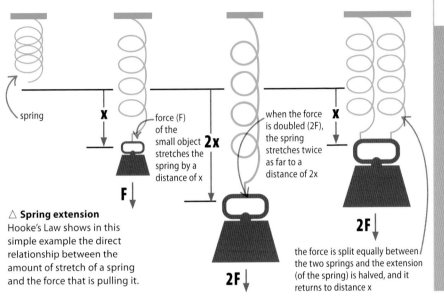

spring

X

2x

F ↓

force (F) of the small object stretches the spring by a distance of x

when the force is doubled (2F), the spring stretches twice as far to a distance of 2x

X

2F ↓

2F ↓

the force is split equally between the two springs and the extension (of the spring) is halved, and it returns to distance x

△ **Spring extension**
Hooke's Law shows in this simple example the direct relationship between the amount of stretch of a spring and the force that is pulling it.

Young's modulus

The elasticity of an object depends on its shape, size, and structure. The English polymath Thomas Young (1773–1829) devised a way of measuring the elasticity of solids—known as Young's modulus—to compare different materials.

Stiffness of select materials	
rubber	0.01–0.1
nylon	3
oak	11
gold	78
glass	80
stainless steel	215.3

△ **Measuring stiffness**
Young's modulus is measured in gigapascals (GPa). The higher the number, the stiffer (less elastic) the material.

Material properties

Many properties of materials are related to the way they deform under stress. They depend partly on the molecules from which materials are made, but also on the shapes and sizes of larger structures inside the material, such as crystals or fibers.

Description of materials under stress	
hard	difficult to scratch or dent
tough	difficult to break or deform
plastic	changes shape permanently when stressed
elastic	returns to original size and shape when stress is removed
brittle	breaks suddenly under stress, with little deformation
ductile	can be drawn out into a wire
malleable	can be hammered into shape

△ **Describing materials**
These terms are used to describe the behavior of materials under stress. Many materials change their behaviors with temperature. For example, warm rubber is very elastic, but very cold rubber is brittle.

Velocity and acceleration

THESE QUANTITIES TELL US HOW QUICKLY SOMETHING IS MOVING.

When motion of an object changes in rate or direction, the motion is described in terms of velocity and acceleration.

Speed and velocity

Speed is a measure of the rate at which a distance is covered. It is commonly measured in kilometers per hour. Velocity is also measured in these units, however, this measure also takes direction into account. Thermodynamicists and nuclear physicists use speed more often than velocity.

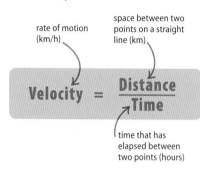

rate of motion (km/h)

space between two points on a straight line (km)

$$\text{Velocity} = \frac{\text{Distance}}{\text{Time}}$$

time that has elapsed between two points (hours)

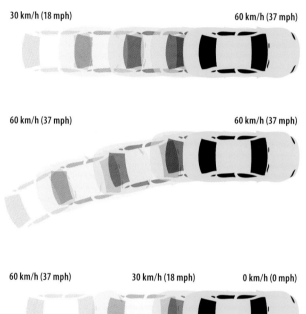

30 km/h (18 mph)   60 km/h (37 mph)

60 km/h (37 mph)   60 km/h (37 mph)

60 km/h (37 mph)   30 km/h (18 mph)   0 km/h (0 mph)

◁ **Increasing velocity**
An increase in the car's velocity is known as acceleration. A constant force gives a constant increase in velocity.

◁ **Changing direction**
The car continues at 60 km/h (37 mph), but then changes lanes. The car's speed is constant, but its velocity is changing.

◁ **Decreasing velocity**
When a car slows, the change in velocity is known as a deceleration.

Relative velocity

The relative velocity compares how fast an object is traveling in comparison to another. If two objects are traveling in the same direction, the relative velocity can be calculated by subtracting the velocity of the slower object from the velocity of the quicker one. Two objects moving in the opposite direction along the same path would be heading for a collision. The relative velocity of these objects would be greater than either of their individual speeds.

▷ **Relative speed zero**
Runner A is moving at the same velocity as B, so their relative velocity is 0 km/h (0 mph).

▷ **Catching up**
Runner A is gaining on B because his velocity is 1 km/h (0.6 mph) faster.

▷ **Heading for collision**
Runner A and B are moving in opposite directions. Their relative velocity is the sum of their two speeds, which is 14 km/h (8.7 mph).

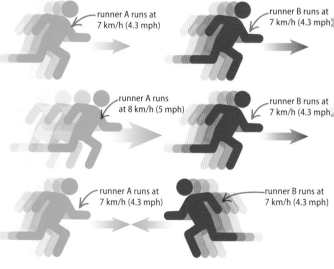

runner A runs at 7 km/h (4.3 mph)   runner B runs at 7 km/h (4.3 mph)

runner A runs at 8 km/h (5 mph)   runner B runs at 7 km/h (4.3 mph)

runner A runs at 7 km/h (4.3 mph)   runner B runs at 7 km/h (4.3 mph)

Changing velocity

Acceleration is a measure of the rate of change in velocity—how long it takes for an object to increase (or decrease) from one velocity to another. Acceleration is calculated by subtracting the starting velocity (V_1) from the final velocity (V_2) to obtain the change in velocity. This figure is then divided by the time that has passed.

rate of change of velocity (meters per second per second)

total change in velocity (meters per second)

time in which this change happened (seconds)

$$Acceleration = \frac{V_2 - V_1}{Time}$$

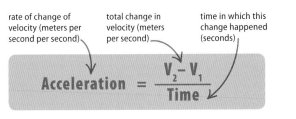

▷ **Motorcycle versus truck**
This graph shows the greater acceleration of the motorcycle compared to the truck, before they both reach the same cruising speed.

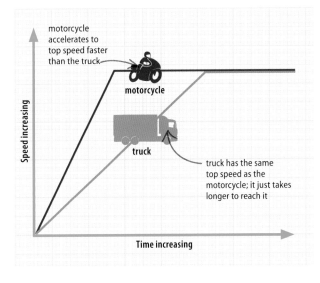

motorcycle accelerates to top speed faster than the truck

motorcycle

truck

truck has the same top speed as the motorcycle; it just takes longer to reach it

Speed increasing

Time increasing

Oscillation

An oscillation is a regular movement about a central point. Whether it is a pendulum swinging from side to side, a weight bouncing on the end of a spring, or the molecules vibrating inside a solid, the motion results from regular accelerations and decelerations. In turn, these produce an average velocity of zero because the object ends up coming back to the same central point. This phenomenon is caused by two opposing forces that accelerate the object to the center, but the resulting velocity moves the same distance in the opposite direction.

REAL WORLD

Timekeeping

Oscillations repeat at a constant rate. The time it takes for the oscillator to complete a full cycle is called the period. A pendulum oscillates with a fixed period, and the long pendulums in grandfather clocks have a period of two seconds. Each swing turns the cogs just enough to keep the hands moving at the right rate.

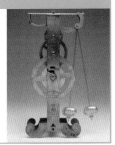

at the starting point the weight has no velocity and downward acceleration is at its maximum

▷ **Ups and downs**
When the weight is at its highest point above the center, or equilibrium point, its velocity is zero, and the downward acceleration is at its maximum. When the weight is at the equilibrium point its velocity is at maximum, while the acceleration has dropped to zero. The weight continues to the lowest point, where its velocity slows to zero and accelerates in the opposite direction.

at the equilibrium point, the weight is at its maximum downward velocity

at its lowest point, the weight stops and acceleratesin the opposite direction

at the equilibrium point, the weight is at its maximum upward velocity

the weight returns to its starting point and process continues

Gravity

THE FORCE OF GRAVITY AFFECTS EVERY OBJECT
IN THE UNIVERSE.

**Gravity is the force that holds planets together and keeps
them orbiting stars, as well as holding us to the Earth.**

Attraction

Gravity is a force of attraction. Although all objects attract all
others, gravity between objects on Earth is usually too small
to notice. This is why it took a genius like Isaac Newton to
understand that gravity does affect all objects, whatever their size.

acceleration of the
apple falling to Earth is
much greater than the
acceleration of Earth
moving toward the apple

apple pulls the Earth with
the same force as the Earth
pulls the apple

▷ **The falling apple**
Due to gravity, an apple
pulls the Earth with the
same force as the Earth
pulls the apple. However,
because the Earth is so
much more massive, it
accelerates far less than
the apple, and moves up
an immeasurably small
distance, while the apple
falls a greater distance.

Physicists think that the force of
gravity is carried by **tiny particles
called gravitons**—but they have
yet to find any.

Universal law

Isaac Newton discovered that gravity affects everything
in the Universe. He explained that the gravitational force
between two objects, such as planets, depends on their
masses and the distance between them. Newton also
showed that spherical objects, such as the Earth, act
as if all their mass is concentrated at their centers.

△ **Attraction**
All objects are attracted to each other by the force
of gravity. Above, two objects of the same mass
are attracted by the pull of this gravitational force.

△ **Double the mass**
Changing the masses of the two objects so that they
are twice as heavy will make the gravitational force
between them four times as strong.

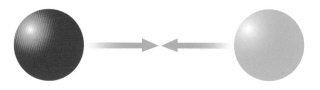

△ **Increase the distance**
Changing the distance between the two objects so
that they are twice as far away from each other will
make the gravitational force four times less.

Weight and mass

Weight is not the same thing as mass. Mass is the amount of matter an object contains, while weight is the force with which the Earth or another body pulls on the object. An object can have the same mass but weigh differently, depending on the gravitational force acting on the object.

Someone who weighs **40 kg (88 lb) on Earth** would weigh **more than one tonne on the Sun**—if it were possible to stand on its surface.

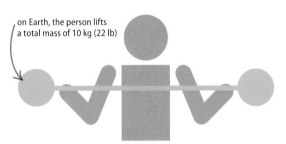

on Earth, the person lifts a total mass of 10 kg (22 lb)

on the Moon, the same person exerting the same amount of effort can lift 60 kg (132 lb)

△ **On Earth**
Lifting a barbell requires this person to exert a greater force than the barbell's weight. Here, a weightlifter is lifting a barbell with weights whose total mass is 10 kg (22 lb).

△ **On the Moon**
On the Moon, the force of gravity is about one sixth that on Earth. So the same effort is required to lift 60 kg (132 lb) on the Moon as it is to lift 10 kg (22 lb) on Earth.

Ballistics

A thrown object is pulled back down to Earth by the force of gravity. At the same time, its sideways motion is decelerated by the drag force applied by the air. If there were no atmosphere, the object would travel a much greater distance.

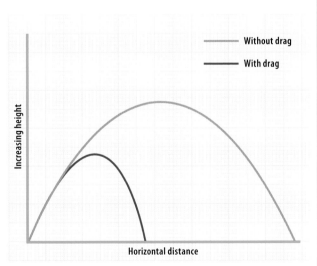

Without drag

With drag

Increasing height

Horizontal distance

△ **Air resistance and motion**
On Earth, air resistance applies a drag force, slowing the object's motion (red line). Where there is no air resistance, such as on the Moon, a thrown object moves in a parabola (green line)—a steady speed in a horizontal direction, while moving up and then down.

Orbit

The harder an object is thrown, the faster and farther it travels before its path takes it back to the Earth's surface. If the object is propelled hard enough, it will gain enough speed to counteract the pull of gravity so it will never land and will orbit the Earth.

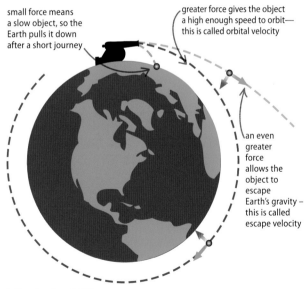

small force means a slow object, so the Earth pulls it down after a short journey

greater force gives the object a high enough speed to orbit—this is called orbital velocity

an even greater force allows the object to escape Earth's gravity – this is called escape velocity

△ **Newton's satellite**
This diagram is based on an illustration by Isaac Newton. He showed that if a cannonball is fired with enough force, its speed will allow it to orbit or escape Earth completely.

Newton's laws of motion

NEWTON'S LAWS EXPLAIN HOW FORCES ACT ON OBJECTS.

When a force acts on an object that is free to move, the object
will move in accordance with Newton's three laws of motion.

SEE ALSO

❰ **38–39** Movement

❰ **172–173** Forces and mass

❰ **176–177** Velocity and acceleration

Understanding motion **182–183** ❱

A new direction for physics

Isaac Newton (1642–1727) published his laws of motion
in 1687, setting the direction for physics over the next two
centuries. He explained that when the forces acting on an
object are balanced, there is no change in the way it moves.
When the forces are unbalanced, there is an overall force in
one direction, which alters the object's speed or the direction
in which it is moving. Newton also emphasized the complicated
relationship between objects and forces, which is due mainly
to the effects of friction and air resistance. Without these effects,
he concluded, the motions of objects are much simpler. So,
his laws apply most obviously to bodies in space, such as
planets and spacecraft.

REAL WORLD

Blast off!

Newton's laws of motion can
be used to explain how a rocket
blasts into space. At the start,
there is no force acting on the
rocket, so it does not move.
Then, when the rocket's engines
fire, the force they produce lifts
the rocket up and off the
launchpad. As the hot gases
shoot down, an equal force
pushes the rocket up.

First law of motion

The first law of motion states that any object will continue to remain stationary, or move
in a straight line at a constant speed, unless an external force acts on it. So, a soccer ball is
stationary until it is kicked, and then it moves until other forces bring it to a halt.

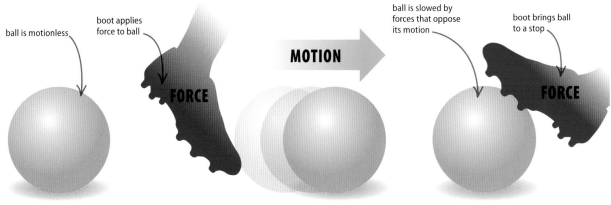

ball is motionless

boot applies
force to ball

MOTION

FORCE

ball is slowed by
forces that oppose
its motion

boot brings ball
to a stop

FORCE

△ **At rest**
Although the force of gravity is
acting on the soccer ball, the ground
below it stops the ball from moving,
so it remains in a state of rest.

△ **Force applied**
The impact of a boot applies a force
to the ball. For as long as the boot is
in contact with the ball, the ball will
be accelerated by it.

△ **Motion is arrested**
The ball immediately begins to slow due
to the resistance of the air and friction with
the ground. It is brought to a stop when it
encounters a stationary object (the boot).

Second law of motion

The second law of motion states that when a force acts on an object, it will tend to move in the direction of the force. The larger the force on an object, the greater its acceleration will be. The more massive an object is, the greater the force needed to accelerate it.

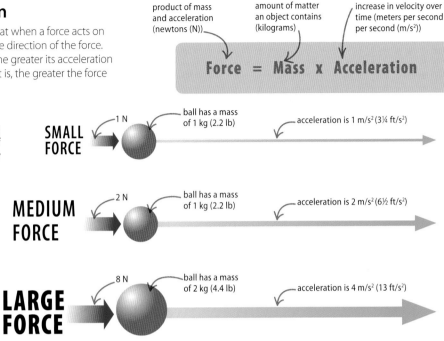

product of mass and acceleration (newtons (N))

amount of matter an object contains (kilograms)

increase in velocity over time (meters per second per second (m/s²))

$$Force = Mass \times Acceleration$$

▷ **Small mass, small force**
A force of 1 N acting on a mass of 1 kg (2.2 lb) will produce an acceleration of 1 m/s² (3¼ ft/s²)—velocity increases by 1 m (3¼ ft) per second every second.

SMALL FORCE — 1 N — ball has a mass of 1 kg (2.2 lb) — acceleration is 1 m/s² (3¼ ft/s²)

▷ **Small mass, medium force**
A force of 2 N acting on a mass of 1 kg (2.2 lb) will produce an acceleration of 2 m/s² (6½ ft/s²).

MEDIUM FORCE — 2 N — ball has a mass of 1 kg (2.2 lb) — acceleration is 2 m/s² (6½ ft/s²)

▷ **Double mass, large force**
A force of 8 N acting on a mass of 2 kg (4.4 lb) will produce an acceleration of 4 m/s² (13 ft/s²).

LARGE FORCE — 8 N — ball has a mass of 2 kg (4.4 lb) — acceleration is 4 m/s² (13 ft/s²)

Third law of motion

The third law of motion states that any object will react to a force applied to it. The force of reaction is equal and acts in an opposite direction to the original force that produces it. In the diagram, two people of the same mass are each standing on a skateboard, which reduces friction. When they push together (action), the result (reaction) will be that they move in the opposite direction from each other.

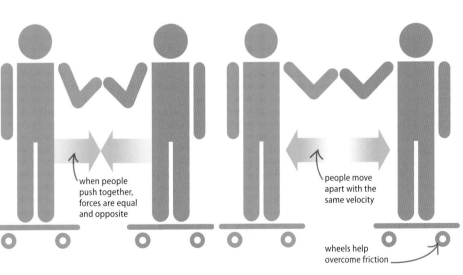

when people push together, forces are equal and opposite

people move apart with the same velocity

wheels help overcome friction

△ **Action**
The third law works when a force acts between two objects. Even if the second person does not make any effort to push back, her or his body will always react to the force in the same way.

△ **Reaction**
The forces between the two people are equal and opposite as they push away from each other on their skateboards. The masses are equal, so they move away from each other at the same velocity.

Understanding motion

FORCES ARE ABLE TO TRANSFER ENERGY TO MAKE
OBJECTS MOVE.

Forces are rarely applied to an object one at a time and in straight
lines. To understand how objects move, some principles are applied.

Momentum

A moving object carries on moving
because it has momentum. It will keep
moving until a force stops it. For example,
when you catch a ball, you must exert a
force on it to remove its momentum and
stop it moving. However, when your hand
and the ball collide, the ball will exert a
force on your hand so that the momentum
of your hand will change. The momentum
gained by your hand is equal to the
momentum lost by the ball. Momentum
is calculated by multiplying the mass of
an object by its velocity—the heavier an
object and the faster it moves, the greater
its momentum. This conservation of
momentum, as shown in the illustration,
is evidence that energy is never created or
destroyed, but transferred between objects.

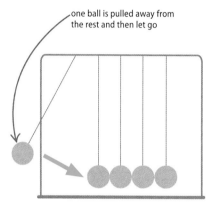

one ball is pulled away from
the rest and then let go

△ **Collision action**
As the left ball hits the line of other
balls, its velocity decreases and its
momentum drops to zero.

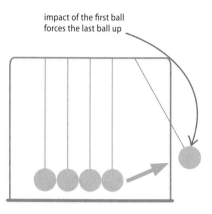

impact of the first ball
forces the last ball up

△ **Motion reaction**
The momentum of the first ball passes
through its neighbors until it reaches
the right ball, which is forced to move.

Kinetic energy

The energy of a moving object is described as kinetic energy.
The more kinetic energy an object has, the faster it moves.
Some objects can have a relatively small mass but great kinetic
energy. For example, the asteroid that is thought to have killed
the dinosaurs 65 million years ago had a huge impact despite
having a relatively small mass. This is because it hit the ground
at around 30 km (19 miles) per second, giving it as much
energy as one million express trains.

The **purpose of an engine** is
to convert the energy in fuel—
or perhaps a battery—into
kinetic energy.

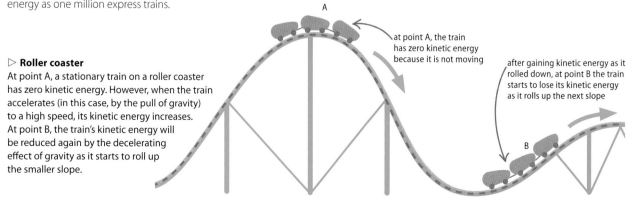

▷ **Roller coaster**
At point A, a stationary train on a roller coaster
has zero kinetic energy. However, when the train
accelerates (in this case, by the pull of gravity)
to a high speed, its kinetic energy increases.
At point B, the train's kinetic energy will
be reduced again by the decelerating
effect of gravity as it starts to roll up
the smaller slope.

A

at point A, the train
has zero kinetic energy
because it is not moving

after gaining kinetic energy as it
rolled down, at point B the train
starts to lose its kinetic energy
as it rolls up the next slope

B

Torque

This term refers to the turning effect of a force—its ability to create rotation rather than linear (straight-line) motion. Torque is dependent on the size of a force and its distance from the turning point, or pivot. Forces applied farther from the pivot result in a larger torque. The torque (or moment) of a force is calculated by multiplying force and distance.

The great Greek mathematician **Archimedes of Syracuse** (c.287 BCE–c.212 BCE) reasoned that if he had a lever long enough, he could generate enough force to **lift up Earth**.

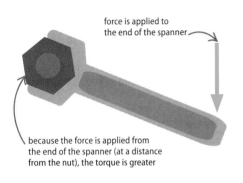

force is applied to the end of the spanner

because the force is applied from the end of the spanner (at a distance from the nut), the torque is greater

△ **Large torque**
Applying a force to the end of a spanner handle maximizes its torque, making it easier to undo a stiff nut.

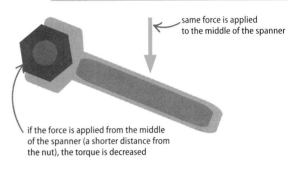

same force is applied to the middle of the spanner

if the force is applied from the middle of the spanner (a shorter distance from the nut), the torque is decreased

△ **Small torque**
Applying the same force to halfway along the handle results in half the torque, so turning the nut requires more effort.

Rotational motion

When an object moves in a circle, it is acted on by two forces. The centripetal force pulls it toward the center, such as gravity on a space satellite or the strength of a string attached to a ball. A second centrifugal force counteracts the centripetal force by pulling the object away from the center.

▷ **In a spin**
The object accelerates toward the center of the circle. This is balanced by a virtual force—called centrifugal force—which reacts to the centripetal force and keeps the object from moving to the center. The result is a continuous curving motion around the center.

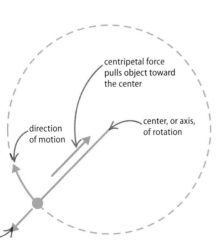

centripetal force pulls object toward the center

center, or axis, of rotation

direction of motion

virtual force, known as centrifugal force, pulls the object away from the center

REAL WORLD
Angular momentum

Any spinning object has what is called angular momentum, which is proportional to the mass of the object, its rotational speed, and the average distance of the mass from the center of the spin. The ice skater uses this phenomenon to control the speed of her spins. When she stretches out her arms, she spreads her mass over a wider area, which creates a relatively slow rate of spin. When she draws them back in, all her mass is centered over the axis of rotation, and she spins faster.

Pressure

PRESSURE IS THE RESULT OF ONE THING
PRESSING ON THE SURFACE OF ANOTHER.

Pressure can be applied to or by any
medium, including air and water.

What is pressure?

Pressure is defined as force per unit area, and is measured
in pascals (Pa), which is equal to one newton per square
meter. To calculate the pressure, divide the force pressing
on the object by the area it is spread across.

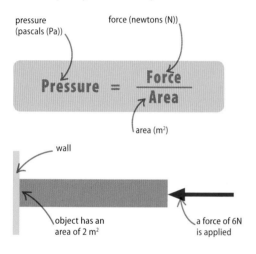

pressure (pascals (Pa))   force (newtons (N))

$$Pressure = \frac{Force}{Area}$$

area (m²)

wall

object has an area of 2 m²   a force of 6N is applied

△ **Larger area, lower pressure**
If a force of 6 N is spread over an area of 2 m², the
applied pressure is 6 divided by 2, which equals 3 Pa.

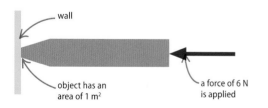

wall

object has an area of 1 m²   a force of 6 N is applied

△ **Smaller area, higher pressure**
If a force of 6 N is spread over an area of 1 m², the
applied pressure is 6 divided by 1, which equals 6 Pa.
This is why drawing pins and nails are pointed: their
small-area tips apply very high pressures, so they
penetrate materials easily.

Atmospheric pressure

At the Earth's surface, the atmosphere applies a pressure of about
about 101,000 pascals to all objects. We cannot feel this pressure
because it is balanced by an equal and opposite pressure inside our
bodies. In different weather conditions and at different heights, the local
atmospheric pressure changes. This can be measured using a barometer.

▽ **Height and pressure**
Gas molecules are constantly on the move and bumping
into each other. When they hit another molecule or the
wall of a container, they exert pressure on it. The air
molecules close to the Earth are at the bottom of the
atmosphere with all the other air molecules on top of
them. Therefore, the pressure is higher and the air is
denser. Air grows thinner higher from the Earth's surface.

high above
the Earth, the
atmospheric
pressure is low,
so air molecules are
widely separated

closer to Earth,
the atmospheric
pressure is high,
so the air molecules
are densely
packed together

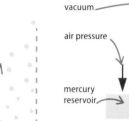

vacuum

air pressure

mercury
reservoir

mercury level
is short because
that is all
the low
atmospheric
pressure can
support

△ **Low atmospheric pressure**
If the air pressure above the reservoir
is low, it will not produce enough force
to make the mercury rise up the tube.

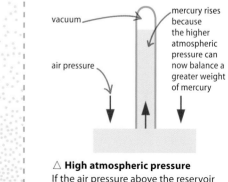

vacuum

air pressure

mercury rises
because
the higher
atmospheric
pressure can
now balance a
greater weight
of mercury

△ **High atmospheric pressure**
If the air pressure above the reservoir
is high, it will push the mercury up
the tube.

Water pressure

Water is far denser than air. This means that as one goes deeper under water the pressure increases rapidly. On Earth, the water pressure at 10 m (33 ft) depth is about one "atmosphere," which is roughly the air pressure at sea level. At a depth of 20 m (66 ft), the water pressure is about two atmospheres, and at 30 m (99 ft) it is about three atmospheres, and so on.

▷ **Under pressure**
In this milk carton, the pressure increases toward the base, so water will squirt out under greater pressure from the bottom hole than from the top one.

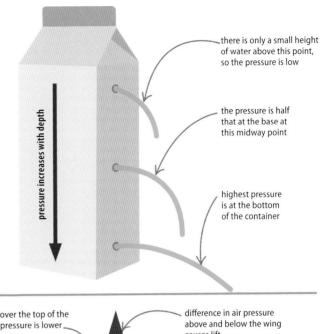

pressure increases with depth

there is only a small height of water above this point, so the pressure is low

the pressure is half that at the base at this midway point

highest pressure is at the bottom of the container

Bernoulli effect

Pressure varies according to the motion of a medium—this is called the Bernoulli effect. In an airplane wing, the top surface of the wing has more camber (longer curve) than the bottom surface, so the air flows faster over the top of the wing than it does underneath.

▷ **Taking flight**
There is less air pressure above the wing than there is beneath the wing. The difference in the air pressure above and below the wing causes lift.

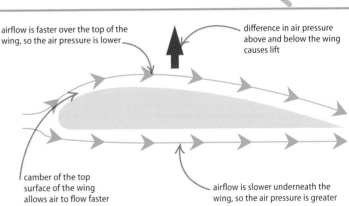

airflow is faster over the top of the wing, so the air pressure is lower

difference in air pressure above and below the wing causes lift

camber of the top surface of the wing allows air to flow faster

airflow is slower underneath the wing, so the air pressure is greater

Hydraulics

In an hydraulic system, a liquid (often oil) is used to transfer force from one place to another. Usually, hydraulic systems also convert a low force at one place to a higher one at another. Hydraulic systems rely on the fact that liquids (unlike gases) are almost incompressible—if they are pressed, rather than reducing in volume they force objects like pistons to move away.

narrow pipe means only a small force is needed, but the piston must be pushed down a long way to move the car up a short distance

◁ **Hydraulic multiplication**
The increase in force created by an hydraulic system can be calculated from the areas of the two sides (shown here in cross-section): if one side has twice the area of the other, the force will be doubled.

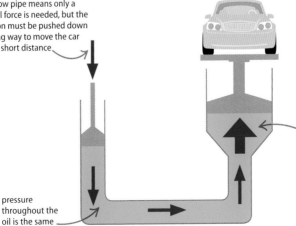

this end of the pipe is wider than the other, so the force applied by the oil pressure is greater

pressure throughout the oil is the same

Machines

MACHINES MAKE WORK EASIER TO DO.

A simple machine is a device that increases the size, or changes the direction, of a force. Machines can use this energy to lift, cut, or move masses.

Simple machines

Even the most complex devices can be broken down into half a dozen simple machines working together. All of these machines have been in use since ancient times, and at first glance some may not seem to be machines at all. However, the way they all multiply the force applied to them, or multiply the distance over which that force acts, makes them machines.

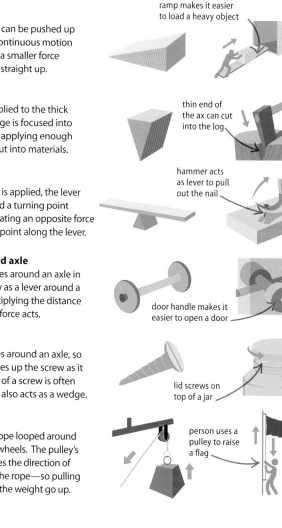

▷ **Ramp**
A heavy load can be pushed up a ramp in a continuous motion that requires a smaller force than lifting it straight up.

ramp makes it easier to load a heavy object

▷ **Wedge**
Any force applied to the thick end of a wedge is focused into the thin end, applying enough pressure to cut into materials.

thin end of the ax can cut into the log

▷ **Lever**
When a force is applied, the lever moves around a turning point (fulcrum), creating an opposite force at a different point along the lever.

hammer acts as lever to pull out the nail

▷ **Wheel and axle**
A wheel moves around an axle in the same way as a lever around a fulcrum, multiplying the distance over which a force acts.

door handle makes it easier to open a door

▷ **Screw**
A screw wraps around an axle, so the load moves up the screw as it turns. The tip of a screw is often pointed, so it also acts as a wedge.

lid screws on top of a jar

▷ **Pulley**
A pulley is a rope looped around one or more wheels. The pulley's wheel changes the direction of the force on the rope—so pulling down makes the weight go up.

person uses a pulley to raise a flag

Levers

Levers magnify a small force into a large force. They work by moving a load around a turning point. There are three types of levers; the difference between them depends on where the effort, load, and fulcrum are positioned.

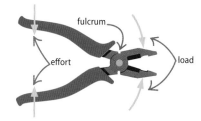

△ **First-class lever**
When you use pliers, the effort is applied on the opposite side of the fulcrum to the load.

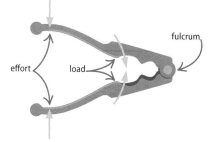

△ **Second-class lever**
When you use a nutcracker, the load is positioned between the effort and the fulcrum.

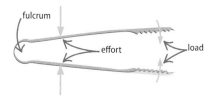

△ **Third-class lever**
When you use tongs, the effort is applied between the load and the fulcrum.

Pulleys

Compound pulleys are good examples of the way machines create a mechanical advantage—amplifying a small effort into a large force capable of lifting big loads. The double pulley is the simplest type. Single pulleys do not create a mechanical advantage, but they allow a force to be applied in a different direction.

The **first machine** ever created was the wedge-shaped handax from the Stone Age.

▷ **Single pulley**
A simple pulley is used to change the direction of a force. The effort and load are equal, so the force needed to lift this object is the same as that required to lift it by hand.

effort required to lift the load

load force is equal to the effort

▷ **Double pulley**
The rope runs around two pulleys, doubling the mechanical advantage. This allows a person to lift the same load with half the effort of a single pulley.

load force is twice that of effort

effort pulls rope twice the distance the load is raised

Gears

When wheels are given interlocking teeth they become cogs or gears, which are used to transmit a turning force, or torque. The magnitude of the transmission depends on the gear ratio, a comparison of the number of teeth on each gear. For example, when the driver gear (moved by the effort force) has twice the number of teeth of the driven gear, this second gear rotates twice as fast and with half the torque.

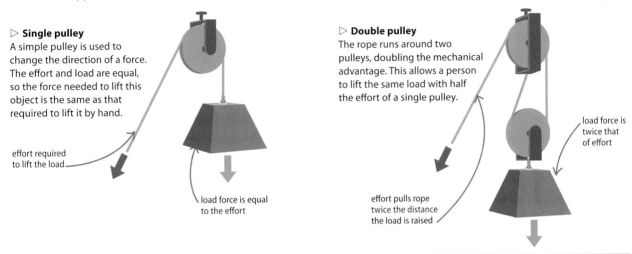

driven gear has seven teeth and rotates counterclockwise

driver gear has 28 teeth and rotates clockwise

◁ **Gear ratio**
To calculate the gear ratio of the gears below, divide the number of teeth of the driver (left; 28 teeth) by that of the driven gear (right; seven teeth). The answer is four, which means that the smaller, driven gear will turn four times faster than the speed of the larger, driver gear.

worm gear turns the screw, transforming the rotation by 90°

rack and pinion gear converts linear motion of toothed rail into rotation

bevel gears interlock at right angles to each other

△ **Transmission**
Several gears together are often called gear trains, or transmissions. They are used in machines to redirect force from one moving part to another.

Excavator

Construction machines, such as this excavator, show how simple machines are combined. The digger moves on tracks, which are driven by wheels acting as pulleys at each end. The shovel uses a wedge to cut into the ground, and moves using hydraulic levers.

Heat transfer

THERMODYNAMICS IS THE STUDY OF THE WAY HEAT MOVES
FROM ONE SUBSTANCE TO ANOTHER.

Heat is the name used for the type of energy that makes the atoms
and molecules move inside a substance. Adding energy makes these
particles move more quickly—and results in the substance heating up.

Measuring heat

Temperature is a measure of the heat in a substance. It is an average figure for the energy contained by every particle. Temperature and energy are not interchangeable. A spark from a fire can have a very high temperature, but it does not cause much of a burn because it contains only a small amount of energy. Temperature is measured using a scale. The difference between the upper and lower points is divided into a fixed number of units, or degrees, and any temperature can be expressed in multiples of degrees.

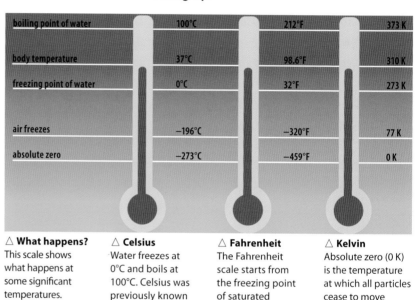

	Celsius	Fahrenheit	Kelvin
boiling point of water	100°C	212°F	373 K
body temperature	37°C	98.6°F	310 K
freezing point of water	0°C	32°F	273 K
air freezes	−196°C	−320°F	77 K
absolute zero	−273°C	−459°F	0 K

△ **What happens?**
This scale shows what happens at some significant temperatures.

△ **Celsius**
Water freezes at 0°C and boils at 100°C. Celsius was previously known as centigrade.

△ **Fahrenheit**
The Fahrenheit scale starts from the freezing point of saturated saltwater (0°F).

△ **Kelvin**
Absolute zero (0 K) is the temperature at which all particles cease to move completely.

Conduction

Heat always moves from areas with high thermal energy to areas with less. In other words, hot things always cool down, while cold things warm up to match their surroundings. Heat moves through a solid by conduction. This is a phenomenon in which the motion of particles in the hot part of the solid gradually transfers to neighboring ones, making them move faster and sending heat energy through the solid. Metals conduct heat better than nonmetals because their electrons are more free to move and pass their energy on.

▽ **Glowing hot**
Metal changes color as it heats up. It glows red and orange at first and then becomes white hot. Experts can judge a metal's temperature just by the color.

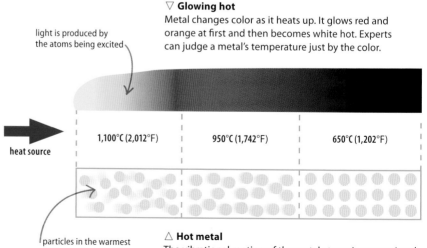

light is produced by the atoms being excited

heat source

1,100°C (2,012°F)   950°C (1,742°F)   650°C (1,202°F)

particles in the warmest region push against those farther along, making the heat flow

△ **Hot metal**
The vibrational motion of the metal atoms is proportional to the heat energy they contain. Those in the warmest part of the bar move faster than those in the cooler areas.

Convection currents

Heat moves through liquids or gases by a process called convection. This works on the basis that hot fluids rise upward while cooler ones sink. As a fluid receives energy its particles move faster and spread out. That results in it becoming less dense and rising upward through the cooler, denser fluid. The cooler fluid sinks, and fills the space left by the warmer fluid. This cool fluid is then exposed to the same heat source, and it is heated and rises up. As a result, heat is transferred around the fluid in a continuous convection current of rising and falling fluids.

Convection currents are responsible for the movement of **tectonic plates** on the Earth's crust.

▽ **Heating water**
The heat from the burner spreads out across the bottom of the metal pan by conduction, but it travels through the water via convection currents.

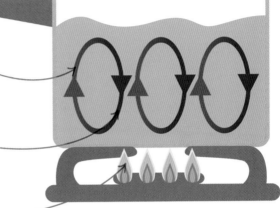

handle is made from plastic, which does not conduct heat well, so it does not get too hot to hold

warm water rises up to the top of the pan, where it begins to cool

cool liquid sinks to the bottom, where it is heated enough to rise again

heat source transfers energy to the water at the bottom of pan

Radiating heat

Heat can travel in the form of electromagnetic radiation—mainly infrared and microwaves. In general, smaller objects radiate away their heat more quickly than larger ones. This is because small objects have large surface areas compared to their volumes: a cube with a surface area of 24 unit^2 (such as the example below, left) has a volume of 8 unit^3 and a cube with a surface area of 6 unit^2 has a volume of only 1 unit^3.

greater surface area is exposed, so this tower radiates heat more quickly

▷ **Comparing surface areas**
This cube has the same volume (8 unit^3) as the tower to the right, but has a smaller surface area (24 unit^2). Therefore, less of its heat energy has access to the surface, so it radiates heat more slowly. The tower has a larger surface area (28 unit^2) than the cube, so more of its heat energy has access to the surface, where it can radiate into space, making the object cool more quickly than the cube.

smaller surface area is exposed, so this cube radiates heats more slowly

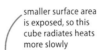

Saving heat

Animals that live in cold parts of the world are larger than their relatives that live in warmer places. For example, polar bears are much larger than the sun bears of southern Asia. The big polar animal loses precious heat more slowly than its tropical cousins because its large body gives it a small surface area to volume ratio.

Using heat

SOME MACHINES HARNESS THE ENERGY IN HEAT TO CREATE
MOTION, WHILE OTHERS TRANSFER HEAT TO WARM FOOD.

Many vehicles are powered by harnessing the heat energy released
from fuels. By contrast, a fridge releases heat to keep the contents
inside cool so they can last longer.

Internal combustion engine

Most road vehicles are powered by internal combustion engines. In external
combustion engines, such as steam locomotives, the burning fuel is kept
separate from the high-pressure steam that drives the engine. In internal
combustion engines the power comes from burning fuel (gasoline or diesel)
inside a cylinder, creating motion in a four-stroke cycle.

The first internal
combustion engine
design was fueled
by **gunpowder**.

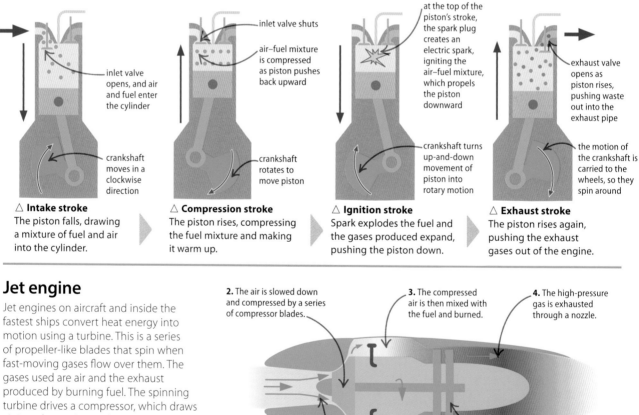

△ **Intake stroke**
The piston falls, drawing
a mixture of fuel and air
into the cylinder.

△ **Compression stroke**
The piston rises, compressing
the fuel mixture and making
it warm up.

△ **Ignition stroke**
Spark explodes the fuel and
the gases produced expand,
pushing the piston down.

△ **Exhaust stroke**
The piston rises again,
pushing the exhaust
gases out of the engine.

Jet engine

Jet engines on aircraft and inside the
fastest ships convert heat energy into
motion using a turbine. This is a series
of propeller-like blades that spin when
fast-moving gases flow over them. The
gases used are air and the exhaust
produced by burning fuel. The spinning
turbine drives a compressor, which draws
air into the engine and squeezes it so it
gets hot. The hot air makes the fuel burn
more quickly, driving the turbine around
faster. The aircraft is thrust forward by the
jet of gas sent backward by the turbine.

2. The air is slowed down
and compressed by a series
of compressor blades.

3. The compressed
air is then mixed with
the fuel and burned.

4. The high-pressure
gas is exhausted
through a nozzle.

the turbine

1. As the turbine spins,
air is sucked into the
engine though an inlet.

△ **How a jet engine works**
Most aircraft have turbine engines, where
a turbine is used to pull air into the engine,
which begins the process described above.

Rocket engine

Rocket engines do not burn their fuels in air. Instead, the fuel is mixed with another chemical, called an oxidizer, which creates a very hot and vigorous reaction. The hot expanding gases produced by the reaction are forced out of a small nozzle. The action of the gas leaving the engines results in a reaction force that drives the rocket forward.

▽ **How a liquid-fueled rocket engine works**
Unlike a jet engine, a rocket engine carries both the fuel and oxidizer on board. Smaller rockets, such as fireworks, use solid fuels, while the largest rockets have liquid fuels.

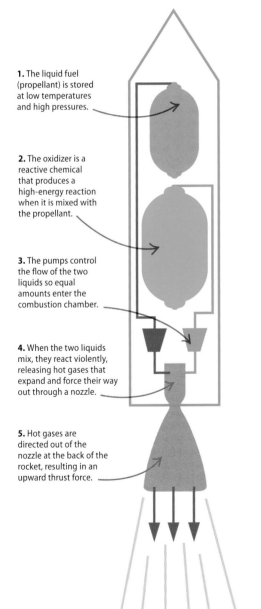

1. The liquid fuel (propellant) is stored at low temperatures and high pressures.

2. The oxidizer is a reactive chemical that produces a high-energy reaction when it is mixed with the propellant.

3. The pumps control the flow of the two liquids so equal amounts enter the combustion chamber.

4. When the two liquids mix, they react violently, releasing hot gases that expand and force their way out through a nozzle.

5. Hot gases are directed out of the nozzle at the back of the rocket, resulting in an upward thrust force.

Refrigeration

Cold is the absence of heat, and a refrigerator chills food by removing heat from the internal storage space. Heat energy always moves from hot places to cold ones. A refrigerator works by passing a cold gas behind the storage space, so heat from the air inside moves to that gas, making the air colder. The cold gas is produced by expanding a liquid very rapidly. The temperature drops as the molecules spread out, thus preserving food and drinks.

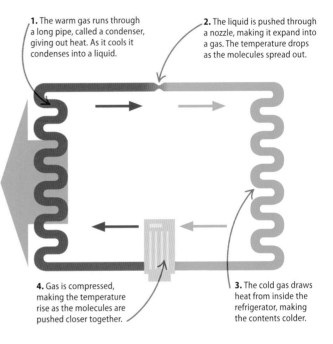

1. The warm gas runs through a long pipe, called a condenser, giving out heat. As it cools it condenses into a liquid.

2. The liquid is pushed through a nozzle, making it expand into a gas. The temperature drops as the molecules spread out.

4. Gas is compressed, making the temperature rise as the molecules are pushed closer together.

3. The cold gas draws heat from inside the refrigerator, making the contents colder.

△ **Refrigeration cycle**
In a refrigerator, a refrigerant (a substance used for cooling) travels around a system of pipes. First, heat radiates from the warm refrigerant. Second, the refrigerant begins to expand and cool. Third, the cold refrigerant cools the refrigerator because thermal energy moves from the refrigerator to the refrigerant. Finally, the compressor squeezes the refrigerant so it gets warmer as it begins to release its thermal energy.

REAL WORLD

Microwave oven

A microwave heats food using high-energy microwaves, which are absorbed by the bonds in water and fat molecules. These vibrate, which causes the food to heat up.

Waves

WAVES ARE VIBRATIONS THAT TRANSFER ENERGY.

Many different types of energy travel in waves. Sound waves carry noises through air, while seismic waves travel inside the Earth and cause earthquakes.

What is a wave?

Waves are vibrations that transfer energy as they travel. Some energy waves—sound waves, for example—need to travel through a medium, such as water or air. The medium does not travel with the wave, but moves back and forth as energy is passed through it, similar to the way a "wave" travels around a sports stadium as people move up and down in their seats. There are two main types of wave: transverse and longitudinal.

REAL WORLD

Seismic waves

Seismic waves are caused by the movement of rocks underground. As the vibrations travel up to the Earth's surface, they can produce earthquakes. The huge amounts of energy in seismic waves can be detected many thousands of kilometers away by sensitive instruments called seismometers.

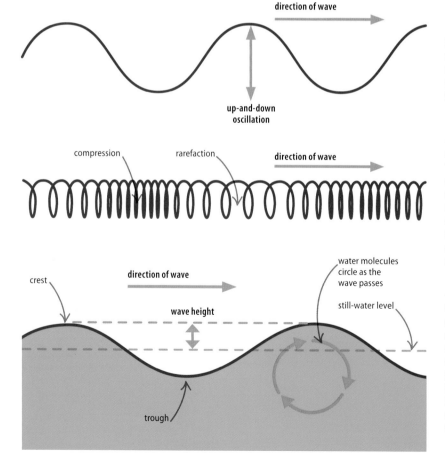

direction of wave

up-and-down oscillation

◁ **Transverse wave**
Light and other electromagnetic waves are transverse waves. This type of wave oscillates (vibrates) up and down at right angles (transversely) to the direction of travel of the wave, following an S-shaped path.

compression   rarefaction   **direction of wave**

◁ **Longitudinal wave**
Sound energy travels in longitudinal waves. The effect is like releasing a stretched spring and watching the energy travel along the coils, squeezing them together (sections called compressions) and stretching them apart (sections called rarefactions). Sound moves through air by pushing and pulling air molecules in a similar pattern of compressions and rarefactions.

crest   **direction of wave**   water molecules circle as the wave passes

wave height   still-water level

trough

◁ **Ocean wave**
Ocean waves are formed by the action of the wind pushing against the surface of the water, and have qualities of both of the wave types above. At the surface, water rises and falls between its highest point (crest) and lowest point (trough), equidistant to the still-water level. As the wave passes, the water molecules below the surface do not move forward but loop in circles.

Measuring waves

Waves have three important measurements. These are wavelength, frequency, and amplitude.

▽ Wavelength

The wavelength is the length of one complete wave cycle. It is the distance measured between any point on a wave and the equivalent point on the next wave.

longer wavelengths take a longer time to complete

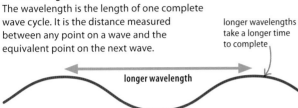

longer wavelength

shorter wavelengths take a shorter time to complete

shorter wavelength

crest

trough

▽ Frequency

This is the number of waves passing any point in one second. The unit of frequency is the hertz (Hz).

waves with a short wavelength have a higher frequency than those with longer wavelengths

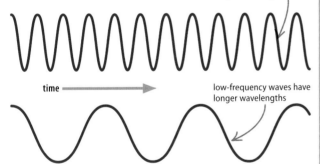

time

low-frequency waves have longer wavelengths

▽ Amplitude

The amplitude is the height of a crest or trough as the wave travels, measured from the central rest postion.

dim light waves have a small amplitude

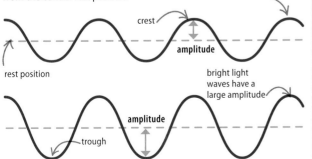

crest

amplitude

rest position

bright light waves have a large amplitude

amplitude

trough

Wave speed

The speed of a wave is related to its frequency and wavelength. They are linked by this equation:

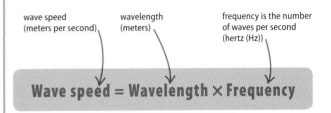

wave speed (meters per second)

wavelength (meters)

frequency is the number of waves per second (hertz (Hz))

Wave speed = Wavelength × Frequency

Calculating wave speed

The waves below are traveling across water. To find the wave speed, multiply the wavelength by the frequency.

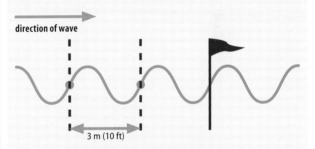

direction of wave

3 m (10 ft)

△ **The wavelength** has been worked out—each wave is 3 m (10 ft) long. A marker is used to help count the number of waves in one second.

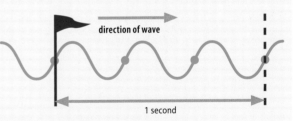

direction of wave

1 second

△ **One second later** Three waves have passed the marker, so the frequency is 3 Hz.

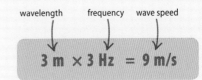

wavelength   frequency   wave speed

3 m × 3 Hz = 9 m/s

Electromagnetic waves

THESE ARE WAVES THAT CARRY ENERGY THROUGH SPACE.

Electromagnetic (EM) waves transfer energy from one place to another. There are different types but they all travel through a vacuum, such as space, at the speed of light.

SEE ALSO

❰ **30–31** Photosynthesis
❰ **168–169** Inside atoms
❰ **170–171** Energy
Light　　　　　　　**196–197** ❱
Astronomy　　　　　**230–231** ❱

Along the spectrum

Visible light is just one type of EM wave; other types are invisible. The full range of waves, called the electromagnetic spectrum, is made up of waves of different frequencies and wavelengths. At one end are radio waves, which have the longest wavelengths and lowest frequencies. At the other end are gamma rays, with the shortest wavelengths and highest frequencies.

▽ **Properties and uses**
The different types of electromagnetic radiation have different properties and uses, depending on their wavelength. Waves with shorter wavelengths, such as gamma rays and X-rays, can carry large amounts of energy, while longer radio waves do not.

REAL WORLD
Snake sense

Some animals can sense infrared radiation well enough to find warm objects in the dark. Some snakes have pits —see the hollow depression on the snout of this viper—that contain heat receptor cells. At night, or when their prey is hiding, these sensors can detect the body heat of warm-blooded prey, such as mice.

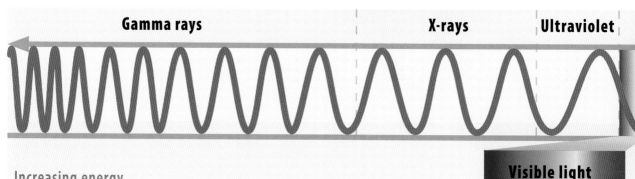

Gamma rays　　　　　　　　　　**X-rays**　　　**Ultraviolet**

Increasing energy　　　　　　　　　　　　　　　　**Visible light**

Gamma rays
These are produced by radioactivity and can carry a lot of energy. They cannot be seen or felt but are very harmful. While they can cause cancer, they also kill cancer cells. Other uses include sterilizing food and surgical instruments.

X-rays
X-rays are used to make images of inside the body because they pass through skin and soft tissue, but are absorbed by harder materials such as bone. In high doses they can be harmful, so X-rays must be used with caution.

Ultraviolet (UV)
UV radiation is found naturally in sunlight. You cannot see it or feel it, although you can experience the effects of too much UV as sunburn. Sunblock and sunglasses should be worn to protect skin and eyes from UV damage.

Visible light
This set of wavelengths is the only one that our eyes can see. The color seen depends on the wavelength of light, with violet and blue having shorter wavelengths than green and yellow. Red has the longest wavelength of all.

The source of EM radiation

EM radiation is associated with the force that holds electrons in place around atoms (see pages 168–169). However, the electrons can move around, jumping between higher and lower energy levels, or shells. These changes result in the atom absorbing or emitting energy in the form of EM radiation.

▽ **Energy in**
For an electron to jump from one shell to the next one up, it needs a specific amount of energy. It cannot make the move in small jumps, nor can it jump beyond the shell and fall back. It will only move if it receives radiation with exactly the right amount of energy.

▽ **Energy out**
When an electron jumps back to its original position nearer to the nucleus, it releases energy in the form of a specific wavelength of radiation. This process is what makes objects glow with visible light, give off heat, or emit other forms of radiation.

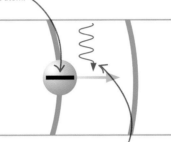

1. The electron is in a low-energy position near the center of the atom.

2. If the electron receives radiation of the correct wavelength, it will jump to the next level.

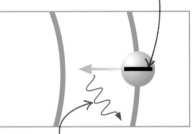

3. The electron is at a high-energy level, farther away from the nucleus than normal.

4. As it drops down to the lower level, the electron gives out radiation of a specific wavelength.

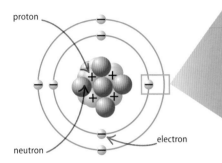

proton

neutron

electron

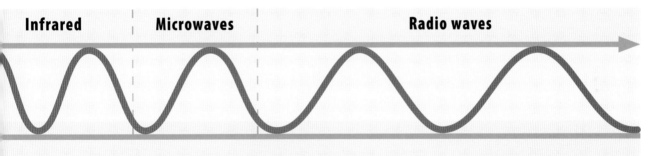

Infrared   **Microwaves**   **Radio waves**

Increasing wavelength

Infrared
Infrared means "below red," and has a lower frequency and longer wavelength than visible red light. We experience infrared waves as heat, and can see it at work in heaters, grills, and toasters. It is also used in television remote controls and in fiber optics.

Microwaves
This band of wavelengths is used in many types of personal communications, including mobile phones, wi-fi, and Bluetooth, as well as in microwave ovens. It is also used in radar technology, as a way of locating airplanes and ships.

Radio waves
Radio waves are the longest in the spectrum. They are used to transmit radio and TV signals around Earth. Television uses higher frequencies than radio. Radio waves from space can be picked up using radio telescopes and used to study the Universe.

Light

LIGHT ENABLES US TO SEE A BRIGHT AND COLORFUL WORLD.

Light is the only type of electromagnetic radiation we can see. We are able to perceive it as a wide range of colors.

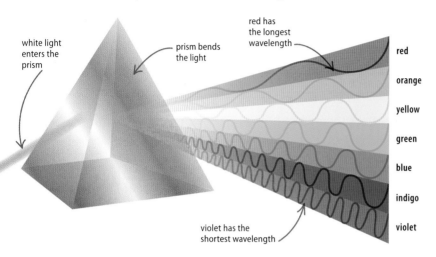

white light enters the prism

prism bends the light

red has the longest wavelength

red

orange

yellow

green

blue

indigo

violet

violet has the shortest wavelength

Color spectrum

If white light is shined into a triangular block of glass, called a prism, the glass refracts (bends) the light. In an effect called dispersion, the light is split into different wavelengths, the band of visible colors known as the spectrum. The spectrum begins with the longest wavelength (red), and ends with the shortest wavelength (violet). Most people see seven distinct colors, but the spectrum is really continuous changing color.

◁ **Splitting light**
A prism bends light by different amounts, according to its wavelength.

Making color

We see color based on information sent to the brain from millions of light-sensitive cells in the eye, called cones. There are three types of cone which respond to either red, green, or blue light. You see all colors as a mix of these three colors, known as primary colors.

▷ **Reflective colors**
Objects either reflect or absorb the different colors in white light. The reflected colors are the ones we see.

white objects reflect all colors in white light

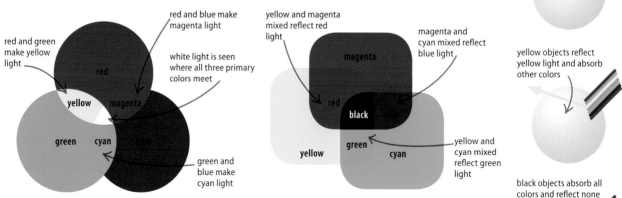

red and green make yellow light

red and blue make magenta light

white light is seen where all three primary colors meet

red

yellow

magenta

green

cyan

blue

green and blue make cyan light

yellow and magenta mixed reflect red light

magenta and cyan mixed reflect blue light

magenta

red

black

blue

yellow

green

cyan

yellow and cyan mixed reflect green light

yellow objects reflect yellow light and absorb other colors

black objects absorb all colors and reflect none

△ **Making colors with light**
If you shine three flashlights, one of each of the primary colors, at a white surface, where they overlap they will create white light. Different combinations will create magenta, yellow, and cyan, known as secondary colors. This effect is used in televisions to create a full-color picture.

△ **Mixing pigments**
Making colors with pigments (inks and paints) is done in a very different way from colored light. The primary pigments are magenta, yellow, and cyan. Each reflects light of a different color. When the pigments are mixed the number of colors they reflect is reduced, and all three together make black.

Reflection

When rays hit a smooth, shiny, flat surface, such as a flat mirror, they are reflected perfectly to give a clear but reversed image. Rough surfaces cause light to bounce off in different directions, so there is no reflected image.

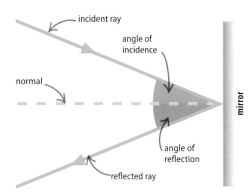

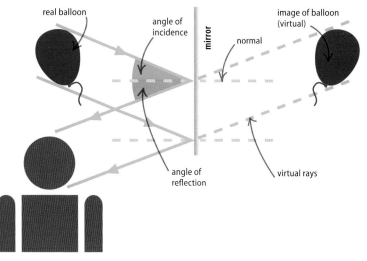

△ Angles of incidence and reflection
A reflection is made up of an incoming ray, called the incident ray, and an outgoing ray, called the reflected ray. The angle of incidence is equal to the angle of reflection, measured from a imaginary line at 90° to the mirror, called the normal.

△ Virtual image
The image in a mirror appears to be behind it—light rays appear to be focused there, but they do not actually meet at that point. This is called a virtual image. The image on a movie screen is called a real image because rays from the projector focus directly on the screen.

Refraction

Light rays usually travel in straight lines, but pass through different media (materials)—such as air, water, or glass—at different speeds. When light moves from one medium to another, the change in speed makes the beam change direction. This effect is known as refraction.

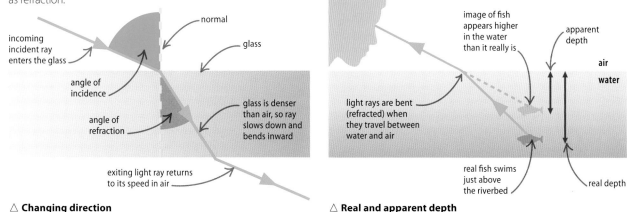

△ Changing direction
If light travels through air and then enters at an angle a more dense medium, such as glass, the rays slow down and refract inward. They travel in a straight line through the glass but at an angle to their original direction. As the rays pass out from the glass to the air, they return to their original path and speed up again.

△ Real and apparent depth
Light rays refract when they pass from water to the lighter medium of air. This means that when you look from an angle at an object in water, it is not in fact where you see it. A fish swimming in the water is actually deeper than it appears to be.

Optics

THE SCIENCE OF OPTICS EXPLAINS AND EXPLORES
THE PROPERTIES AND BEHAVIOR OF LIGHT.

SEE ALSO	
❰ **64** Vision	
❰ **196–197** Light	
Telescopes	**230** ❭
The Sun	**232–233** ❭

**Light is a type of electromagnetic radiation. It is carried by
a stream of particles that can also behave like a wave.**

Light sources

The Sun, lights, and TV screens all emit (send out) light—they
are luminous. But most objects reflect and/or absorb light that
bounces off them. Transparent materials, such as glass and
water, let light pass right through them. They transmit light.

Features of light	
form of radiation	Light is a form of electromagnetic radiation (see pages 194–195). It radiates (spreads out) from its source.
light rays travel in straight lines	You can see this in the beams from lighthouses, flashlights, and lasers. Because light rays are straight, if an object blocks them you get a dark region of shadow.
transfers energy	Energy is needed to produce light. All materials gain energy when they absorb light—solar cells use the energy in sunlight to produce electricity.
stream of particles that can behave like a wave	Light is carried by a stream of particles, called photons, but in some situations this stream can also behave like a wave.
can travel through empty space	Electromagnetic waves do not need to travel through a medium (a material such as water or air). The light from the Sun and stars, for example, reaches us through empty space.
travels fast	Light is the fastest thing in the Universe. In a vacuum, such as space, its speed is exactly 299,792 km per second (roughly 186,282 miles per second).

△ **Understanding light**
The most important source of light on Earth is the Sun. Sunlight is
produced by the energy generated deep in its core (see pages 232–233).
In contrast, the Moon simply reflects the light of the Sun and shines
much less brightly. The table above gives the main features of light.

Lenses

A lens is a piece of transparent glass or plastic that uses
refraction (see page 197) to change the directions of light
rays. Lenses are used to focus light in glasses, cameras, and
telescopes. There are two main types—convex and concave.

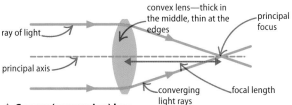

△ **Convex (converging) lens**
When rays pass through a convex lens they converge (bend inward)
and meet at a point behind the lens, called the principal focus. The
distance from the center of the lens to the principal focus is called
the focal length.

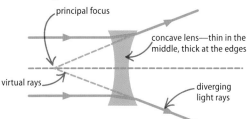

△ **Concave (diverging) lens**
A concave lens makes light diverge (spread out). When parallel rays
pass through a concave lens they spread out as if they came from
a focal point in front of the lens, called the principal focus.

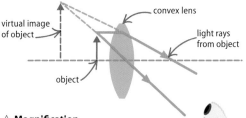

△ **Magnification**
If an object is placed between the center of a convex lens and the
principal focus, the rays never converge. Instead they appear to come
from a position behind the lens as a magnified image. It is a virtual
image (see page 197).

Interference

Where two rays of light meet, they affect each other, a phenomemon known as interference. If the waves are in phase (in step), they reinforce each other. This is called constructive interference. If they are out of phase (out of step), they cancel each other out. This is called destructive interference. Astronomers use the interference between light beams from different parts of stars to image them.

Bubble colors

When light is reflected from a soap bubble, some is reflected from the inner surface of the bubble, and some from the outer surface. The light rays from the two surfaces interfere to produce new wavelengths, seen as different colors.

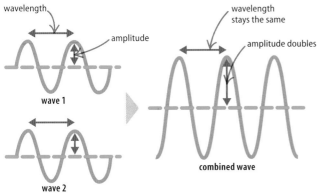

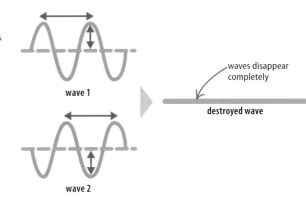

△ **In phase**
When two waves that are in phase meet, their amplitudes add together to make a single wave with double the amplitude. This is called constructive interference.

△ **Out of phase**
If the waves are out of phase, the interference is destructive. As the two waves come together, their amplitudes cancel each other out and the wave is destroyed.

Diffraction

Experiments with light have helped scientists to understand its properties. We know that light can travel as waves, because it behaves like other types of waves, such as sound. For example, both light and sound waves are reflected and refracted (see page 197). Another feature of waves is that, when they pass through a gap, or around an obstacle, they spread out. This effect is called diffraction.

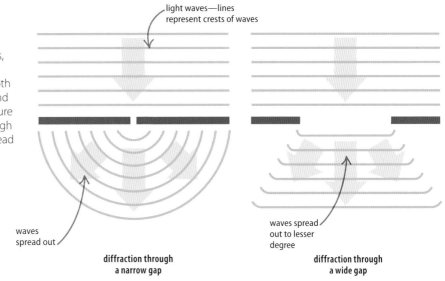

light waves—lines represent crests of waves

▷ **Spreading out**
Waves spread out like ripples as they pass through a narrow gap. Wider gaps cause less diffraction.

waves spread out

waves spread out to lesser degree

diffraction through a narrow gap

diffraction through a wide gap

Sound

SOUNDS ARE VIBRATIONS, CARRIED EITHER BY SOLIDS,
LIQUIDS, OR GASES.

SEE ALSO
❮ **64** Hearing
❮ **184** Atmospheric pressure
❮ **192–193** Waves

Sounds are of great benefit for communication and can also be
harnessed for medical or industrial use. However, unwanted
sound is a serious pollutant that damages health and well-being.

Pitch and loudness

The characteristics of sound that we experience as pitch and
loudness are closely related to the physical properties of sound
waves. Generally, the higher the frequency of a wave—the
number of peaks and troughs that pass a point each second—
the higher its pitch; the larger the amplitude of the wave, the
louder it sounds.

We can hear sounds so
quiet that they make our
eardrums move **less than
the width of an atom**.

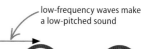
high-amplitude waves
results in a loud sound

high-frequency waves make
a high-pitched sound

low-amplitude waves
results in a quiet sound

low-frequency waves make
a low-pitched sound

△ **Loudness**
These sound waves have the same frequency but different
amplitudes. A higher amplitude indicates there is a larger
variation in air pressure, and greater volume.

△ **Pitch**
These sound waves have the same amplitude but different
frequencies. A higher frequency creates a more rapid variation
in air pressure and results in a higher pitch.

Echoes

Sound waves reflect from surfaces,
especially hard, smooth ones. If the
surface is far enough from the sound
source for an adequate time to pass
before the reflection returns, it can
be heard or detected as an echo.
Underwater echoes are used by ships
to scan the sea floor. The return time
depends on the depth of the bed,
so maps of the seafloor can be made
in this way.

▷ **Mapping the seabed**
This diagram illustrates how a ship uses
echoes to map the sea floor.

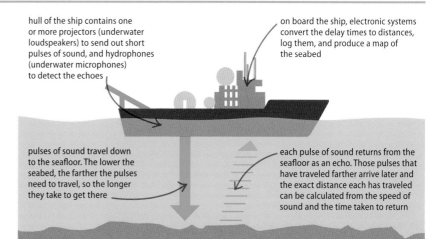

hull of the ship contains one
or more projectors (underwater
loudspeakers) to send out short
pulses of sound, and hydrophones
(underwater microphones)
to detect the echoes

on board the ship, electronic systems
convert the delay times to distances,
log them, and produce a map of
the seabed

pulses of sound travel down
to the seafloor. The lower the
seabed, the farther the pulses
need to travel, so the longer
they take to get there

each pulse of sound returns from the
seafloor as an echo. Those pulses that
have traveled farther arrive later and
the exact distance each has traveled
can be calculated from the speed of
sound and the time taken to return

Doppler effect

If a sound source is moving toward a listener, the pulses of pressure that make up the sound waves get closer together because the source is moving a little closer to each one before sending out the next. This means that the sound's frequency is higher than if the source were stationary. If the source is receding, the pulses become farther apart and the frequency lowers. This is called the Doppler effect.

Sound underwater

Sound travels better in water than in air. Marine animals use sound for a wide range of tasks. Some use it to communicate over huge distances, others to probe their surroundings, while some even use it to stun their prey. Dolphins and some whale species are especially dependent on sound for communication, which makes them particularly vulnerable to underwater noise pollution.

▽ **Police siren**
The Doppler effect explains why a siren on a police car changes in pitch as the vehicle drives by.

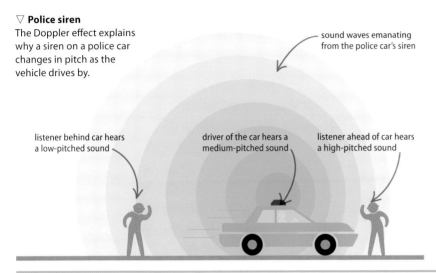

sound waves emanating from the police car's siren

listener behind car hears a low-pitched sound

driver of the car hears a medium-pitched sound

listener ahead of car hears a high-pitched sound

Supersonic motion

Sound travels at a speed of around 343 m (1,340 ft) per second through air. However, when an object travels faster than sound, it overtakes the sound waves ahead of it. An example of this is the supersonic jet, which flies faster than the speed of sound, so a person cannot hear it coming toward him or her—the jet passes before the sound arrives. However, when the sound catches up, it arrives suddenly as a shock wave, which is heard on the ground as a sonic boom.

High-power, high-frequency sound can be used to **smash kidney stones apart**, avoiding the need for surgery.

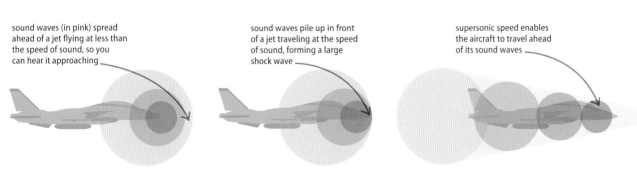

sound waves (in pink) spread ahead of a jet flying at less than the speed of sound, so you can hear it approaching

sound waves pile up in front of a jet traveling at the speed of sound, forming a large shock wave

supersonic speed enables the aircraft to travel ahead of its sound waves

△ **Subsonic flight**
The sound waves ahead of an aircraft flying slower than the speed of sound have higher frequencies than those behind them.

△ **The shock front**
When the speed of sound is reached, the sound waves can no longer spread ahead of the plane, creating a shock front.

△ **Supersonic flight**
The shock front of a supersonic plane is heard as a sonic boom by anyone it passes over.

Electricity

ELECTRICITY IS THE PHENOMENON ASSOCIATED
WITH EITHER MOVING OR STATIC ELECTRIC CHARGE.

Atoms contain tiny particles called electrons that carry negative electrical charge.
These orbit the positively charged atomic nucleus, but can become detached.

Static electricity

When an object contains an excess of electrons, it is said to be negatively charged. It will repel other negatively charged objects. Objects containing many atoms that have lost electrons are positively charged. Such objects attract negatively charged objects, and repel other positively charged objects. Since the electrons are not flowing to or from such objects, this type of electricity is called static. Objects with static charge also attract neutral objects, by repelling electrons within them to leave an area of positive charge.

Static discharge

In stormy weather, electrons gradually move from the Earth to low clouds. Charges also separate within clouds. The ground and the upper parts of clouds become strongly positively charged, while the lower parts of clouds become strongly negative. Eventually, the clouds discharge as the charges neutralize each other. Discharges within clouds are seen as sheet lightning, while forked lightning is a cloud-to-ground discharge. The lightning can travel at a staggering 209,200 km/h (130,000 mph) with an electric current of around 300,000 amperes.

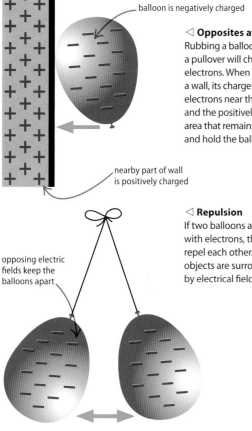

balloon is negatively charged

◁ **Opposites attract**
Rubbing a balloon against a pullover will charge it with electrons. When it touches a wall, its charge repels electrons near the surface and the positively charged area that remains will attract and hold the balloon.

nearby part of wall is positively charged

◁ **Repulsion**
If two balloons are charged with electrons, they will repel each other. Charged objects are surrounded by electrical fields.

opposing electric fields keep the balloons apart

△ **Dangerous places**
When lightning strikes, it takes the shortest route to the ground. A lone tree is likely to be struck in a thunderstorm. High buildings are often struck, too, so they are fitted with lightning rods to conduct any lightning safely down to the ground.

Lightning can heat the air surrounding it to a temperature that is **more than five times hotter than the Sun's surface!**

Electric currents

When electric charge flows through a material, it is called an electric current. It is caused by a drift of electrons through a material called a conductor (see below). In an electrical circuit (see page 206), a power source, such as a battery, gives the electrons energy so that the charge flows from the negative terminal (connection) of the power source around the circuit to the positive terminal. Current only flows when an electric circuit is complete, with no gaps. In a circuit, individual electrons actually travel extremely slowly (less than 1 mm (0.04 in) per second), but because they are closely situated to one another they are able to pass electrical energy around the circuit at more than 100 million m (328 million ft) per second.

Copper is a very good conductor of electricity, and is used to make a variety of cooking utensils and pipes for carrying hot water, both in homes and industry.

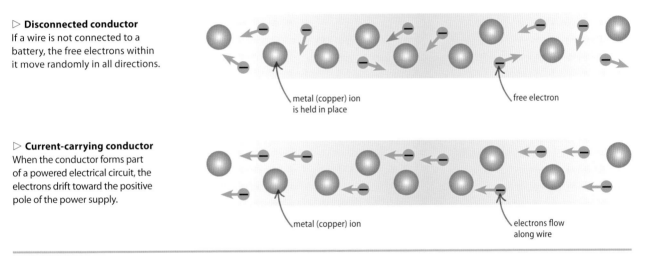

▷ **Disconnected conductor**
If a wire is not connected to a battery, the free electrons within it move randomly in all directions.

metal (copper) ion is held in place

free electron

▷ **Current-carrying conductor**
When the conductor forms part of a powered electrical circuit, the electrons drift toward the positive pole of the power supply.

metal (copper) ion

electrons flow along wire

Controlling electricity

Some materials are better at carrying an electrical current than others and are called conductors. Many metals make good conductors, as their atoms easily release electrons to carry the current. Materials such as glass, rubber, and most plastics are made of atoms that do not easily release their electrons. As a result, these conduct electricity poorly, cannot carry a current, and are called insulators.

neutral wire completes the circuit, connecting an appliance back to the supply

earth wire sends the current to the ground if the circuit fails

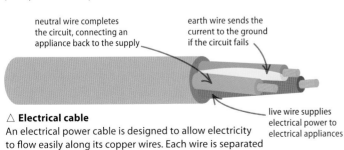

△ **Electrical cable**
An electrical power cable is designed to allow electricity to flow easily along its copper wires. Each wire is separated by a plastic sleeve—a good insulator. The colors of the sleeves vary between countries.

live wire supplies electrical power to electrical appliances

REAL WORLD
Amber

Amber is the dried resin of certain trees, and it quickly collects a static charge when it is rubbed. A piece of charged amber will attract light objects, such as feathers. The ancient Greeks were aware of these effects, and the words "electron" and "electricity" come from the Greek word for amber.

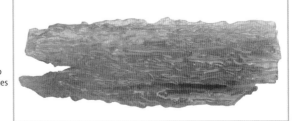

Current, voltage, and resistance

THESE ARE THE FACTORS THAT DETERMINE HOW ELECTRICITY FLOWS THROUGH A CIRCUIT.

There are two variables that control the amount of current that flows around a circuit: voltage and electrical resistance.

What is voltage?

Voltage is a measure in volts (V) of potential difference—the difference in electrical energy between two points, such as the difference in potential energy at two different points of a circuit. A voltage is required to make electrons move and an electric current to flow. Batteries are labeled in terms of their voltage. A typical car battery is 12 volts, while a flashlight battery may be 1.5 volts.

Volts, amperes, and ohms are named after three scientists who helped develop **the science of electricity**: Alessandro Volta, André-Marie Ampère, and Georg Simon Ohm.

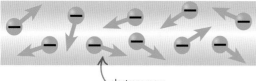

electrons move in all directions

voltage is applied

electrons drift in the same direction

△ **No voltage**
If a conductor's ends are not connected to a battery or other power source, the free electrons within it drift randomly in all directions.

△ **Voltage**
If the ends of a conductor are connected to a battery, the battery's voltage makes electrons drift along, creating electrical current.

Resistance

Any piece of wire and any component in a circuit holds back the flow of electricity through it to some extent. This is usually because electrons moving around the circuit are scattered ("bounced") by the ions (charged atoms) of the material, which slows the electrons down and makes them lose energy. This "holding back" is called electrical resistance. The lost energy appears in the form of heat, sound, or light. The resistance of a wire depends on factors such as its length and diameter.

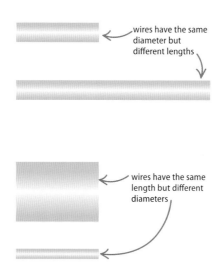

wires have the same diameter but different lengths

wires have the same length but different diameters

◁ **Length**
A shorter wire has less resistance than a longer wire of the same diameter. This is because electrons have less distance to travel and suffer fewer collisions and energy loss. In a longer wire, they have farther to go, so they encounter more collisions, greater resistance, and greater energy loss.

◁ **Diameter**
A thinner wire has a greater resistance than a thicker wire of the same length, because it has less room for electrons to move through. In the thicker wire, more electrons can travel side by side (like a crowd in a wide corridor), so the electron flow is greater.

Ohm's Law

Ohm's Law is a formula that shows the relationship between voltage, current, and resistance. Changing the value of one of these three variables will affect the other two. The resistance in a circuit, for example, can be increased by adding an extra component, such as a lamp or a resistor—a device designed to resist the current.

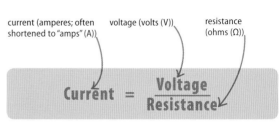

current (amperes; often shortened to "amps" (A)) voltage (volts (V)) resistance (ohms (Ω))

$$\text{Current} = \frac{\text{Voltage}}{\text{Resistance}}$$

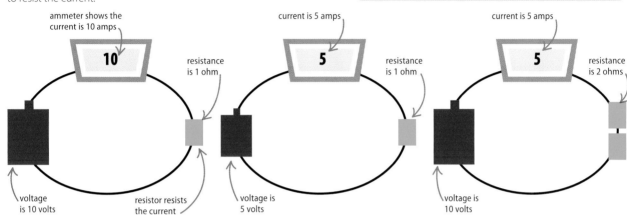

ammeter shows the current is 10 amps

10

resistance is 1 ohm

voltage is 10 volts

resistor resists the current

current is 5 amps

5

resistance is 1 ohm

voltage is 5 volts

current is 5 amps

5

resistance is 2 ohms

voltage is 10 volts

△ **Circuit 1**
In this circuit, the battery provides 10 volts and there is a resistance of just 1 ohm, so the current is 10 amps.

△ **Circuit 2**
In this second circuit, there is still 1 ohm of resistance, but the voltage has been halved, which reduces the current to 5 amps.

△ **Circuit 3**
Here, the voltage is again 10 volts, but another 1 ohm resistor has been added, which reduces the current to 5 amps.

Electric heat and light

When electricity flows along a conductor, the resistance that occurs converts some of the electrical energy into heat and sometimes light. The amount of resistance and heat produced can be increased by using a high-resistance wire.

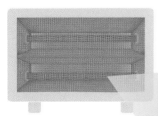

tightly coiled thin wire converts electrical energy to heat energy

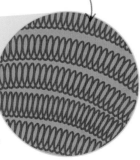

△ **Electric heater**
An electric heater uses long lengths of high-resistance wire coiled tightly, so that more wire can be fitted into the heater, generating more heat.

REAL WORLD

Superconductors

Certain materials lose practically all of their electrical resistance at very low temperatures. This phenomenon, called superconductivity, can be used to create very efficient electromagnets. These powerful superconducting electromagnets are used in Magnetic Resonance Imaging (MRI) scanners in medicine, in large particle colliders, and in some magnetic levitation (Maglev) rail vehicles, including this Japanese train.

Circuits

ALL ELECTRONIC AND ELECTRICAL SYSTEMS
AND EQUIPMENT ARE BUILT FROM CIRCUITS.

Circuits are composed of power sources, conductors, and
electronic or electrical components that carry out specific tasks.

Circuit basics

In any circuit, a power source—such
as a cell—pushes electrical current
along one or more conductors, often
wires. When the current passes through
a component, such as a light bulb, the
component changes the electricity
and also changes itself in response.
For example, a resistor controls the
flow of current to protect the device
from overload. Similarly, a light bulb
opposes the current and lights up.
If the circuit is broken—by means
of a switch, for example—the current
ceases to flow.

△ **Switch**
This allows or halts
the flow of current.

△ **Light bulb**
This will light up when
current flows through it.

△ **Resistor**
The purpose of this is to
resist the flow of current.

△ **Cell**
This causes current to
flow around the circuit.

△ **Voltmeter**
This device measures
the voltage, in volts.

△ **Variable resistor**
This device controls
the amount of current.

△ **Capacitor**
This device stores
electrical charge.

△ **Ammeter**
This component measures
the current, in amps.

△ **Motor**
A motor moves when
current flows through it.

In series

When components are connected in series, they share the voltage
of the power source, such as a cell. If there are two identical
components, then each will receive half the voltage. The current
will stop flowing around the circuit if there is a break at any point.

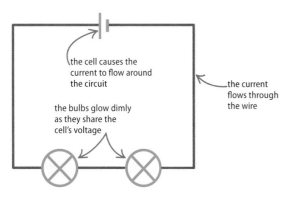

the cell causes the
current to flow around
the circuit

the current
flows through
the wire

the bulbs glow dimly
as they share the
cell's voltage

△ **Series circuit**
These two bulbs are arranged in series so they have
to share the voltage. They glow dimly as a result.

In parallel

When components are connected in parallel, they are
each subject to the whole voltage from the power source.
The current will continue to flow through one bulb if the
wire leading to the other is broken.

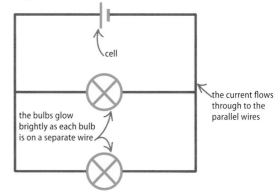

cell

the bulbs glow
brightly as each bulb
is on a separate wire

the current flows
through to the
parallel wires

△ **Parallel circuit**
These two bulbs are arranged in parallel so they both
receive the full voltage from the cell and glow brightly.

Capacitor

A capacitor is a component used in many circuits to store and release electric charge. There are many different types and sizes of capacitor, many of which are used in circuits to smooth out a varying electric current. At its simplest, a capacitor may consist of two plates of electrically conductive material separated by an insulator called a dielectric. In a direct current (DC) circuit (see page 216), the capacitor stops the flow of current once it has been fully charged.

Supercapacitors that store and release large electrical charges are being developed to replace electric vehicle batteries, since they can be **recharged far more quickly** and more often.

REAL WORLD

Camera flash

Some capacitors are used because they can release their entire charge in just a fraction of a second. Most digital cameras use capacitors, which are charged up by the camera's battery, to power their flash function. The capacitor releases all of its charge almost instantly to enable the flash to fire brightly so that it lights up a dim scene as a photo is taken.

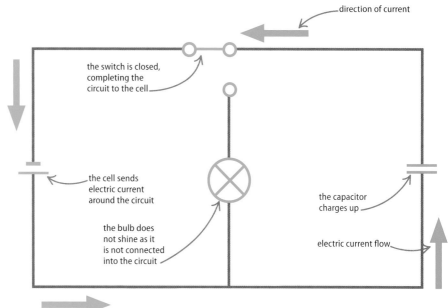

direction of current

the switch is closed, completing the circuit to the cell

the cell sends electric current around the circuit

the bulb does not shine as it is not connected into the circuit

the capacitor charges up

electric current flow

△ **Capacitor stores charge**

In this direct current (DC) circuit, charge flows from the cell to the capacitor. The electric charge builds up on the capacitor's plate while some current continues to flow across the capacitor and around the circuit. As its charge builds, the capacitor resists the flow of current. Once fully charged, it completely stops the flow of current through it.

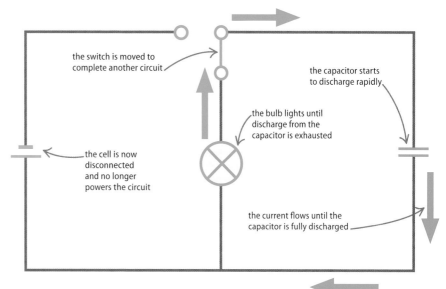

the switch is moved to complete another circuit

the cell is now disconnected and no longer powers the circuit

the bulb lights until discharge from the capacitor is exhausted

the capacitor starts to discharge rapidly

the current flows until the capacitor is fully discharged

△ **Capacitor releases charge**

Moving the switch disconnects the cell from the circuit but closes and completes another circuit that still contains the capacitor. The capacitor discharges (releases its electrical charge) and the bulb lights up. The bulb will only shine for a short while and will stop once the capacitor is fully discharged.

Electronics

IN ELECTRONIC SYSTEMS, INFORMATION FLOWS IN THE FORM OF
PRECISELY CONTROLLED ELECTRICAL SIGNALS THROUGH CIRCUITS.

Almost all modern machines, from computers and phones to washing
machines and cars, contain electronic devices of many kinds.

Electronic components

These are components designed
to handle, control, and change the
amount of electric current flowing
through circuits in a device. The
current acts like an electrical signal,
instructing the circuit and device to
perform specific tasks, from adding
up numbers on a calculator to
displaying a word on screen. When
first invented, these devices were large
and bulky, and individually built and
wired together. Now they have been
miniaturized so that thousands
can exist together on a tiny silicon
microchip. When electronic circuits
are designed, each component is
represented by a special symbol,
including the ones on the right.

△ **Diode**
This makes current
flow in one direction.

△ **Connected wires**
The symbol for wires
that are connected.

△ **Overlapping wires**
These wires cross but
are not connected.

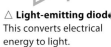

△ **Light-emitting diode**
This converts electrical
energy to light.

△ **Amplifier**
This device Increases
electrical power.

△ **Transistor**
This device controls
the size of current.

△ **Piezo transducer**
Converts electrical
energy to sound.

△ **Fuse**
This component burns
out if the circuit shorts.

△ **Thermistor**
This device converts
heat to electricity.

△ **Generator**
This generates
electrical voltage.

△ **AC power supply**
This supplies energy as
an alternating current.

△ **DC power supply**
This supplies energy
as a direct current.

△ **Inductor**
This is a type of
electromagnet.

△ **Transformer**
This varies the
current and voltage.

△ **Microphone**
This changes sound
into electrical energy.

△ **Aerial**
This device sends or
receives radio waves.

Integrated circuits

Modern electronic circuits are built
onto tiny rectangles of silicon to make
microchips. They are called integrated
circuits because the components are
all constructed together. An integrated
circuit is built up from layers of different
materials. Some of these layers are
insulators, some are conductors,
and some are semiconductors, which
allow electricity to flow, but only in
certain conditions. Patterns etched in
the layers produce the components
and their interconnections.

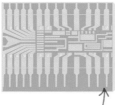

components and
connections are etched
onto layers of semi-
conductor material

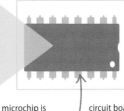

microchip is
encased in
a body with pins
to connect to
a circuit board

circuit board
is fitted with
microchips and
other components

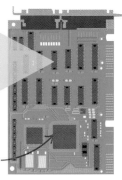

△ **Integrated circuit**
Electronic components are
so small they are only visible
under a microscope.

△ **Microchip**
This is constructed from a tiny
wafer of silicon, and contains
many integrated circuits.

△ **Circuit board**
Containing many microchips
and other components, this
forms a key part of many devices.

Using codes

We use numbers made up of ten numerals (0, 1, 2, 3, 4, 5, 6, 7, 8, and 9), but computers use only two numerals: 0 and 1. This is because computer circuits store data in the form of switches. Each switch holds a single "bit" of information. If the switch is on, this information is a 1; if the switch is off, it is a 0. This means that all information must be coded for the computer as 1s and 0s. This leads to very long numbers, so, to make it easier for humans to handle, binary is often converted into hexadecimal (base 16) numbers.

Decimal	Binary	Hexadecimal
0	0000	0
1	0001	1
2	0010	2
3	0011	3
4	0100	4
5	0101	5
6	0110	6
7	0111	7
8	1000	8
9	1001	9
10	1010	A
11	1011	B
12	1100	C
13	1101	D
14	1110	E
15	1111	F

△ **Conversion table**
This table shows the decimal (base 10) number system we use converted to binary (base 2) numbers and hexadecimal (base 16) numbers.

The first electronic component was the **diode**, invented in 1904 by English scientist Ambrose Fleming.

Logic gates

A logic gate is used to make a simple decision. It accepts an electrical signal from its inputs (it can have one or two) and then outputs either an "on," high-voltage signal (representing 1) or an "off," low-voltage signal (representing 0). In computers and many other electronic devices, large numbers of logic gates are linked together to form complex circuits. The image below shows three commonly found logic gates and their possible inputs and outputs.

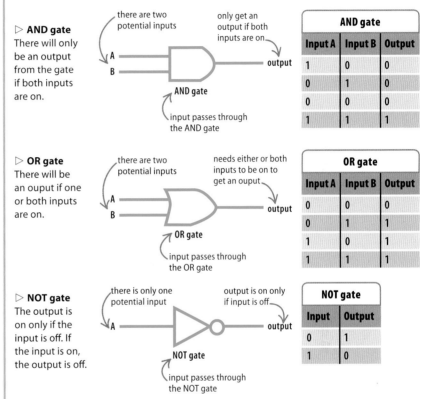

▷ **AND gate**
There will only be an output from the gate if both inputs are on.

there are two potential inputs

only get an output if both inputs are on

output

AND gate

input passes through the AND gate

AND gate		
Input A	Input B	Output
1	0	0
0	1	0
0	0	0
1	1	1

▷ **OR gate**
There will be an ouput if one or both inputs are on.

there are two potential inputs

needs either or both inputs to be on to get an ouput

output

OR gate

input passes through the OR gate

OR gate		
Input A	Input B	Output
0	0	0
0	1	1
1	0	1
1	1	1

▷ **NOT gate**
The output is on only if the input is off. If the input is on, the output is off.

there is only one potential input

output is on only if input is off

output

NOT gate

input passes through the NOT gate

NOT gate	
Input	Output
0	1
1	0

Retinal implant

Modern electronic devices can be so small, reliable, and sensitive that they can be implanted in the human retina to help some partially sighted people see. Light falling onto the implant is converted into electrical signals that stimulate the optic nerve. The brain interprets these signals as patterns of dark and light, and allows the patient to see objects.

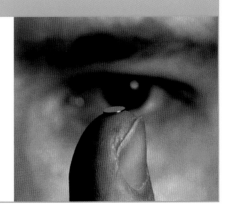

Magnets

MAGNETS PRODUCE A MAGNETIC FIELD, WHICH ATTRACTS SOME
MATERIALS AND CAN ATTRACT OR REPEL OTHER MAGNETS.

SEE ALSO	
❬ **124–125** Transition metals	
❬ **172–173** Forces and mass	
❬ **203** Electric currents	
Electric motors	**212–213** ❭
Electricity generators	**214–215** ❭

Some magnets occur naturally, while some materials can
be made magnetic by passing an electric current through
them. Some materials can be permanently magnetized.

Magnetic force

In magnetic materials, areas called domains behave like tiny
magnets. When not magnetized, these are all jumbled up and
point in different directions, but when placed in a magnetic field
or stroked repeatedly by a magnet, the domains all line up so
that all their north poles point in one direction and the south
poles in the opposite direction, making the material magnetic.

▷ **Single bar magnet**
The area around the magnet
where its magnetism can
affect other materials is
called its magnetic field.
A bar magnet has a north
pole at one end and a
south pole at the other.
Cutting a bar magnet in
two creates two magnets,
each with their own north
and south poles.

the closer the lines
of force, the stronger
the magnetic field

▷ **Horseshoe magnet**
Magnets come in all kinds of
shapes, such as the horseshoe
magnet. This type of magnet
also has a north and south
pole, but it is curved, so the
poles are close together.

magnetism is
mostly confined
to between
the poles

▽ **Attract or repel**
Two magnets will be attracted
to each other if unlike poles
(one north and one south) face
each other. However, like poles
repel, pushing each other away.

magnetic fields surrounding
the south poles of two magnets
repel each other

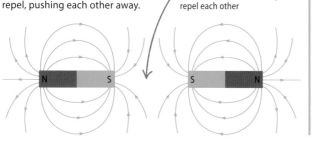

Permanent magnets

Some materials, including iron, nickel, cobalt, and their alloys (metals
combined with metals or nonmetals), are ferromagnetic. These
can be magnetized by an electric current or by stroking another
magnet. Once magnetized, these materials stay magnetic unless
demagnetized by a shock, excess heat, or a variable magnetic field.

▽ **Magnetic objects**
Steel is an alloy of iron,
and is used to make cans
and paper clips. "Copper"
coins actually contain nickel.

▽ **Nonmagnetic objects**
Common plastics are not
magnetic, nor are aluminum
beer and soda cans, or brass
musical instruments.

REAL WORLD

Lodestone compass

Lodestone is a naturally occurring magnetic mineral that was
used thousands of years ago to make the first compasses. If
a piece of lodestone is allowed to spin freely, it will align itself
with the Earth's magnetic field, pointing in a north–south
direction. The word "magnet" comes from "Magnesia," the
area in Greece where lodestones and manesium were found.

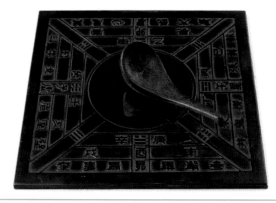

Earth's magnetic field

The Earth can be thought of as one big powerful magnet whose magnetic field, called the magnetosphere, stretches tens of thousands of kilometers out into space. The planet's magnetism is caused by the motion of liquid metals in its outer core. For an unknown reason, the direction of the Earth's field reverses suddenly, about once every million years.

▷ **Magnetic Earth**
The magnetic pole at Earth's north is a south pole, because the north poles of compasses are attracted by it. Confusingly, it is often called the South Magnetic Pole. There is a difference of a few degrees between the direction in which a compass points and the North Geographic Pole. This difference is called the angle of declination.

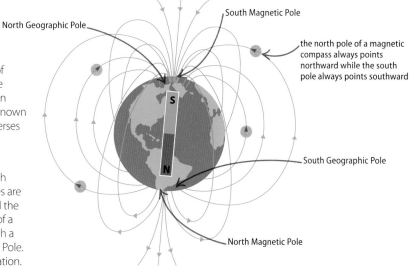

North Geographic Pole

South Magnetic Pole

the north pole of a magnetic compass always points northward while the south pole always points southward

South Geographic Pole

North Magnetic Pole

Electromagnet

Magnets are not the only source of magnetic fields. An electric current flowing through a conductor produces a circular magnetic field at right angles to the conductor. The current creates an electromagnet—a device that is extremely useful since its magnetism can be controlled and switched on and off. The poles of an electromagnet will be reversed if the direction of the current is reversed.

▽ **Field direction**
The direction of the magnetic field can be remembered by making a loose fist with your fingers of your right hand as if grasping the conductor. Sticking your thumb up in the direction of the current, your fingers follow a curving path in the direction of the magnetic field.

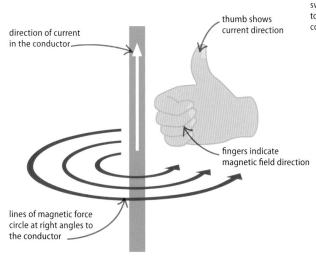

direction of current in the conductor

thumb shows current direction

fingers indicate magnetic field direction

lines of magnetic force circle at right angles to the conductor

▽ **Solenoid**
A solenoid is a common form of electromagnet. It consists of a coil of wire through which an electric current is passed to produce a magnetic field. The soft iron core in the middle of this solenoid helps produce a stronger magnetic field and does not retain its magnetism after the current is switched off.

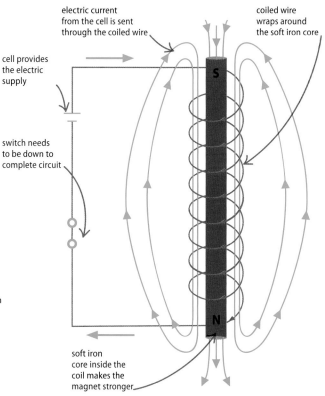

electric current from the cell is sent through the coiled wire

coiled wire wraps around the soft iron core

cell provides the electric supply

switch needs to be down to complete circuit

soft iron core inside the coil makes the magnet stronger

Electric motors

AN ELECTRIC CURRENT AND THE FORCES IN A MAGNETIC
FIELD CAN COMBINE TO CREATE MOTION.

An electric motor turns because of the forces of attraction and
repulsion between a permanent magnet and an electromagnet.

Inside a motor

A wire coil sits between the opposite poles of one or more permanent magnets.
When an electric current is passed through the wire coil, it generates a magnetic
field, which interacts with the magnetic field of the surrounding permanent
magnets, repelling like poles and attracting unlike poles, which make the wire
coil rotate half a turn. The electric current is then reversed to switch the wire coil's
magnetic poles, so that it moves another half-turn. Repeating this process results
in the coil spinning around.

▽ **Left-hand rule**
This rule can be used to work out
the direction an electric motor turns.

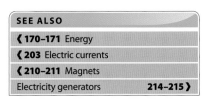

thumb shows the
direction of thrust
applied to the wire

first finger shows
the direction of the
magnetic field

middle finger shows the
direction of the current
(positive to negative)

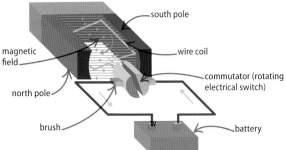

south pole

magnetic
field

wire coil

north pole

commutator (rotating
electrical switch)

brush

battery

△ **Stage 1**
In this simple DC electric motor, current flows from the battery
through the commutator and into the wire coil. This turns it into
an electromagnet and generates a magnetic field, which interacts
with the field of the permanent magnet.

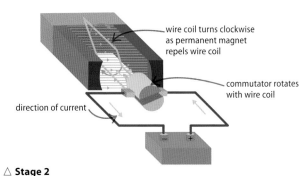

wire coil turns clockwise
as permanent magnet
repels wire coil

commutator rotates
with wire coil

direction of current

△ **Stage 2**
Repelled by the permanent magnet's like poles, the wire coil
starts turning. After a quarter-turn, the permanent magnets
also begin attracting the opposite pole of the wire coil, helping
to complete the half-turn.

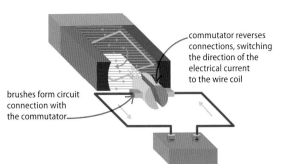

commutator reverses
connections, switching
the direction of the
electrical current
to the wire coil

brushes form circuit
connection with
the commutator

△ **Stage 3**
With the poles of the wire coil and permanent magnet now
lining up, the commutator reverses the direction of the current
in the wire coil. This switches the polarity of the wire coil's
magnetic field.

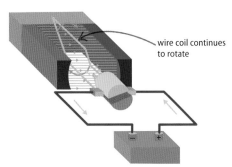

wire coil continues
to rotate

△ **Stage 4**
With the coil's current reversed, the like poles of the coil and
permanent magnet repel again. The coil continues to rotate.
When it completes another half-turn, the commutator will
reverse the current again to keep the coil spinning.

Loudspeaker

A loudspeaker uses the motion generated by the forces between a permanent magnet and an electromagnet to reproduce sound. Fluctuating electric current enters the coil, producing a fluctuating magnetic field. The forces between this field and that of the permanent magnet move the coil rapidly in and out. The coil moves the cone and these movements generate sound waves.

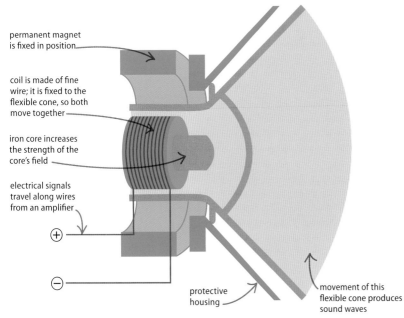

permanent magnet is fixed in position

coil is made of fine wire; it is fixed to the flexible cone, so both move together

iron core increases the strength of the core's field

electrical signals travel along wires from an amplifier

(+)

(−)

protective housing

movement of this flexible cone produces sound waves

△ **Electromagnetism in action**
The forces acting on the moving parts of a loudspeaker are electromagnetic, produced by the interaction of the permanent magnet and the coil electromagnet.

Robotic arm

The joints and parts of an industrial robot arm, such as this car welding robot, are powered by electric stepper motors. A central rotor can be turned in steps by the magnets, making the motor capable of very precise movements.

The **world's smallest electric motor** is just 1nm (1 nanometer) across.

Linear motor

This type of electric motor creates a force in a straight line rather than the turning force of a traditional rotary motor. It achieves this by a continuous sequence of magnetic attraction and repulsion between electromagnets along a track and magnets attached to a sled, train, or some other object running along the track. The electromagnets repeatedly switch their polarity to move the object down the track without the need for wheels.

▷ **Magnetic motion**
Maglev (magnetic levitation) trains use powerful magnets to float above a track and are propelled forward at great speed by a linear motor.

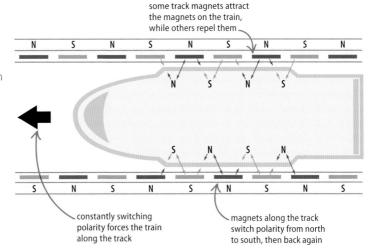

some track magnets attract the magnets on the train, while others repel them

constantly switching polarity forces the train along the track

magnets along the track switch polarity from north to south, then back again

Electricity generators

GENERATORS USE INDUCTION TO CHANGE MOTION
INTO ELECTRICAL POWER.

Generators, also called dynamos, are vital in many areas of
technology. For example, turbines use them to change the kinetic
energy of moving wind, water, or steam into electrical energy.

Electromagnetic induction

In 1831, English scientist Michael Faraday (1797–1867) discovered
that when a magnet was moved in or out of a coil of wire, an
electric current was produced. A voltage and current is produced
in a conductor (the coil of wire) when it cuts across a magnetic
field because the magnetic field lines of force act on the free
electrons in the conductor, causing them to move. This principle,
known as induction, is the basis on which all generators work.

Built in 1871, the **Gramme
dynamo** was the first
electricity generator to
generate power commercially.

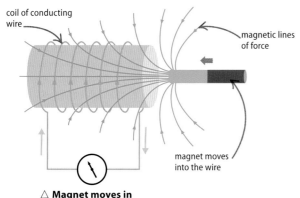

coil of conducting
wire

magnetic lines
of force

magnet moves
into the wire

△ **Magnet moves in**
A generator works when the magnet moves
into the wire. The induction effect is stronger
if the conductor is coiled.

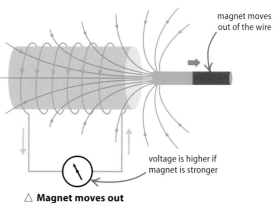

magnet moves
out of the wire

voltage is higher if
magnet is stronger

△ **Magnet moves out**
When the magnet moves out of the
conductor, current is induced in the
opposite direction.

Bicycle dynamo

A bicycle dynamo contains a permanent
magnet fitted to a shaft. As the bicycle
wheel turns, the dynamo shaft turns,
rotating the permanent magnet inside
a coil of wire wrapped around a soft
iron core. The changing magnetic field
of the turning permanent magnet
induces a current in the coil, which
flows from the dynamo to power
the bicycle's front and rear lights.

▷ **Electromagnetic
induction in action**
In this bicycle
dynamo, the wire
coil is fixed in place
and the permanent
magnet rotates
inside it. Friction
between the
grooved dynamo
wheel and the tire
wall causes the shaft
holding the magnet
to rotate when the
bicycle wheel turns.

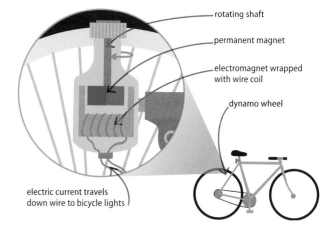

rotating shaft

permanent magnet

electromagnet wrapped
with wire coil

dynamo wheel

electric current travels
down wire to bicycle lights

Direct current generator

Generators can be built to produce either direct current (DC) or alternating current (AC) (see page 216). A DC generator has the same parts as a DC electric motor but works in reverse (see page 212). The wire coil of the conducting wire is turned inside a magnetic field that is generated by a large permanent magnet. As the wire in the coil cuts across the magnetic field lines, a voltage and current are created in the coil.

▷ **Right-hand rule**
This rule shows the direction in which a current will flow in a wire when the wire moves in a magnetic field.

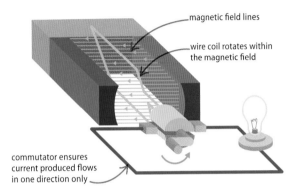

thumb shows the direction of thrust applied to the wire

first finger points in the direction of magnetic field

middle finger shows the direction of the current

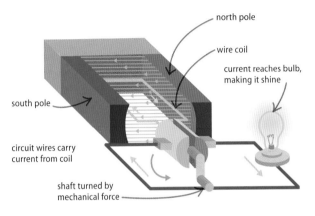

north pole
wire coil
current reaches bulb, making it shine
south pole
circuit wires carry current from coil
shaft turned by mechanical force

△ **Stage 1**
An experimental direct current generator sees the wire coil turned by a hand crank. As it passes through the magnetic field of the permanent magnet, a current is induced in the wire coil.

magnetic field lines
wire coil rotates within the magnetic field
commutator ensures current produced flows in one direction only

△ **Stage 2**
An electric current is only generated when the wire coil is cutting the horizontal magnetic field lines. When the wire coil is vertical, no current is produced and the bulb does not light up.

Alternating current generator

An alternating current (AC) generator, known as an alternator, does not use a commutator. As a result, the current produced changes direction twice for every complete 360° turn of the coil. Individual slip rings are fitted to each of the two ends of the coil to provide a path for the current to leave. Brushes contact the slip rings and complete the path for the current into the circuit to which the generator is attached.

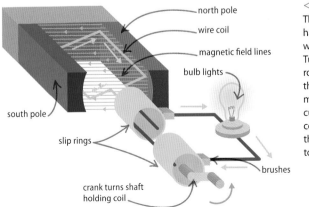

north pole
wire coil
magnetic field lines
bulb lights
south pole
slip rings
crank turns shaft holding coil
brushes

◁ **Alternator in action**
This simple alternator has a single loop of wire acting as the coil. Turning a hand crank rotates the coil between the poles of a permanent magnet. An alternating current is induced in the coil, which flows through the slip rings and brushes to light the bulb.

REAL WORLD
Wind-up electrics

In parts of the world where electricity is unreliable or absent, and batteries are expensive, radios (as below), laptops, and other electronic devices can be powered by hand. A small generator inside the device is turned by a hand crank to charge up the rechargeable batteries inside.

Transformers

TRANSFORMERS CHANGE THE VOLTAGE OF AC POWER.

Alternating current can be changed to a higher or lower voltage by a device called a transformer. For example, high voltage from a power station needs to be transformed to a lower voltage for use in homes.

SEE ALSO	
❰ **186–187** Machines	
❰ **200** Pitch and loudness	
❰ **203** Electric currents	
❰ **208–209** Electronics	
❰ **214** Electromagnetic induction	
Power grid	**220** ❱

Direct and alternating current

There are two kinds of current: direct and alternating. Direct current (DC) is usually produced by batteries and it flows one way around a circuit. Electricity—as used in homes—has an alternating current (AC), in which the direction of the flow of electricity reverses dozens of times a second. Transformers are devices that can be used to change the voltages and currents of AC (see pages 204–205). They make it easy to change AC to a high-voltage form for transmission over long distances, and to a low-voltage form for domestic use.

▷ **AC and DC voltage**
In this graph, the green line represents DC voltage and the green area is the energy transferred by this voltage. To transfer the same energy in the same time (as shown by the blue areas), the AC voltage must rise higher than the DC voltage at some parts of its cycle, as the orange line shows.

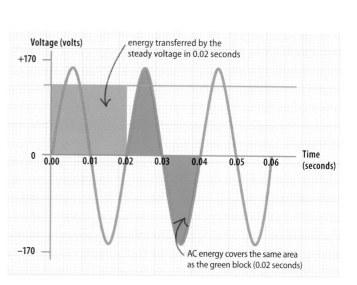

Transformers

An inductor is a coil of wire that stores energy in a magnetic field. A transformer is two inductors in one: two coils share the same core. When an alternating current passes through one coil, the core sets up currents in the other coil. If this second coil has more turns than the first, then the voltage across it is higher.

▷ **Voltage and coils**
The ratio of the number of volts that pass through the primary and secondary transformers is equal to the ratio of the number of turns in the primary and secondary coils.

voltage (volts) across primary coil

number of turns of primary coil

$$\frac{V_p}{V_s} = \frac{N_p}{N_s}$$

voltage (volts) across secondary coil

number of turns of secondary coil

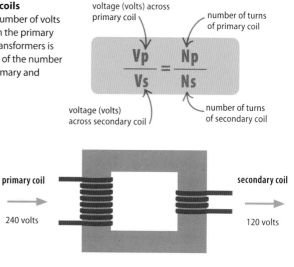

△ **Step-up transformer**
The second inductor has twice as many turns as the first, so its voltage is twice that in the first.

△ **Step-down transformer**
The second inductor has half as many turns as the first, so the voltage is halved as a result.

Induction in action

Induction is the production of an electrical current by a changing magnetic field. Many devices, from microphones (see below) to microcomputer parts, rely on it. Computer hard disks, for instance, store data magnetically on a stack of disklike platters. The surface of each platter contains billions of individual areas, each of which can be magnetized. The patterns of magnetism store data as binary digits (1s for magnetized areas and 0s for demagnetized areas) and are produced by induction caused by a tiny moving electromagnet.

▷ **Hard disk**
A hard disk drive consists of a set of disks called platters. Data is written, read, or deleted by an electromagnet on a moving arm.

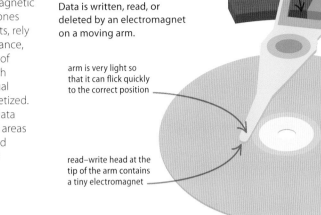

magnet moves the arm from side to side

arm is very light so that it can flick quickly to the correct position

read–write head at the tip of the arm contains a tiny electromagnet

disk spins at a high speed, and the head reads and writes as the surface passes beneath

a metal mesh protects the delicate diaphragm. Outdoors, special shields are used to reduce the noise of the wind blowing across it

sound waves make the diaphragm vibrate

coil's movement within the magnetic field generates a current in the coil

diaphragm's movements move coil up and down

wires transfer current's signals to an amplifier, recorder, or loudspeaker

magnet

◁ **Take a look inside**
In a microphone, sound waves vibrate a diaphragm, which is attached to a wire coil. As the coil moves, it induces currents, which form a changing electrical signal.

◁ **Microphone**
A microphone is carefully designed so that it mimics the sounds it receives, and does not overemphasize particular frequencies.

Electromagnets are used to **lift heavy loads** of steel. The most powerful can lift single loads **weighing more than 250 tons**.

REAL WORLD

Induction cooking

The electromagnet in an induction stovetop generates a magnetic field. Some of its energy transfers to the metal pan via the process of induction, as circulating electric currents. The electrical resistance of the metal means that some of this electrical energy is converted to heat, warming the pan but not the surface.

Power generation

ELECTRICITY IS PRODUCED IN DIFFERENT WAYS.

Electricity is generated on a large scale in power stations. They work in different ways, but they all harness a source of energy and use it to power giant electricity generators.

SEE ALSO

❰ 28 Respiration
❰ 126–127 Radioactivity
❰ 156–157 Carbon and fossil fuels
❰ 214–215 Electricity generators
Renewable energy **224–225 ❱**

Thermal power station

This is the most common type of power station. Its source of energy is a fossil fuel, generally natural gas, coal, or oil. The heat released by burning the fuel boils water into steam. The steam is forced under high pressure through the turbine, making it spin. This rotation is transmitted to the generator.

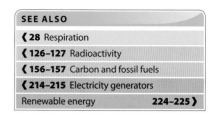

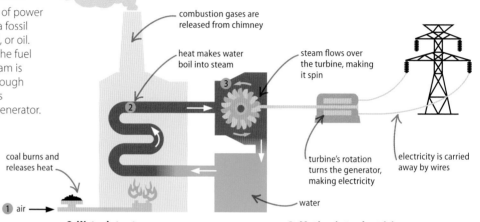

combustion gases are released from chimney

heat makes water boil into steam

steam flows over the turbine, making it spin

coal burns and releases heat

turbine's rotation turns the generator, making electricity

electricity is carried away by wires

water

① air ➝

1. Heat from fuel
Solid fuels, like coal, are crushed into small particles, increasing its surface area so that it burns faster and hotter. The gases released by the combustion are released from a chimney.

2. Water into steam
The water boils in the furnace, turns into steam, and passes over the propeller-like blades. The steam then condenses back into water to begin the process again.

3. Motion into electricity
The rotational motion of the turbine is passed to the generator, where a conductor is spun around in a magnetic field, inducing an electric current.

Hydroelectricity

In a hydroelectric power station, the energy of falling water is used to generate electricity. A dam built across a river builds up a large reservoir of water behind it. The water is released through a pipe, or penstock, to form a high-pressure flow that spins a turbine at the bottom. A water-driven turbine has cup-shaped blades, unlike the wing shapes on a gas or steam turbine.

▷ **Power station**
The turbine inside the dam is connected to a generator in a power station on the downstream side of the dam.

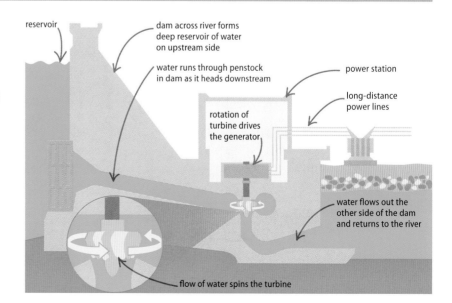

reservoir

dam across river forms deep reservoir of water on upstream side

water runs through penstock in dam as it heads downstream

power station

long-distance power lines

rotation of turbine drives the generator

water flows out the other side of the dam and returns to the river

flow of water spins the turbine

Energy from atoms

Nuclear power stations use radioactive materials, such as uranium or thorium, as a source of heat. Radioactive elements produce heat as they decay, but a great deal more heat is produced by a process called nuclear fission. The fuel is refined to contain large amounts of a particular radioactive isotope (see pages 126–127) that can be split into two smaller atoms. Uncontrolled fission causes the explosion of a nuclear bomb, but the process is slowed down in nuclear reactors.

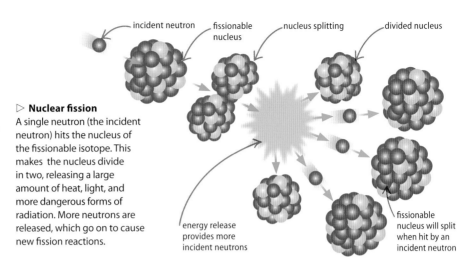

incident neutron   fissionable nucleus   nucleus splitting   divided nucleus

▷ **Nuclear fission**
A single neutron (the incident neutron) hits the nucleus of the fissionable isotope. This makes the nucleus divide in two, releasing a large amount of heat, light, and more dangerous forms of radiation. More neutrons are released, which go on to cause new fission reactions.

energy release provides more incident neutrons

fissionable nucleus will split when hit by an incident neutron

Nuclear reactor

The fission reaction takes place inside a reactor filled with water or gas. The reactor has a core containing fuel rods made of radioactive material. The reaction heats the water, which is pumped through a heat exchanger, where the superheated water makes steam that drives the turbines. There are also control rods, made largely of boron, which soak up some of the free neutrons, limiting the number of fissions that occur and so controlling the process.

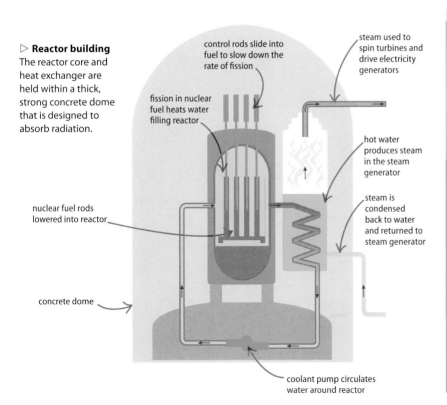

▷ **Reactor building**
The reactor core and heat exchanger are held within a thick, strong concrete dome that is designed to absorb radiation.

control rods slide into fuel to slow down the rate of fission

fission in nuclear fuel heats water filling reactor

nuclear fuel rods lowered into reactor

concrete dome

steam used to spin turbines and drive electricity generators

hot water produces steam in the steam generator

steam is condensed back to water and returned to steam generator

coolant pump circulates water around reactor

Cherenkov radiation

The water surrounding a nuclear reactor has an eerie blue color, which is caused by Cherenkov radiation, named after the Russian scientist Pavel Cherenkov (1904–1990). This happens because charged particles move through the water at an extremely high velocity.

Electricity supplies

ELECTRICITY IS SENT FROM POWER STATIONS FOR USE IN
HOMES AND WORKPLACES VIA A HUGE NETWORK OF CABLES.

Almost all the electricity used in homes, offices, and factories is generated at large
power stations far from where people live. It is sent across country in a power grid,
before being transformed into a usable current suitable for domestic use.

Power grid

Electricity is generated as an alternating current (AC). This is boosted to several hundred
thousand volts by a transformer before it enters the power grid. The high voltage reduces
the amount of energy lost as heat as currents travel along wires hundreds of kilometers
long. Burying high-voltage lines is very expensive, so most of the power grid is made up
of lightweight aluminum cables suspended from pylons, high in the air for safety.

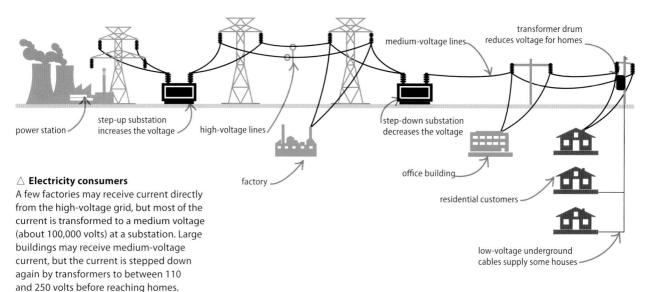

transformer drum
reduces voltage for homes

medium-voltage lines

power station

step-up substation
increases the voltage

high-voltage lines

step-down substation
decreases the voltage

factory

office building

residential customers

low-voltage underground
cables supply some houses

△ **Electricity consumers**
A few factories may receive current directly
from the high-voltage grid, but most of the
current is transformed to a medium voltage
(about 100,000 volts) at a substation. Large
buildings may receive medium-voltage
current, but the current is stepped down
again by transformers to between 110
and 250 volts before reaching homes.

Future power grids may
use superconductors to
carry **ten times as much
current** as today's cables.

REAL WORLD

Power cuts

When the power grid fails, it results in
a power cut. No current arrives at homes
and offices, and the lights—and
everything else—go off. A power cut
can be caused by a simple failure of a
transformer or a cable being damaged
by a storm. However, huge power cuts
have also been caused by solar storms,
where a surge of charged particles from
the Sun overloads the grid, causing
it to shut down.

Domestic circuits

A domestic electricity supply connects to the grid at the consumer unit, or fuse box. Powerful electrical appliances, such as an oven, have a direct connection to the fuse box. Others are connected by ring or radial circuits. Ring circuits can use thinner wires to supply the same power as a radial circuit, but radial circuits can be extended easily and can carry smaller amounts of current. So, radial circuits are often used for lighting and ring circuits are used for power sockets.

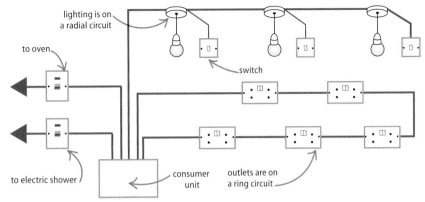

lighting is on a radial circuit

to oven

switch

to electric shower

consumer unit

outlets are on a ring circuit

△ **Wiring the house**
This simplified diagram shows the wiring in a house. Normally each floor of a house has two circuits: one for the lighting, another for the electrical outlets.

Protecting circuits

If too much current runs through domestic circuits, the wiring or appliances connected to it may get very hot and cause a fire. The consumer unit contains automatic switches called circuit breakers that cut off the supply if dangerous electrical surges occur. The circuit breaker also responds to short circuits, where faulty wiring or a damaged appliance results in the circuit drawing much more current than is normal. Fuses in plugs will also cut dangerous currents.

▽ **Electrical plug**
Most appliances connect to the electrical supply via a plug that fits into an outlet in the wall. Every plug has a live wire that delivers the current to it. The neutral wire carries the current back to the main circuit.

plug is unearthed, meaning it can expose someone to an electric shock

▽ **Earth wire**
A third wire is sometimes used in plugs. The earth wire connects the appliance to the ground via the domestic circuit. If a fault damages the insulation in the plug, any leaking current will flow safely to the ground via the earth wire.

plug has a longer middle prong, which opens socket for the shorter ones either side

▽ **Fused plugs**
The plugs in many countries are fitted with fuses—thin wires through which the current passes. If too much current passes through, the fuse wire gets hot and melts, breaking the circuit before other components get too hot.

fuse

Adaptor

Many devices come equipped with an oversized plug, known as an adaptor. The current flowing through domestic circuits is AC, which is fine for simple devices such as light bulbs and heaters. However, the back-and-forth surges of an AC supply would damage sensitive electronics, such as microchips, so an adaptor is used to filter the AC into a direct current (DC), which only travels in one direction.

▷ **Rectifier**
The main component in an adaptor is the rectifier. This is a type of diode (D) that only lets current pass in one direction.

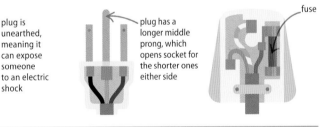
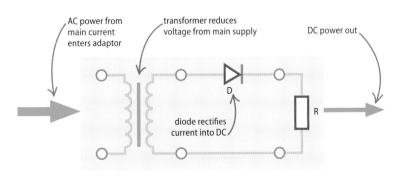

AC power from main current enters adaptor

transformer reduces voltage from main supply

DC power out

D

diode rectifies current into DC

R

Energy efficiency

ENERGY IS LOST AS HEAT BY ALL MACHINES AND PROCESSES.

SEE ALSO
❮ **170–171** Energy
❮ **188–189** Heat transfer
❮ **216–217** Transformers

When a machine or activity is designed or planned, it should be as efficient as possible. This means that as much as possible of the energy output should be used for work.

Lost energy

At a subatomic scale, some processes occur with 100 percent transfer of energy from one form to another, but on a larger scale, this never happens. Some energy is always turned into heat, which is usually unwanted. Other types of unwanted energy may also be produced: many machines make a lot of noise, which is a wasteful and unpleasant form of acoustic energy.

▽ **Energy loss**
Only a third of the energy consumed by thermal power plants reaches customers as electricity.

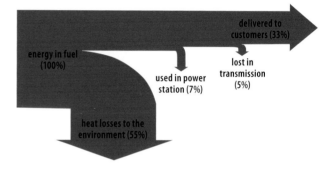

delivered to customers (33%)

energy in fuel (100%)

lost in transmission (5%)

used in power station (7%)

heat losses to the environment (55%)

▽ **Energy conversion**
Every type of energy conversion has a maximum possible efficiency. Some processes are wasteful, while others convert a very high proportion of one form of energy into another.

CONVERSION EFFICIENCIES		
Energy process	**Conversion taking place**	**Maximum possible efficiency**
photosynthesis	radiant energy from the Sun to chemical energy in the plant	6%
solar cell	radiant energy from the Sun to electrical energy, often produced by silicon crystals	28%
muscle	chemical energy from chemicals in the blood to kinetic energy as the muscle contracts	30%
coal-fired power station	chemical energy from coal to electrical energy from turbines	40%
internal combustion	chemical energy from gasoline or diesel to kinetic energy used to make vehicle move	50%
wind turbine	kinetic energy of the wind to electrical energy from a generator	60%
electric heater	electrical energy to thermal energy, produced by electrical resistance of the element	100%

REAL WORLD
Fiberoptic cable

Until a few decades ago, telephone and computer signals were usually sent in the form of electric currents that traveled down copper wires. Although copper conducts electricity very well, some of the electrical energy is lost in the form of heat. Now, optical fibers have replaced many copper wires. In an optical fiber, signals travel in the form of light, and only a tiny amount of the light energy is changed to heat, making it a far more efficient system.

Many **household appliances** are not energy efficent—the only 100 percent efficient device is the **electric heater**. However, they can be very expensive to use.

Heat loss and insulation

Sometimes we wish to produce as much heat as we can, rather than as little as possible. When we do, it is important to prevent the heat escaping. The main ways to reduce heat loss from buildings are to keep doors and windows closed and to install heat insulation.

▷ **Keeping warm for less**
Heating a well-insulated home costs only a small fraction of the amount required to heat an uninsulated one. Here are some ideas to help keep a house warm and save money.

Cavity wall insulation
Most houses are built with hollow outer walls. The gaps can be filled with foam, which sets hard and provides effective insulation.

Doors and windows
In an uninsulated house, gaps around doors and windows can account for 11 percent of heat loss.

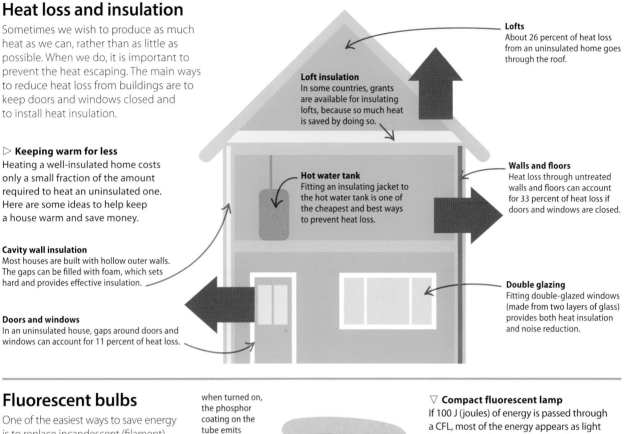

Loft insulation
In some countries, grants are available for insulating lofts, because so much heat is saved by doing so.

Lofts
About 26 percent of heat loss from an uninsulated home goes through the roof.

Hot water tank
Fitting an insulating jacket to the hot water tank is one of the cheapest and best ways to prevent heat loss.

Walls and floors
Heat loss through untreated walls and floors can account for 33 percent of heat loss if doors and windows are closed.

Double glazing
Fitting double-glazed windows (made from two layers of glass) provides both heat insulation and noise reduction.

Fluorescent bulbs

One of the easiest ways to save energy is to replace incandescent (filament) bulbs, with energy-efficient fluorescent versions. An incandescent bulb glows because the current passes through a high-resistance bare wire (the filament). The resistance means that enough electrical energy is converted to heat to make it glow with light.

▷ **Compact fluorescent lamps (CFL)**
Compact fluorescent lamps (CFLs) are gradually replacing domestic, incandescent bulbs because they are more energy efficient and last longer. CFLs feature a spiraling glass tube full of gases, which emit ultraviolet light when an electric current passes through them. This triggers a phosphor coating on the tube to shine brightly.

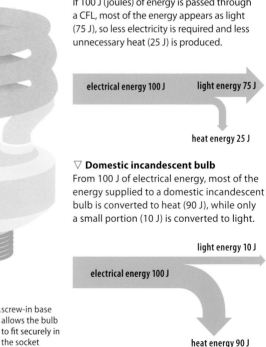

when turned on, the phosphor coating on the tube emits photons of light

ultraviolet light

argon gas and mercury vapor

electric current passed through to the glass tube

screw-in base allows the bulb to fit securely in the socket

▽ **Compact fluorescent lamp**
If 100 J (joules) of energy is passed through a CFL, most of the energy appears as light (75 J), so less electricity is required and less unnecessary heat (25 J) is produced.

electrical energy 100 J   light energy 75 J

heat energy 25 J

▽ **Domestic incandescent bulb**
From 100 J of electrical energy, most of the energy supplied to a domestic incandescent bulb is converted to heat (90 J), while only a small portion (10 J) is converted to light.

light energy 10 J

electrical energy 100 J

heat energy 90 J

Renewable energy

RENEWABLE ENERGY SOURCES ARE AN ACTIVE
AREA OF RESEARCH WORLDWIDE.

Fossil fuels (coal, oil, and natural gas) will not
last forever, and they cause serious pollution.
Nuclear energy produces waste that remains
dangerous for many centuries. So, alternative
sources of energy are being developed.

Solar energy

There are two main ways to convert
sunlight into usable energy. The first
way is in a solar thermal collector, where a
liquid is heated by being pumped through
sunlit pipes, and then used to heat a
boiler. The second way is to turn it into
electrical power by means of photovoltaic
cells. These contain material such as the
element selenium, which produces an
electrical voltage when light falls on it.

▷ **Heat from the Sun**
Solar thermal collectors can be used to heat water
for warming a house, as shown in this illustration.

1. The solar thermal
collectors contain
water-filled pipes,
which are heated
by the Sun.

2. Heated water is pushed from the
top of the collectors down the pipe
by the cooler water entering the
collectors at the base.

3. The electronic controller controls the
pump and checks water temperature in
solar thermal collectors and tank.

4. The water passes through
pipes in the tank, warming
the water in the tank. This
cools the water in the pipes.

9. The water has
lost heat to the
water in the tank.
The pump pushes
it up the pipe to
the solar thermal
collectors, to be
heated again.

5. The pipe takes the
hot water to the hot
faucets in the house.

6. The gas or electric
boiler provides extra
heat to the water tank
if solar-heated water
is too cold.

8. A pump keeps
the water moving
through the pipes.

7. Cold water from main
supply replaces hot water
used in the house, so the
water tank is always full.

Wind turbines

A wind turbine converts the motion of the wind into electrical
power. Many wind turbines are grouped into arrays and these
are often offshore, where conditions are windier. As winds do
not always blow, the amount of energy from wind farms is
variable. Wind farms can be unpopular
because of their appearance and the
noise they make, so careful siting
is essential.

▷ **Inside a wind turbine**
When the wind turbine blades
rotate, a generator (dynamo)
produces electricity. A system
of gears converts the relatively
slow spin of the blades into a
more rapid rotation in the dynamo,
producing more electrical power.

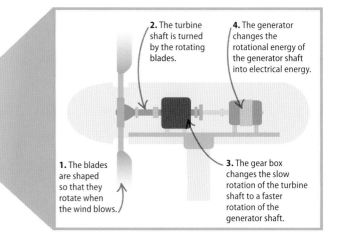

2. The turbine
shaft is turned
by the rotating
blades.

4. The generator
changes the
rotational energy of
the generator shaft
into electrical energy.

1. The blades
are shaped
so that they
rotate when
the wind blows.

3. The gear box
changes the slow
rotation of the turbine
shaft to a faster
rotation of the
generator shaft.

Tidal power

The movements of the sea can be converted into usable power in a number of ways. In one type of tidal power system, both the incoming and outgoing tides produce electricity by turning turbines.

▽ Inward tide
The incoming tide passes through a gap in a sea wall and turns a turbine mounted in the gap.

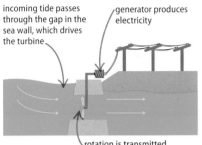

incoming tide passes through the gap in the sea wall, which drives the turbine

generator produces electricity

rotation is transmitted to the generator

▽ Outward tide
When the tide goes out, the water flows back through the gap, turning the turbines again, generating more electricity.

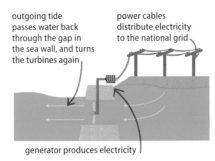

outgoing tide passes water back through the gap in the sea wall, and turns the turbines again

power cables distribute electricity to the national grid

generator produces electricity

REAL WORLD

Energy from the waves

The Pelamis wave energy converter draws power from sea waves. The converter is made of floating sections that are hinged together. As they bend with the waves, the "hinges" force fluid along pipes, and the pressure of the fluid is used to rotate turbines and generate electricity.

Geothermal energy

The interior of Earth is hotter than the surface—on average, the temperature increases by an interval of around 30°C (50°F) for every kilometer (0.6 miles) of depth. The difference between the surface and underground temperatures can be exploited to generate electricity, or simply to heat water for domestic use.

▷ Electricity from underground
In this geothermal system, water is pumped underground and pushes groundwater up to the surface. This water contains mineral salts and is referred to as geothermal brine. The brine is hot enough to boil and the high-pressure steam produced is used to turn turbines in a power station, generating electricity. The power station then distributes this electricity to the power grid, so it can be used in homes and buildings.

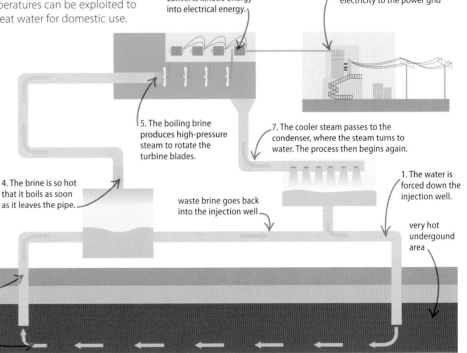

6. An array of turbines and electrical generators converts kinetic energy into electrical energy.

the power station adjusts voltage and distributes electricity to the power grid

5. The boiling brine produces high-pressure steam to rotate the turbine blades.

7. The cooler steam passes to the condenser, where the steam turns to water. The process then begins again.

1. The water is forced down the injection well.

4. The brine is so hot that it boils as soon as it leaves the pipe.

waste brine goes back into the injection well

very hot undergound area

3. The hot brine is forced up the production well.

2. The water from the injection well heats up, and, when chemicals dissolve in it, it turns into brine.

The Earth

OUR PLANET IS THE THIRD FROM THE SUN AND ONE OF FOUR
IN THE SOLAR SYSTEM MADE MAINLY OF ROCK AND METAL.

Earth is the only planet where liquid water is known to exist.
It is also the only place in the Universe known to support life.

Inside the Earth

Earth is a mixture of rock and metal.
The solid inner core is made of iron
and nickel, and the outer core is a
mix of molten iron and nickel.
The surrounding mantle is
a thick layer of solid and
semimolten rock. A thin,
rocky outer shell consists
of thick continental crust
(land) and thinner
oceanic crust (seafloor).

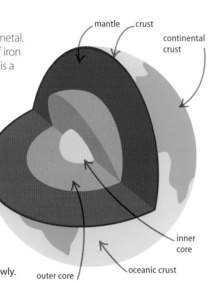

▷ **Inner heat**
Inside, the Earth is very hot.
The temperature at the inner
core reaches 4,700°C (8,500°F).
The heat causes the semimolten
rock in the mantle to circulate slowly.

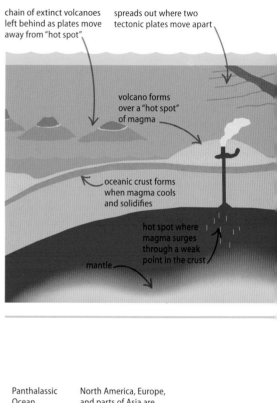

The seasons

Earth experiences seasons because it rotates on a tilted axis as it travels
around the Sun, an orbit that takes one year. As different areas of the
planet face toward or away from the Sun, the length of the day and
temperature change, which affect plant growth and animal behavior.

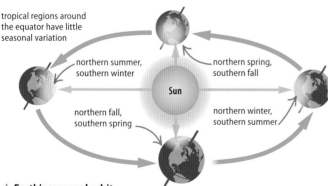

△ **Earth's seasonal orbit**
When the North Pole turns to face the Sun it is summer in the Northern
Hemisphere and winter in the south. Six months later, when the South Pole
tilts toward the Sun, it is summer in the south and winter in the north.

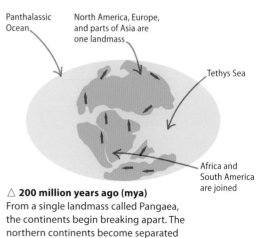

△ **200 million years ago (mya)**
From a single landmass called Pangaea,
the continents begin breaking apart. The
northern continents become separated
from those in the south by the Tethys Sea.

Plate tectonics

Earth's crust is broken up into sections called tectonic plates. The plates drift on the mantle as it is slowly churned by currents caused by heat from the core. Where two plates move together, at a convergent boundary, one plate dives under another to form a mountain range. At a divergent boundary the plates move apart and molten material from the mantle, known as magma, erupts at the surface as a volcano. Where plates grind alongside each other, earthquakes occur as the rocks catch and then jerk free.

▽ **Violent Earth**
As plates constantly move, oceans are pulled apart, and continents may either crash into each other or break away.

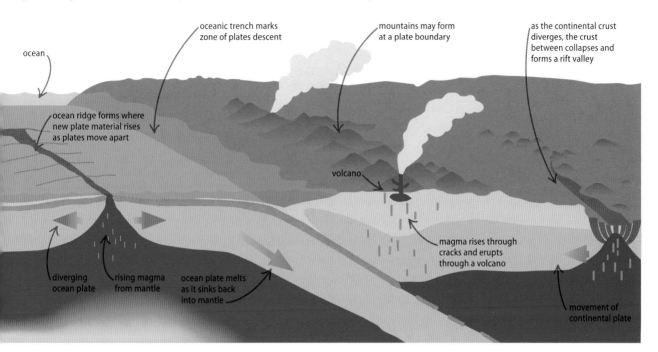

ocean

oceanic trench marks zone of plates descent

mountains may form at a plate boundary

as the continental crust diverges, the crust between collapses and forms a rift valley

ocean ridge forms where new plate material rises as plates move apart

volcano

magma rises through cracks and erupts through a volcano

diverging ocean plate

rising magma from mantle

ocean plate melts as it sinks back into mantle

movement of continental plate

Continental drift

Over millions of years, the motion of Earth's plates has caused the continents to drift apart. If you could put them together, they would all fit, like a jigsaw puzzle. This idea is supported by matching patterns of rocks and fossils on lands now separated, and may explain why similar animals are found on opposite sides of the world.

India separates from the southern continents

North America begins to break away from Eurasia

Antarctica and Australia are the same continent

South America separates from Africa

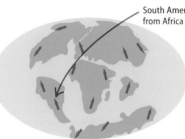

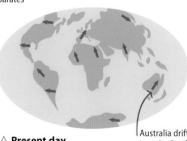

Australia drifts into the Pacific Ocean

△ **130 mya**
At this point, North America begins to break up from Eurasia (the landmass comprising Europe and Asia). Australia and Antarctica are joined together.

△ **70 mya**
Divergent plates continue to open up the Atlantic Ocean. South America drifts west, Antarctica heads for the South Pole, and India creeps towards Asia.

△ **Present day**
India is in place after colliding with the Eurasian mainland. Greenland separates from North America.

Weather

CHANGES IN CONDITIONS IN THE ATMOSPHERE
PRODUCE DIFFERENT WEATHER EVENTS.

Weather changes occur when sections of atmosphere with
different temperatures, pressures, and humidities (water content)
come into contact.

SEE ALSO	
❮ **74–75**	Ecosystems
❮ **100–101**	Changing states
❮ **102–103**	Gas laws
❮ **184**	Atmospheric pressure
❮ **202**	Static discharge
❮ **226–227**	The Earth

Precipitation

Rain is an example of precipitation, where water vapor
in the atmosphere condenses to a liquid and falls to the
ground. Warm air can hold more water vapor than cool air.
Precipitation occurs when air saturated with water vapor
is forced to cool and the excess falls as raindrops. Hail and
snow are also forms of precipitation. Hailstones are formed
when rain is repeatedly blown upward into colder areas of
the atmosphere, while snowflakes form when water vapor
condenses in already freezing air. Precipitation occurs at
weather fronts, where air masses of different temperatures
meet. There are three types, shown below.

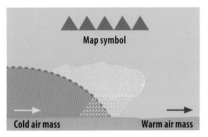

Cold air mass | Warm air mass

◁ **Cold front**
Cold air moves
under warmer,
wetter air. As the
warm air rises, its
pressure falls and
it cools, dropping
its water as rain.

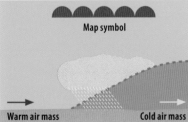

Warm air mass | Cold air mass

◁ **Warm front**
A mass of warm air
flows over a block
of cold air, forming
rain and clouds.
Warm fronts move
more slowly
than cold fronts,
resulting in
sustained rain.

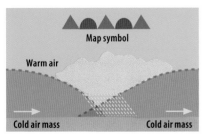

Warm air

Cold air mass | Cold air mass

◁ **Occluded front**
This occurs when
the warm air mass
is pushed off the
ground completely
by cooler air.
Occluded fronts
also produce rain.

Wind

Wind forms when air rushes from an area of high pressure to an
area of low pressure. The bigger the difference between the two
pressures, the stronger the wind.

▽ **Beaufort wind scale**
This scale describes the strength of wind by its effects, so
people can judge wind speeds without a measuring device.

Scale	Wind speed km/h (mph)	Strength	Observation
0	0–2 (0–1)	calm	smoke rises vertically
1	3–6 (2–3)	light air	smoke drifts slowly
2	7–11 (4–7)	light breeze	leaves rustle
3	12–19 (8–12)	gentle breeze	small flags fly
4	20–29 (13–18)	moderate breeze	trees toss, dust flies
5	30–39 (19–24)	fresh breeze	small branches sway
6	40–50 (25–31)	strong breeze	large branches sway
7	51–61 (32–38)	near gale	trees in motion
8	62–74 (39–46)	gale	twigs break
9	75–87 (47–54)	strong gale	branches break
10	88–101 (55–63)	storm	trees snap
11	102–119 (64–74)	violent storm	widespread damage
12	120+ (75+)	hurricane	extreme damage

REAL WORLD

Tornado

The fastest winds on Earth are
inside tornadoes. They form
when a column of spinning
air inside a thunderstorm cloud
makes contact with the ground.
An average tornado is about
80 m (260 ft) across and the
air in it moves at 170 km/h
(110 mph), sucking objects
off the ground, high into the air.

Clouds

Clouds are made of minute water droplets or ice condensed around tiny specks of dust that are blown in the air. Clouds are mostly white because their water droplets scatter a lot of light. When the cloud is filled with water and close to raining, it looks dark gray because it absorbs a lot of light.

▽ **Cloud types**
Below are types of cloud that are defined by their height and appearance.

1. Cirrostratus
These flat and wispy clouds form at high altitudes from ice crystals.

2. Cirrus
Cirrus are high, wispy clouds, and indicate that stormy weather is likely.

3. Cirrocumulus
These high, fluffy clouds means rain is on its way.

4. Altostratus
Sheets of cloud at medium altitudes suggest gray skies and light rain are likely.

5. Cumulonimbus
The largest cloud of all produces thunderstorms.

6. Altocumulus
These fluffy clouds form at a medium height, and indicate a cold front is coming.

7. Stratocumulus
The wide, fluffy clouds are closer to the ground than altocumulus.

8. Stratus
This cloud is characterized by its horizontal shape.

9. Nimbostratus
This is a stratus cloud with rain.

10. Cumulus
These low-level fluffy clouds form in mild weather.

11. Fog
This stratus cloud can touch the ground.

Weather maps

Meteorologists (people who study weather) show the current atmospheric conditions on weather maps. This is a useful tool for forecasting the weather. The map shows the weather fronts and areas of low and high pressure. An expert meteorologist can predict how the front will move, and so figure out what the weather will be like over a particular region.

▷ **Pressure gradients**
Isobars (the black lines on the map) link places where the atmospheric pressure is the same. The isobars form rings around centers of low and high pressure. The closer the rings are to each other, the stronger the wind.

Key

Warm front   Cold front   Occluded front

low-pressure zones, where there are usually strong winds and rainfall, are marked by rings of isobars with the lowest pressure near the center

high-pressure zones, where it is usually cloudless and sunny, are marked by rings of isobars with the highest pressure near the center

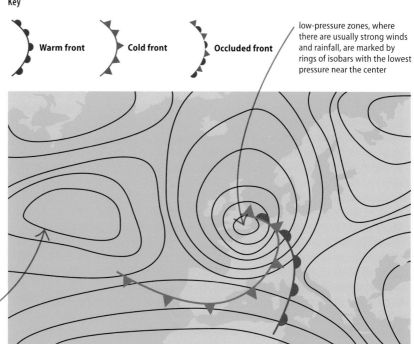

Astronomy

ASTRONOMY IS THE SCIENTIFIC STUDY
OF STARS AND OTHER OBJECTS IN SPACE.

People have been mapping the stars and tracking
the movements of planets for thousands of years.

Telescopes

The first telescopes, developed in
the early 17th century, gathered light
from a distant source and magnified
the image using either lenses (in
refracting telescopes) or mirrors
(in reflecting telescopes). These
are called optical telescopes, because
they focus light. Today, there also
telescopes that reveal other types of
radiation invisible to the human eye,
such as gamma rays and radio waves.
These have led to many important
discoveries in astronomy, such as
active galaxies and the Big Bang.

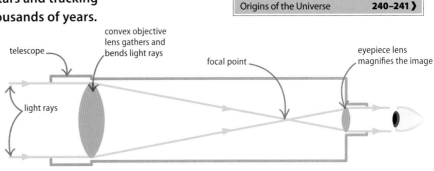

telescope

convex objective
lens gathers and
bends light rays

focal point

eyepiece lens
magnifies the image

light rays

△ **Refracting telescope**
The large objective lens focuses the rays of
light into a small image inside the device.
Then an eyepiece lens magnifies the image.

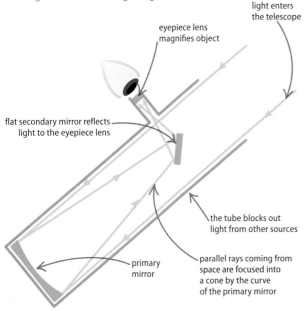

eyepiece lens
magnifies object

light enters
the telescope

flat secondary mirror reflects
light to the eyepiece lens

the tube blocks out
light from other sources

primary
mirror

parallel rays coming from
space are focused into
a cone by the curve
of the primary mirror

△ **Reflecting telescope**
This telescope collects light using a curved primary mirror, which
reflects and focuses the light onto a flat secondary mirror. This shines
the light toward the eyepiece lens, which magnifies the object.
The world's most powerful astronomical telescopes are reflecting
telescopes, some with mirrors up to 10 m (33 ft) across.

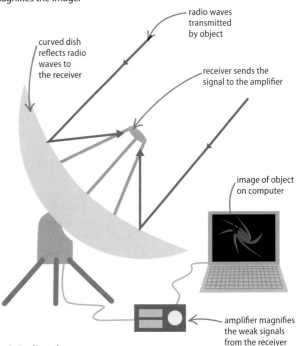

radio waves
transmitted
by object

curved dish
reflects radio
waves to
the receiver

receiver sends the
signal to the amplifier

image of object
on computer

amplifier magnifies
the weak signals
from the receiver

△ **Radio telescope**
A radio telescope is a huge antenna that picks up the longer wavelengths
of radiation coming from space. The radio signals from stars are weak, so
a large dish is used to reflect them onto the central receiver. The signals
are amplified electronically, and a computer processes them to produce
pictures, called radio images.

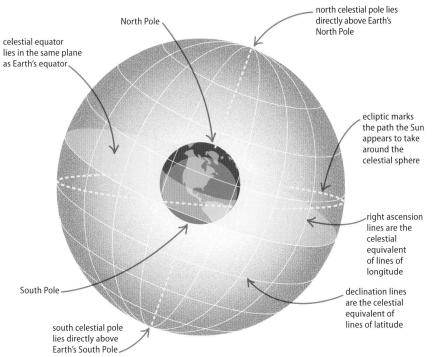

celestial equator lies in the same plane as Earth's equator

North Pole

north celestial pole lies directly above Earth's North Pole

ecliptic marks the path the Sun appears to take around the celestial sphere

right ascension lines are the celestial equivalent of lines of longitude

declination lines are the celestial equivalent of lines of latitude

South Pole

south celestial pole lies directly above Earth's South Pole

Celestial sphere

The objects seen in the night's sky are not all the same distance from Earth. The Moon is obviously much closer than Jupiter, but astronomers plot their movements and the positions of all the stars on an imaginary sphere that surrounds Earth. The view of the celestial sphere, as it is called, changes as Earth rotates within it, so stars appear to rise in the east and set in the west, just like the Sun. An observer on Earth can see a maximum of half of the sphere at one time.

◁ **Plotting the stars**
The Earth's poles and equator are projected onto the celestial sphere. A system of grid lines through the poles, called lines of right ascension, and lines parallel to the equator, called declination lines, means stars can be located by their coordinates.

Spectroscopy

In the laboratory, scientists investigate the chemical elements in hot gases using a technique called spectroscopy. Observing the gases through a spectroscope reveals the different wavelengths of light (see page 196). White light produces a continuous band of colors, but if atoms are present they affect the light and lines of color appear. The atoms of each element have their own unique pattern of lines, called an emission spectrum. Astronomers use spectroscopy to find out what materials are present many light-years away.

emission spectrum of carbon

emission spectrum of hydrogen

emission spectrum of mercury

Light-years

Distances in space are so immense that they are measured in light-years, or the distance light travels in a year—slightly more than 9 trillion km (6 trillion miles). The Sun is eight light-minutes away. Voyager 1 (right), the most distant space probe, is 16 light-hours away, while our next nearest star is four light-years out.

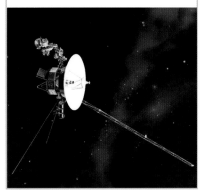

The Sun

THE SUN IS OUR NEAREST STAR. ITS HEAT AND
LIGHT MAKE ALL LIFE ON EARTH POSSIBLE.

Although 100 times wider than Earth, the Sun is an average
star in terms of its size and age. Studying the Sun has helped
us understand how other stars in the Universe work.

Inside the Sun

The Sun is an immense ball of gas 1.4 million km
(870,000 miles) wide. It is made up of almost
three-quarters hydrogen, about a quarter helium,
and small amounts of 65 or so other elements, all
held together by gravity. The temperature, density,
and pressure of the gas increase toward the center.
At the core, nuclear reactions that convert hydrogen
to helium are the source of the Sun's energy. The
energy radiates out, taking many thousands of
years to reach the surface, where it is released
into space as light and heat.

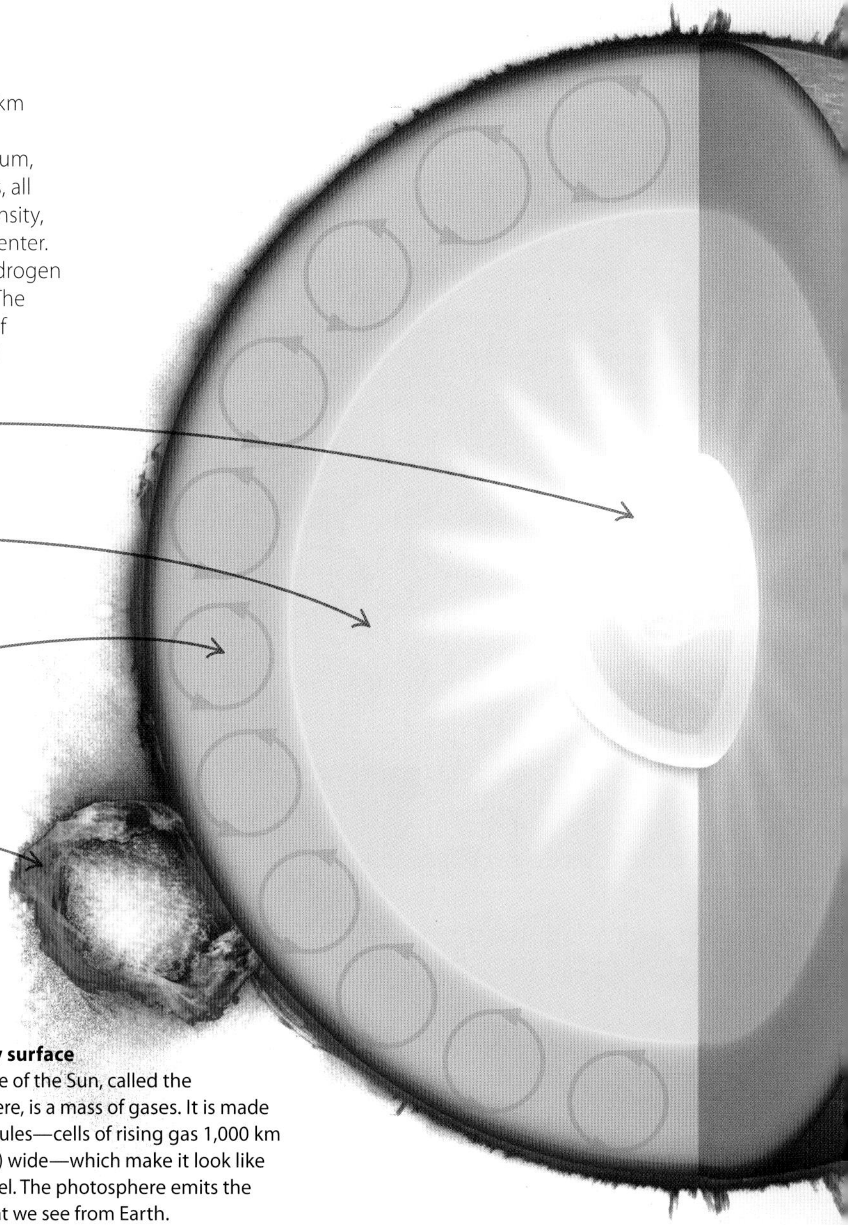

Core
The temperature at
the center of the Sun is
15.7 million°C (28 million°F).

Radiative zone
In this region, energy
slowly radiates from the core
towards the convective zone.

Convective zone
Swirling currents of gas
carry heat from the top of
the radiative zone toward
the surface, where they cool
and then sink back.

Prominence
These looping clouds of gas
can shoot out more than
100,000 km (62,000 miles).

The Sun's mass is
about **750 times
greater** than all
the other objects
in the Solar System
put together.

▷ **Stormy surface**
The surface of the Sun, called the
photosphere, is a mass of gases. It is made
up of granules—cells of rising gas 1,000 km
(620 miles) wide—which make it look like
orange peel. The photosphere emits the
visible light we see from Earth.

Chromosphere
The Sun is surrounded by layers of gas, forming an atmosphere. The inner layer is called the chromosphere. The outer layer is called the corona, and extends into space for millions of kilometers.

Spicules
Jets of gas called spicules shoot up to 10,000 km (6,200 miles) from the photosphere for bursts of up to 10 minutes.

Nuclear fusion

Nuclear fusion occurs when the nuclei of two atoms fuse (join) together to make a large nucleus and energy is released. Every element is made up of atoms with a different number of protons (positive particles) in the nucleus and neutrons (no charge). Hydrogen, the most common element in the Sun, has one proton and usually no neutrons. However, the heat and pressure in the Sun's core increases the chance for hydrogen isotopes to form, with one or two neutrons. They fuse to form helium.

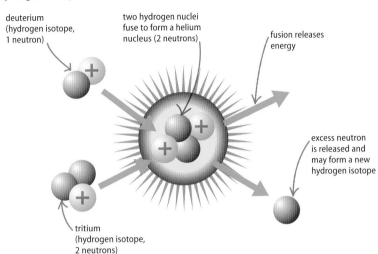

deuterium (hydrogen isotope, 1 neutron)

two hydrogen nuclei fuse to form a helium nucleus (2 neutrons)

fusion releases energy

excess neutron is released and may form a new hydrogen isotope

tritium (hydrogen isotope, 2 neutrons)

The **sunspots** on the Sun's surface may last from a few hours to **several weeks.**

△ **Activity in the Sun's core**
At the Sun's core hydrogen nuclei collide at great speed. The fusion process is complex, but one of the reactions that takes place is shown above. Here, two different hydrogen isotopes (see page 169) fuse to form helium, releasing energy and an excess neutron.

REAL WORLD
Little Ice Age

The number of sunspots rises and falls over an 11-year cycle. It is believed that sunspot activity may affect the climate on Earth. In the late 1600s, there was a long period when few sunspots were recorded, which coincided with a series of very cold winters in Europe that became known as the Little Ice Age. For about 100 years the Thames River, in London, England, froze almost every winter, and frost fairs were held on the thick ice.

Sunspots
These dark areas are around 1,500°C (2,700°F) cooler than the rest of the surface. They occur where magnetism prevents hot gas from reaching the surface.

The Solar System I

THE SUN AND THE OBJECTS THAT ORBIT IT, INCLUDING THE
PLANETS AND SMALLER BODIES, MAKE UP OUR SOLAR SYSTEM.

SEE ALSO

❰ **178–179** Gravity
❰ **226–227** The Earth
❰ **232–233** The Sun
The Solar System II   **236–237** ❱

At the center of the Solar System, the powerful gravitational
force of the Sun holds the eight planets in orbit around it.

The word "planet" comes
from the Greek word
"planetos," which
means **"wanderer."**

Scale of the Solar System

Distances in space are so vast that they are hard to
imagine. Earth is about 150 million km (93 million miles)
from the Sun. To simplify things, astronomers call this
distance an astronomical unit (AU)—so Earth is 1 AU
from the Sun. Using this scale, Neptune, the furthest
planet, is 30 AU from the Sun.

3. Earth
Diameter: 12,756 km (7,926 miles)
Distance from Sun: 1 AU
Year: 365 days
Day: 24 hours
Number of moons: 1
Average surface temperature: 15°C
(59°F)

4. Mars
Diameter: 6,786 km (4,217 miles)
Distance from Sun: 1.5 AU
Year: 687 days
Day: 24.5 hours
Number of moons: 2
Average surface temperature: −63°C
(−81°F)

▷ **The planets**
Each planet in the Solar System has its own features,
such as distance from the Sun and number of moons.
Every planet also has a different year (the time it takes
to orbit the Sun), and length of day (the time it takes
to rotate once about its axis).

2. Venus
Diameter: 12,104 km (7,521 miles)
Distance from Sun: 0.7 AU
Year: 225 days
Day: 243 days
Number of moons: 0
Average surface temperature: 464°C
(867°F)

1. Mercury
Diameter: 4,879 km (3,031 miles)
Distance from Sun: 0.4 AU
Year: 88 days
Day: 58 days
Number of moons: 0
Average surface temperature: 167°C
(333°F)

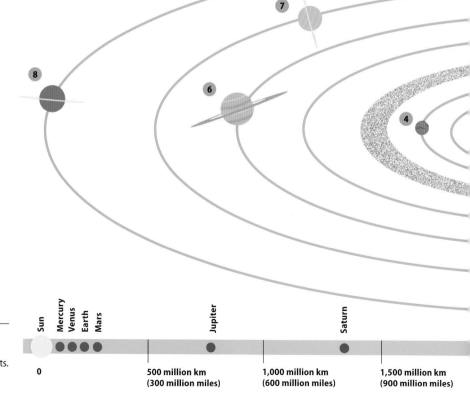

▷ **Inner and outer Solar System**
The first four planets—all small and rocky—
form the inner Solar System. Beyond the
Main Belt, in the outer Solar System,
lie the four planets known as the gas giants.

Sun	Mercury	Venus	Earth	Mars	Jupiter	Saturn

0	500 million km (300 million miles)	1,000 million km (600 million miles)	1,500 million km (900 million miles)

Beyond Neptune

Surrounding the planets is the Kuiper Belt, a region of mainly icy-rocky bodies and a small number of dwarf planets, such as Pluto. Beyond this lies the Oort Cloud, a sphere of yet more ice bodies left over from the formation of the Solar System.

▷ **The Oort Cloud**
More than one trillion comets make up the Oort Cloud. Its outer edge marks the end of the Solar System.

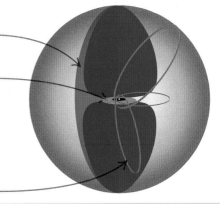

The outer limit
The Oort Cloud reaches 50,000 AU from the Sun.

Kuiper Belt
The Kuiper Belt merges with the Oort Cloud.

Comet orbits
Many comets are ice bodies from the Oort Cloud that have been pushed into closer orbits around the Sun. They travel in all directions, as shown by the orbits in pink.

5. Jupiter
Diameter: 142,984 km (88,846 miles)
Distance from Sun: 5.2 AU
Year: 11.9 years
Day: 10 hours
Number of moons: 63
Cloud-top temperature: −108°C (−162°F)

6. Saturn
Diameter: 120,536 km (74,897 miles)
Distance from Sun: 9.6 AU
Year: 29.5 years
Day: 10.5 hours
Number of moons: 62
Cloud-top temperature: −139°C (−218°F)

7. Uranus
Diameter: 51,118 km (31,763 miles)
Distance from Sun: 19.2 AU
Year: 84 years
Day: 17 hours
Number of moons: 27
Cloud-top temperature: −197°C (−323°F)

8. Neptune
Diameter: 49,528 km (30,775 miles)
Distance from Sun: 30 AU
Year: 165 years
Day: 16 hours
Number of moons: 13
Cloud-top temperature: −201°C (−330°F)

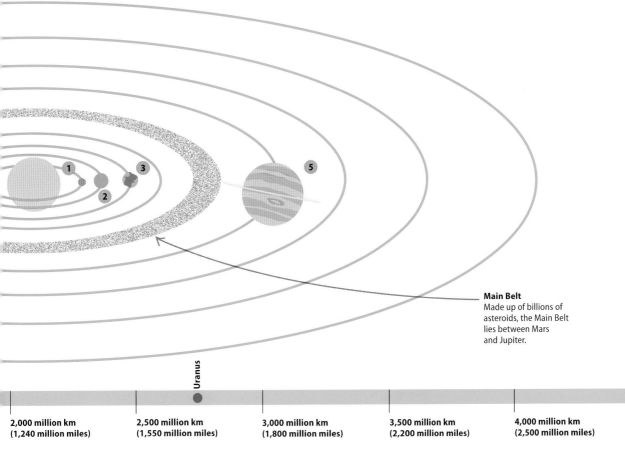

Main Belt
Made up of billions of asteroids, the Main Belt lies between Mars and Jupiter.

Uranus

Neptune

2,000 million km
(1,240 million miles)

2,500 million km
(1,550 million miles)

3,000 million km
(1,800 million miles)

3,500 million km
(2,200 million miles)

4,000 million km
(2,500 million miles)

The Solar System II

AS WELL AS THE PLANETS, THE GRAVITY OF THE SUN ATTRACTS A HUGE NUMBER OF SMALLER OBJECTS.

SEE ALSO
‹ **232–233** The Sun
‹ **234–235** The Solar System I
Stars and galaxies **238–239** ›

Moons orbit most of the planets, including Earth, while smaller bodies, such as comets, follow independent paths.

The Moon

A moon is a body that orbits a planet and there are more than 160 in our Solar System. Earth has just one moon, a cratered ball of rock, which spins as it orbits. The Moon formed when a large asteroid collided with Earth during its formation and some of the debris went into orbit around the Earth, becoming the Moon.

Orbit of the Moon ▷
When we look at the Moon, only the sunlit part is visible. The Moon takes the same amount of time to orbit our planet as to spin once on its axis, so we only ever see one side from Earth. The side we don't see is called the far side of the Moon.

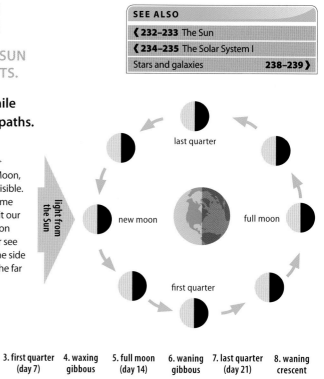

last quarter

light from the Sun

new moon

full moon

first quarter

The lunar cycle ▷
Every month the face of the Moon appears to change from a dark shadow, called a new moon, to become a thin, shining crescent, to a full moon, and then back again. This is because, as the Moon orbits Earth, we see more or less of the half of the Moon that is lit by the Sun.

1. new moon (day 0)
2. waxing crescent
3. first quarter (day 7)
4. waxing gibbous
5. full moon (day 14)
6. waning gibbous
7. last quarter (day 21)
8. waning crescent

side facing the Earth is in darkness

we see the whole sunlit surface

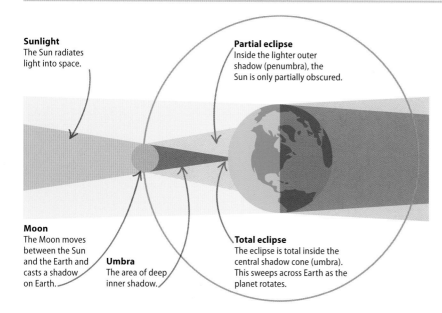

Sunlight
The Sun radiates light into space.

Partial eclipse
Inside the lighter outer shadow (penumbra), the Sun is only partially obscured.

Moon
The Moon moves between the Sun and the Earth and casts a shadow on Earth.

Umbra
The area of deep inner shadow.

Total eclipse
The eclipse is total inside the central shadow cone (umbra). This sweeps across Earth as the planet rotates.

Eclipses

Sometimes, as the Moon orbits Earth, it passes directly in front of the Sun and blocks out the light. This is called a solar eclipse. Sometimes the Moon's path takes it into Earth's shadow, causing a lunar eclipse. When this happens the Moon dims and appears red because the light is bent as it passes through Earth's atmosphere.

◁ **Solar eclipse**
There are only about three total solar eclipses each year, and each one is only seen from the narrow band of deep shadow, called the umbra, on the Earth's surface. More frequent is a partial solar eclipse, when the Moon's shadow covers just part of the Sun.

Dwarf planets

In 1930, a new planet was found beyond the orbit of Neptune. The new body was named Pluto, and it was found to be by far the smallest planet, even smaller than Earth's Moon. By 2005, improved survey techniques had found several bodies similar in size to Pluto in the same area of the Solar System, and one (Ceres) in the Main Belt of asteroids. It was then decided to name objects of this size range dwarf planets, and Pluto became one of them.

▷ **Independent bodies**
Dwarf planets are independent bodies large enough to have become almost spherical because of their internal gravity, but are too small to be called planets.

Ceres
This is the largest body in the Main Belt of asteroids. It is made of rock and metal.

Haumea
This misshapen body in the Kuiper Belt is largely made of ice.

Makemake
A very cold ball of ice, this dwarf planet comprises frozen methane, ammonia, and water.

Pluto
Pluto has five moons. The largest is Charon, about half the size of Pluto.

Eris
Discovered in 2005, Eris is currently the largest dwarf planet known.

Charon
Some scientists consider Charon to be part of a two-planet system with Pluto, rather than Pluto's moon.

Comets

Comets are "dirty snowballs" of dust and ice formed at the birth of the Solar System. They have highly elliptical (oval-shaped) orbits. Some travel far beyond Neptune, taking thousands of years to circle the Sun, while others have short paths of just a few years. As a comet nears the Sun it heats up and dust and gas stream out, forming tails millions of kilometers long.

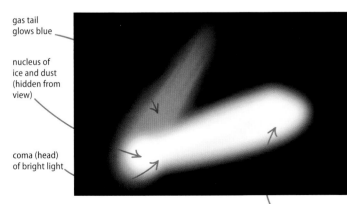

gas tail glows blue

nucleus of ice and dust (hidden from view)

coma (head) of bright light

dust tail reflects white sunlight

△ **Comet tails**
When a comet gets closer to the Sun, beyond the orbit of Mars, it becomes active, developing a coma and tails. When the comet travels out of the inner Solar System these disappear again.

Meteoroids and meteorites

Meteoroids are chunks of space rock that produce streaks of light, called meteors, as they burn up in Earth's atmosphere. Most are small and around 3,000 tons of space rock hits Earth every year as dust. A meteoroid that survives the atmosphere to land on Earth's surface is called a meteorite and these can form large impact craters. Roter Kamm in Namibia, for example, is more than five million years old, 2.5 km (1.5 miles) wide, and 130 m (425 ft) deep.

Stars and galaxies

GALAXIES ARE HUGE STAR SYSTEMS, MADE UP OF
STARS AND LARGE AMOUNTS OF GAS AND DUST.

SEE ALSO

❰ **178–179** Gravity
❰ **194–195** Electromagnetic waves
❰ **230–231** Astronomy
❰ **232–233** The Sun

Our Solar System is just one of billions in our local area,
or galaxy. Our galaxy, the Milky Way, is one of hundreds
of billions in the Universe.

Astronomical objects

Astronomers estimate that about 6,000 objects can
be seen from Earth with the naked eye. Most appear
as points of light, but a closer look through a powerful
telescope reveals that there is a lot more than just
stars and planets out there.

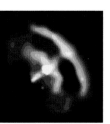

Pulsar
Some dying stars blow apart in a
massive explosion called a supernova.
The core of the star may collapse to form
a small, dense neutron star that emits
beams of energy and rotates at amazing
speeds. If the beams are detected on
Earth, the star is known as a pulsar.

The Sun
Our local star is the source of almost all
light and heat reaching Earth. However,
it is a very average star, in terms of size
and temperature. The next nearest star
to Earth is Proxima Centauri, which is
4.2 light-years away.

Galaxy
Galaxies are vast star systems that
exist in a range of shapes and sizes.
The smallest have a few million stars,
and the largest, several trillion. Around
half of galaxies are spiral-shaped disks,
with a central bulge and arms spiraling
away from it.

Constellation
Ancient astronomers organized the
visible stars into patterns called
constellations, most based on images
from Greek mythology, such as Ursa
Major. Although the stars in a
constellation look close together, they are
at vastly differing distances from Earth.

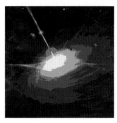

Quasar
Among the most distant of all objects
are quasars—young galaxies seething
with energy as billions of stars form.
Light from distant objects takes many
billions of years to reach Earth, so we
see quasars as they were when the
light left them all that time ago.

Star cluster
Stars are seldom found alone; most
travel either with one or more
companions. Pairs are called binary
stars. Small groups of stars are called
clusters and are made up of stars that
formed at the same time, held
together in groups by gravity.

REAL WORLD
The Milky Way

Our galaxy is called the Milky Way. It is a spiral galaxy, but from
Earth we see it as a pale strip across the night sky. The ancient
Greeks called this the *galaktikos*, which means "milky path." The
Solar System lies in an outer arm.

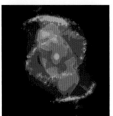

Planetary nebula
As some stars grow older they eject
their outer layers to become planetary
nebulae. The glowing, colored rings of
hot gas and dust make stars such as
the Cat's Eye Nebula some of the most
stunning objects in space. The faint star
at the center is known as a white dwarf.

Star types

Stars have a life cycle and, as they age, their characteristics—size, color, luminosity (energy output)—change. Astronomers group stars according to the light they emit. The color of a star's light identifies how hot it is, with blue as the hottest. However, hot stars are not always the brightest. Arranging stars by luminosity and temperature shows they fall into certain groups.

▷ **Star groups**
When a star first produces energy by nuclear fusion—like our Sun—it is called a main sequence star.

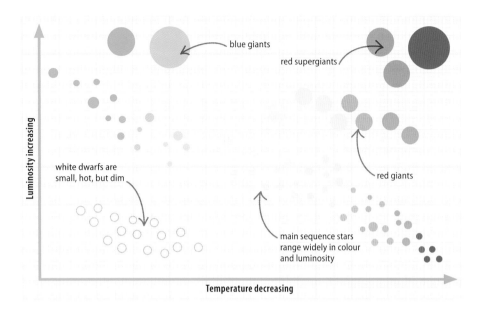

blue giants

red supergiants

red giants

white dwarfs are small, hot, but dim

Luminosity increasing

main sequence stars range widely in colour and luminosity

Temperature decreasing

Star life cycle

Stars are born inside great clouds of gas and dust called nebulae. As the clouds collapse, the temperature and pressure rise, and eventually nuclear fusion begins in the star's core. The star then stabilizes on the main sequence of its life. When the fuel runs out, the star dies.

▽ **Birth, life, and death**
Stars spend around 90 percent of their lives in the main sequence phase. The mass of a star is crucial and will determine what happens to it and when it dies.

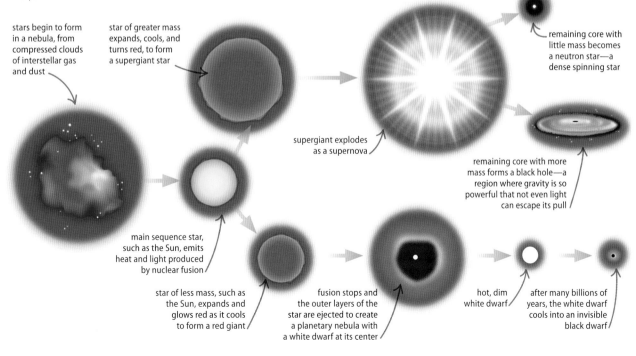

stars begin to form in a nebula, from compressed clouds of interstellar gas and dust

star of greater mass expands, cools, and turns red, to form a supergiant star

supergiant explodes as a supernova

remaining core with little mass becomes a neutron star—a dense spinning star

remaining core with more mass forms a black hole—a region where gravity is so powerful that not even light can escape its pull

main sequence star, such as the Sun, emits heat and light produced by nuclear fusion

star of less mass, such as the Sun, expands and glows red as it cools to form a red giant

fusion stops and the outer layers of the star are ejected to create a planetary nebula with a white dwarf at its center

hot, dim white dwarf

after many billions of years, the white dwarf cools into an invisible black dwarf

Origins of the Universe

THE BIG BANG THEORY EXPLAINS HOW THE UNIVERSE DEVELOPED.

Although nobody knows how the Universe began, evidence suggests that it started with an ancient burst of energy and is still expanding.

SEE ALSO
❮ 168–169 Inside atoms
❮ 170–171 Energy
❮ 178–179 Gravity
❮ 194–195 Electromagnetic waves
❮ 230 Telescopes
❮ 231 Spectroscopy
❮ 238–239 Stars and galaxies

The Big Bang theory

In the 1920s, it was discovered that our galaxy and the millions of galaxies that surround it are all moving away from each other because the Universe is expanding. This implies that the galaxies all began close together, billions of years ago. The idea that the Universe started as a hot burst of energy in the distant past was widely accepted once the remains of that energy were observed in the 1960s. In the 13.7 billion years that have passed since the Universe began, that energy has cooled to −270°C (−454°F), and is now known as cosmic microwave background radiation.

There are two theories that predict the way in which the Universe may end: it might be **torn to pieces** in a "**Big Rip**," or become cold and black in a "**Big Chill**."

▽ **The history of everything**
This diagram shows the story of the Universe. Moving from left to right, the intervals of time become longer: the halfway mark on the picture is 500,000 years, which is only 0.04 percent of the age of the Universe.

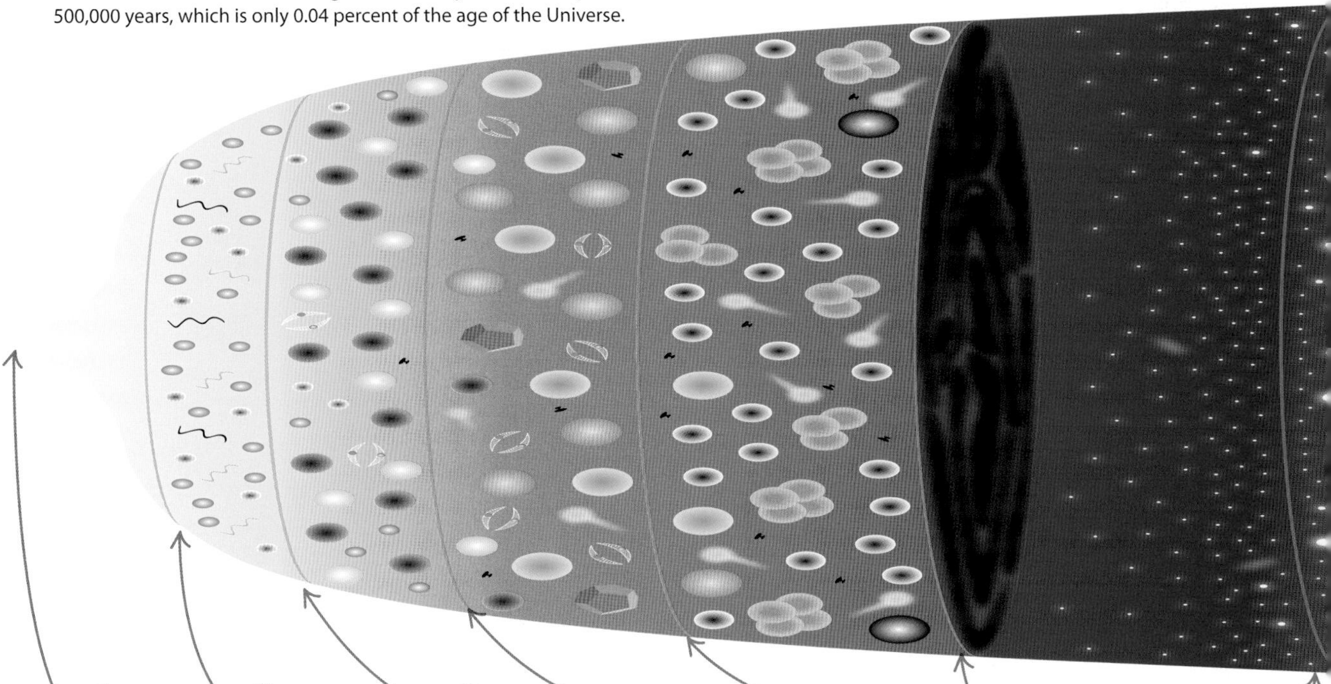

10^{-43} second Nothing is known of the events that led to the Big Bang and the start of our Universe.

10^{-38} second The Universe undergoes an enormous increase in its rate of expansion, called inflation, and emits a huge amount of heat and radiation.

10^{-10} second Electromagnetic and weak forces become distinct. The Universe is cooling rapidly and forming a soup of primitive particles.

0.001 seconds Matter is formed, as subatomic particles, and most of these destroy each other.

3 minutes As the Universe continues to cool, the remaining particles are mostly protons, neutrons, electrons, and neutrinos.

380,000 years The Universe is cool enough for atoms to form. Space becomes transparent, because there are fewer particles to obstruct photons of light.

1 billion years Stars and galaxies form.

Cosmic microwave

Microwave radiation from the Big Bang was discovered by accident by radio astronomers who thought the radiation was background noise. Today, satellites have made very detailed maps of the radiation, which show slight temperature variations (shown as different colors below), revealing slight variations in the density of the young Universe.

Red shift

Galaxies exist in clusters, and the key piece of evidence for the Big Bang is that these clusters are all moving apart—in other words, the Universe is expanding. Astronomers know this because they can split the light from the galaxies into spectra, which are like rainbows containing lines of light or dark that show the substances present in them. In spectra from distant galaxies, the positions of the lines are all shifted to longer wavelengths—that is, toward the red end of the spectrum.

▷ **Red shift**
The light from a star, galaxy, or other bright object is reddened very slightly if it is moving away from the observer.

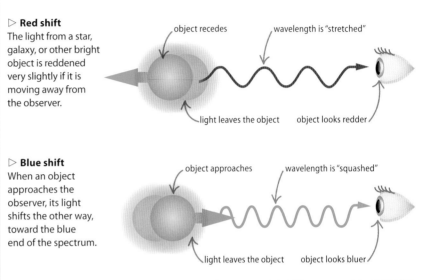

object recedes   wavelength is "stretched"

light leaves the object   object looks redder

▷ **Blue shift**
When an object approaches the observer, its light shifts the other way, toward the blue end of the spectrum.

object approaches   wavelength is "squashed"

light leaves the object   object looks bluer

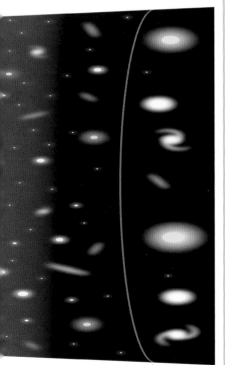

Present
The Universe as it is today, 13.7 billion years old.

What is out there?

Stars and galaxies are just a tiny part of the total mass of the Universe. Most is made up of forms of matter and energy that cannot be seen. Galaxies contain invisible dark matter, which scientists know is there by observing galaxy behavior. There is also an unknown force accelerating the expansion of the Universe, known as dark energy.

▽ **Universal matter**
These pie charts show the make-up of the Universe. The atoms that make the stars and galaxies are mainly hydrogen and helium. The additional heavier elements are made within stars. Those beyond iron (see pages 116–117) form when massive stars explode.

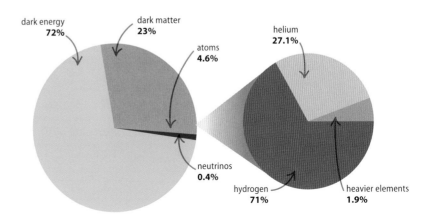

dark energy
72%

dark matter
23%

atoms
4.6%

neutrinos
0.4%

helium
27.1%

hydrogen
71%

heavier elements
1.9%

Biology reference

The plant kingdom

With the scientific name Plantae, this kingdom contains around 300,000 species, ranging from simple mosses to immense trees. The great majority of plants are photosynthetic and manufacture their own supply of sugars using the energy in sunlight.

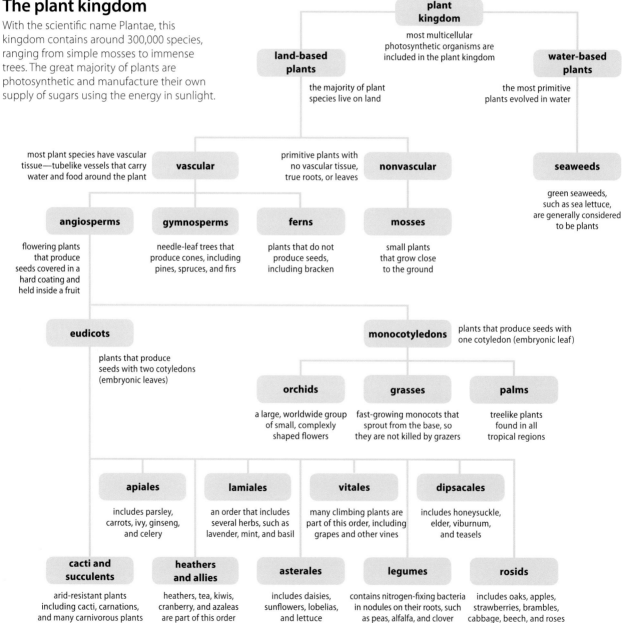

plant kingdom
most multicellular photosynthetic organisms are included in the plant kingdom

land-based plants
the majority of plant species live on land

water-based plants
the most primitive plants evolved in water

most plant species have vascular tissue—tubelike vessels that carry water and food around the plant

vascular

primitive plants with no vascular tissue, true roots, or leaves

nonvascular

seaweeds
green seaweeds, such as sea lettuce, are generally considered to be plants

angiosperms
flowering plants that produce seeds covered in a hard coating and held inside a fruit

gymnosperms
needle-leaf trees that produce cones, including pines, spruces, and firs

ferns
plants that do not produce seeds, including bracken

mosses
small plants that grow close to the ground

eudicots
plants that produce seeds with two cotyledons (embryonic leaves)

monocotyledons
plants that produce seeds with one cotyledon (embryonic leaf)

orchids
a large, worldwide group of small, complexly shaped flowers

grasses
fast-growing monocots that sprout from the base, so they are not killed by grazers

palms
treelike plants found in all tropical regions

apiales
includes parsley, carrots, ivy, ginseng, and celery

lamiales
an order that includes several herbs, such as lavender, mint, and basil

vitales
many climbing plants are part of this order, including grapes and other vines

dipsacales
includes honeysuckle, elder, viburnum, and teasels

cacti and succulents
arid-resistant plants including cacti, carnations, and many carnivorous plants

heathers and allies
heathers, tea, kiwis, cranberry, and azaleas are part of this order

asterales
includes daisies, sunflowers, lobelias, and lettuce

legumes
contains nitrogen-fixing bacteria in nodules on their roots, such as peas, alfalfa, and clover

rosids
includes oaks, apples, strawberries, brambles, cabbage, beech, and roses

The animal kingdom

With the scientific name of Animalia, this kingdom contains more than a million species—the precise number is unclear. All animals are heterotrophs—they survive by feeding on other living things, using this food for fuel or as a source of raw materials.

animal kingdom
includes multicellular heterotrophic organisms that generally have an internal digestive system and head

invertebrates
all animals that do not have backbones (which are called vertebrates)

roundworms
also known as nematodes, found in soils and living as parasites

segmented worms
including marine worms, earthworms, and leeches

sponges
primitive, mainly aquatic animals that absorb food through their outer surface

mollusks
the second-largest phylum, including snails, squid, and clams

small phyla
includes microscopic creatures, such as worms and aquatic life

flatworms
mostly parasitic worms, including flukes and tapeworms

cnidarians
jellyfish, anemones, and corals

echinoderms
a widespread group of marine creatures, including starfish and sea urchins

bryozoans
small filter-feeding animals that frequently grow in colonies

arthropods
the largest phylum of invertebrates, which all have jointed limbs

millipedes
plant-eating arthropods with two pairs of legs on each body segment

centipedes
predatory arthropods with many body segments, each with a single pair of eggs

crustaceans
mainly aquatic arthropods including crabs, lobsters, but also woodlice

arachnids
eight-legged arthropods including spiders, scorpions, mites, and ticks

insects
largest class of arthropods, have six legs and are the only flying invertebrates

contains all the vertebrates, and other animals that have a flexible supporting rod called a notochord

chordates (backbone)

jawless fish
primitive fish that have a spiral of teeth, used to scrape mouthfuls of food

cartilaginous fish
sharks and rays that have skeletons made of cartilage and not bone

bony fish
the largest group of fish, mostly with fins supported by thin rays of bone

amphibians
the first land vertebrates, including frogs, toads, newts, and salamanders

reptiles
a large group of land animals that includes snakes, turtles, and crocodiles

birds
egg-laying animals with feathers; most are able to fly

hair-covered animals that produce milk to feed their young

mammals

monotremes
primitive egg-laying mammals from Australia and New Guinea, including the echidna and platypus

marsupials
pouched mammals from Australia and the Americas, including opossums, kangaroos, and wombats

placentals
carry developing young in an internal chamber, nourishing them with an organ called a placenta

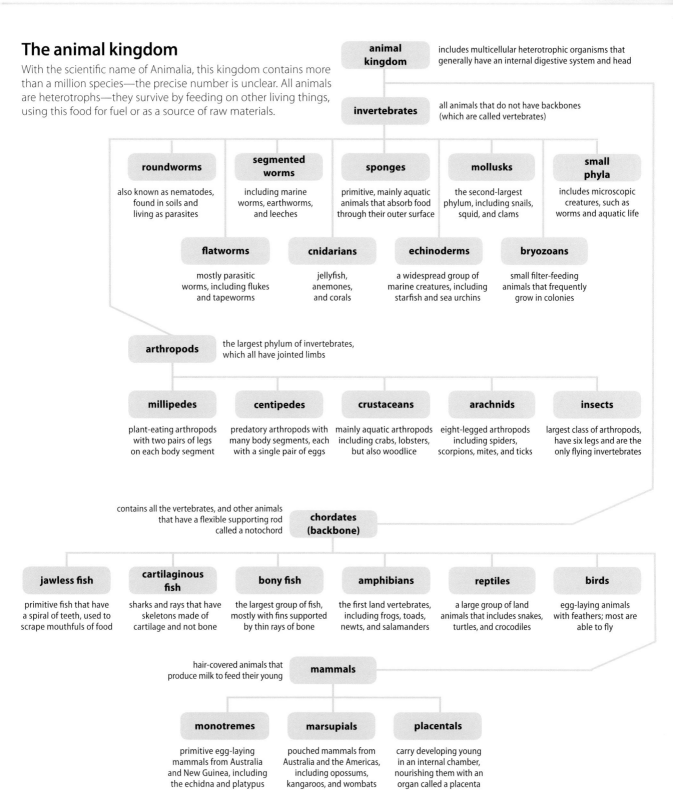

Chemistry reference

Melting and boiling points

Every element has a specific melting and boiling point. This is the temperature at which a solid changes into a liquid or a gas respectively. All temperatures are measured at atmospheric pressure. Metals tend to have high melting points, while simple gases have boiling points below room temperature. However, carbon is a nonmetal, but has the highest melting point of all.

LIST OF ELEMENTS						LIST OF ELEMENTS					
Atomic number	**Name/Symbol**	**Melting point** °C	°F	**Boiling point** °C	°F	**Atomic number**	**Name/Symbol**	**Melting point** °C	°F	**Boiling point** °C	°F
1	hydrogen (H)	−259	−434	−253	−423	29	copper (Cu)	1,083	1,981	2,582	4,680
2	helium (He)	−272	−458	−269	−452	30	zinc (Zn)	420	788	907	1,665
3	lithium (Li)	179	354	1,340	2,440	31	gallium (Ga)	30	86	2,403	4,357
4	beryllium (Be)	1,283	2,341	2,990	5,400	32	germanium (Ge)	937	1,719	2,355	4,271
5	boron (B)	2,300	4,170	3,660	6,620	33	arsenic (As)	817	1,503	613	1,135
6	carbon (C)	3,500	6,332	4,827	8,721	34	selenium (Se)	217	423	685	1,265
7	nitrogen (N)	−210	−346	−196	−321	35	bromine (Br)	−7	19	59	138
8	oxygen (O)	−219	−362	−183	−297	36	krypton (Kr)	−157	−251	−152	−242
9	fluorine (F)	−220	−364	−188	−306	37	rubidium (Rb)	39	102	688	1,270
10	neon (Ne)	−249	−416	−246	−410	38	strontium (Sr)	769	1,416	1,384	2,523
11	sodium (Na)	98	208	890	1,634	39	yttrium (Y)	1,522	2,772	3,338	6,040
12	magnesium (Mg)	650	1,202	1,105	2,021	40	zirconium (Zr)	1,852	3,366	4,377	7,911
13	aluminum (Al)	660	1,220	2,467	4,473	41	niobium (Nb)	2,467	4,473	4,742	8,568
14	silicon (Si)	1,420	2,588	2,355	4,271	42	molybdenum (Mo)	2,610	4,730	5,560	10,040
15	phosphorus (P)	44	111	280	536	43	technetium (Tc)	2,172	3,942	4,877	8,811
16	sulfur (S)	113	235	445	832	44	ruthenium (Ru)	2,310	4,190	3,900	7,052
17	chlorine (Cl)	−101	−150	−34	−29	45	rhodium (Rh)	1,966	3,571	3,727	6,741
18	argon (Ar)	−189	−308	−186	−303	46	palladium (Pd)	1,554	2,829	2,970	5,378
19	potassium (K)	64	147	754	1,389	47	silver (Ag)	962	1,764	2,212	4,014
20	calcium (Ca)	848	1,558	1,487	2,709	48	cadmium (Cd)	321	610	767	1,413
21	scandium (Sc)	1,541	2,806	2,831	5,128	49	indium (In)	156	313	2,028	3,680
22	titanium (Ti)	1,677	3,051	3,277	5,931	50	tin (Sn)	232	450	2,270	4,118
23	vanadium (V)	1,917	3,483	3,377	6,111	51	antimony (Sb)	631	1,168	1,635	2,975
24	chromium (Cr)	1,903	3,457	2,642	4,788	52	tellurium (Te)	450	842	990	1,814
25	manganese (Mn)	1,244	2,271	2,041	3,706	53	iodine (I)	114	237	184	363
26	iron (Fe)	1,539	2,802	2,750	4,980	54	xenon (Xe)	−112	−170	−107	−161
27	cobalt (Co)	1,495	2,723	2,877	5,211	55	cesium (Cs)	29	84	671	1,240
28	nickel (Ni)	1,455	2,641	2,730	4,950	56	barium (Ba)	725	1,337	1,640	2,984

LIST OF ELEMENTS					
Atomic number	Name/Symbol	Melting point °C	°F	Boiling point °C	°F
57	lanthanum (La)	921	1,690	3,457	6,255
58	cerium (Ce)	799	1,470	3,426	6,199
59	praseodymium (Pr)	931	1,708	3,512	6,354
60	neodymium (Nd)	1,021	1,870	3,068	5,554
61	promethiium (Pm)	1,168	2,134	2,700	4,892
62	samarium (Sm)	1,077	1,971	1,791	3,256
63	europium (Eu)	822	1,512	1,597	2,907
64	gadolinium (Gd)	1,313	2,395	3,266	5,911
65	terbium (Tb)	1,356	2,473	3,123	5,653
66	dysprosium (Dy)	1,412	2,574	2,562	4,644
67	holmium (Ho)	1,474	2,685	2,695	4,883
68	erbium (Er)	1,529	2,784	2,863	5,185
69	thulium (Tm)	1,545	2,813	1,947	3,537
70	ytterbium (Yb)	819	1,506	1,194	2,181
71	lutetium (Lu)	1,663	3,025	3,395	6,143
72	hafnium (Hf)	2,227	4,041	4,602	8,316
73	tantalum (Ta)	2,996	5,425	5,427	9,801
74	tungsten (W)	3,410	6,170	5,660	10,220
75	rhenium (Re)	3,180	5,756	5,627	10,161
76	osmium (Os)	3,045	5,510	5,090	9,190
77	iridium (Ir)	2,410	4,370	4,130	7,466
78	platinum (Pt)	1,772	3,222	3,827	6,921
79	gold (Au)	1,064	1,947	2,807	5,080
80	mercury (Hg)	−39	−38	357	675
81	thallium (Tl)	303	577	1,457	2,655
82	lead (Pb)	328	622	1,744	3,171
83	bismuth (Bi)	271	520	1,560	2,840
84	polonium (Po)	254	489	962	1,764

LIST OF ELEMENTS					
Atomic number	Name/Symbol	Melting point °C	°F	Boiling point °C	°F
85	astatine (At)	300	572	370	698
86	radon (Rn)	−71	−96	−62	−80
87	francium (Fr)	27	81	677	1,251
88	radium (Ra)	700	1,292	1,200	2,190
89	actinium (Ac)	1,050	1,922	3,200	5,792
90	thorium (Th)	1,750	3,182	4,787	8,649
91	protactinium (Pa)	1,597	2,907	4,027	7,281
92	uranium (U)	1,132	2,070	3,818	6,904
93	neptunium (Np)	637	1,179	4,090	7,394
94	plutonium (Pu)	640	1,184	3,230	5,850
95	americium (Am)	994	1,821	2,607	4,724
96	curium (Cm)	1,340	2,444	3,190	5,774
97	berkelium (Bk)	1,050	1,922	710	1,310
98	californium (Cf)	900	1,652	1,470	2,678
99	einstienium (Es)	860	1,580	996	1,825
100	fermium (Fm)	unknown		unknown	
101	mendelevium (Md)	unknown		unknown	
102	nobelium (No)	unknown		unknown	
103	lawrencium (Lr)	unknown		unknown	
104	rutherfordium (Rf)	unknown		unknown	
105	dubnium (Db)	unknown		unknown	
106	seaborgium (Sg)	unknown		unknown	
107	bohrium (Bh)	unknown		unknown	
108	hassium (Hs)	unknown		unknown	
109	meitnerium (Mt)	unknown		unknown	
110	darmstadtium (Ds)	unknown		unknown	
111	roentgenium (Rg)	unknown		unknown	
112	copernicum (Cn)	unknown		unknown	

Human elements

The human body contains 25 different chemical elements. Most are found in just tiny amounts. About two-thirds of the body is made of water (H_2O), and almost all of the rest is made up of carbon, nitrogen, calcium, and phosphorus atoms.

▷ **Human elements**
This chart shows the proportion of elements in the body by their mass—so 65 percent of body weight is made up of oxygen atoms, and so on.

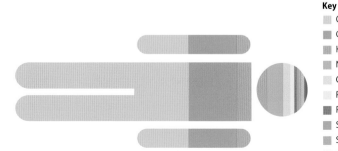

Key
- Oxygen 65%
- Carbon 18%
- Hydrogen 10%
- Nitrogen 3%
- Calcium 1.5%
- Phosphorus 1%
- Potassium 0.25%
- Sulfur 0.25%
- Sodium 0.15%
- Chlorine 0.15%
- Others 0.7%

Physics reference

SI units

All scientists use seven basic units of measurement, known as the SI base units, listed below. "SI" stands for "Système International." The units are maintained by experts in the headquarters, located in Paris, France.

SI UNITS		
Unit	**Symbol**	**Quantity measured**
meter	m	unit of length, defined as the distance light travels through a vacuum in 1/299,792,458th of a second
kilogram	kg	unit of mass, defined by the International Standard Kilogram made of a platinum-iridium alloy in Paris, France
second	s	unit of time, defined in terms of the frequency of a type of light radiated by a cesium atom
ampere	A	unit of electrical current, defined by the attraction force between two parallel conductors that are conducting one ampere
kelvin	K	unit on a scale of temperature that begins at absolute zero: 0 Kelvin or 459.67°F (-273.15°C)
candela	cd	a measure of luminous intensity (how powerful a light source is); one candle has a luminous intensity of one candela
mole	mol	a unit of quantity of a substance (generally very small particles such as atoms and molecules); one mole is made up of 6.02×10^{23} objects (atoms or molecules)

Derived SI units

This table contains just a few units that are derived from combinations of the seven base SI units. Nevertheless these units are very widely used and have been given their own names.

SI UNITS		
Unit	**Symbol**	**Quantity measured**
becquerel	Bq	unit of radioactive decay; the quantity of material in which one nucleus decays per second
Celsius	°C	unit of temperature, with the same magnitude as a Kelvin, but zero is at water's freezing point
coulomb	C	closely related to an ampere, this is the quantity of charge carried each second by a current of one ampere
farad	F	unit of capacitance, which is a capacitor's ability to store charge
hertz	Hz	unit of frequency; the number of cycles or repeating events per second
joule	J	amount of energy transferred when a force of one newton is applied over one meter
newton	N	unit of force required to increase the velocity of a mass by 1 kg by 1 m per second every second
ohm	Ω	unit of resistance; a one ohm resistor allows a current of one ampere to flow when one volt is applied across it
pascal	Pa	unit of pressure; a pascal is a force of one newton applied across an area of one square meter
volt	V	unit of potential difference and the force that pushes electric current
watt	W	unit of power (the rate at which energy is expended); calculated as joules per second

Formulas

Physicists calculate unknown quantities using formulas, in which known quantities are combined in specific ways. Formulas can be rearranged according to which quantity needs to be calculated. Here are some of the main formulas.

PHYSICS FORMULAS		
Quantity	**Description**	**Formula**
Current	voltage / resistance	$I = \dfrac{V}{R}$
Voltage	current x resistance	$V = IR$
Resistance	voltage / current	$R = \dfrac{V}{I}$
Power	work / time	$P = \dfrac{W}{t}$
Time	distance / velocity	$t = \dfrac{d}{v}$
Distance	velocity x time	$d = vt$
Velocity	displacement (distance in a given direction) / time	$v = \dfrac{d}{t}$
Acceleration	final velocity – initial velocity / time	$a = \dfrac{v2-v1}{t}$
Force	mass x acceleration	$F = ma$
Momentum	mass x velocity	$p = mv$
Pressure	force / area	$P = \dfrac{F}{A}$
Density	mass / volume	$\rho = \dfrac{m}{V}$
Volume	mass / density	$V = \dfrac{m}{\rho}$
Mass	volume x density	$m = V\rho$
Area	length x width	$A = lw$
Kinetic energy	½ mass x square of velocity	$E_k = \tfrac{1}{2} mv^2$
Weight	mass x acceleration due to gravity	$W = mg$
Work done	force x distance moved in direction of force	$W = Fs$

The planets

This table gives some basic information on the planets of the Solar System plus the number of observed moons that orbit them. The inner planets have rocky surfaces, while the larger outer planets are mainly made of gases and ice.

PLANETS AND MOONS		
Planet	**Description**	**Number of known moons**
Mercury	rock, metal	0
Venus	rock, metal	0
Earth	rock, metal	1
Mars	rock, metal	2
Jupiter	gas, ice, rock	63
Saturn	gas, ice, rock	62
Uranus	gas, ice, rock	27
Neptune	gas, ice, rock	13

Earth's vital statistics

Our planet is the largest rocky planet in the Solar System. Many of the units scientists use to measure the Universe are based on the size and motion of the planet.

Average diameter	12,756 km (7,928 miles)
Average distance from Sun: km (miles)	149.6 million (93 million)
Average orbital speed around Sun: km (miles)	29.8 km/s (18.5 mps)
Sunrise to sunrise (at the Equator)	24 hours
Mass	5.98×10^{24} kg
Volume	1.08321×10^{12} km³
Average density (water = 1)	5.52 g/cm³
Surface gravity	9.81 m/s²
Average surface temperature	15°C (59°F)
Ratio of water to land	70:30

Glossary

AC (alternating current)
AC is an electrical current that repeatedly changes in direction.

acceleration
An increase or decrease in an object's velocity (speed) due to a force being applied to it.

acid
A compound that breaks up into a negative ion and one or more positive hydrogen ions, which react easily with other substances.

activation energy
The energy needed to start a chemical reaction.

air resistance
A force that pushes against an object that is moving through the air, slowing it down; also called drag.

algae
Plantlike organisms that live in water or damp habitats; in general, they are single-celled.

alkali
A compound that dissociates into negative hydroxide (OH) ions and a positive ion; alkalis react easily with acids.

allotrope
A variant form of an element; for example, carbon can occur as graphite or diamond; while allotropes look different and have various physical properties, they all have identical chemical properties.

alloy
A mixture of two or more metals, or a metal and a nonmetal.

amplitude
The height of a wave.

anatomy
The science that studies the structure of living bodies to discover how they work.

anion
A negatively charged ion formed when an atom or group of atoms gains one or more electrons.

arthropod
A member of the largest animal phylum, which includes spiders, insects, and crustaceans.

atmosphere
A blanket of gases that surrounds a planet, moon, or star.

atom
The smallest unit of an element.

atomic number
The number of protons located in the nucleus of an atom; every element has atoms with a unique atomic number.

attraction
A force that pulls things together; opposite of repulsion.

bacteria (singular: bacterium)
Single-celled organisms that form a distinct kingdom of life; compared to other cells, bacterial cells are small and lack organelles.

base
An ionic compound that reacts with an acid.

biomass
A way of measuring the total mass of living things in a certain region; a useful way of comparing different types of organism in an ecosystem.

boiling point
The temperature at which a heated substance changes from a liquid into a gas; when the gas is cooled, it will condense into a liquid at this same temperature.

buoyancy
The tendency of a solid to float or sink in liquids.

capillary
A small blood vessel that delivers oxygen to body cells.

catalyst
A substance that lowers the activation energy of a chemical reaction, making the reaction occur much more rapidly.

cations
Positively charged ions, which form from atoms (or molecules) that lose one or more electrons.

cell
The smallest unit of a living body.

cellulose
A complex carbohydrate that makes up the wall that surrounds all plant cells.

chemical
A pure substance that has distinct properties.

chlorophyll
The green-colored compound that collects the energy in sunlight so it can be used to react with carbon dioxide and water to make sugar during photosynthesis.

chromosome
A structure in the nucleus of cells that is used to store coils of DNA.

circuit
A series of components (such as light bulbs) connected between the poles of a battery or other power source so an electric current runs through them.

combustion
A chemical process in which a substance reacts with oxygen, releasing heat and flames.

compound
A chemical that is made up of the atoms of two or more elements bonded together.

compression
Squeezing or pushing a substance into a smaller space.

concave
Having a curved surface that resembles the inside of a circle or sphere.

concentration
The amount of one substance mixed into a known volume of another.

condense
To turn from a gas to a liquid; for example, steam condenses into water.

conduction
The process by which energy is transferred through a substance. The energy being transferred is thermal (heat), acoustic, or electrical.

convection
A process that transfers heat through a liquid or gas, with warm areas rising and cooler ones sinking, thus creating a circulating current.

convex
Having a curved surface that resembles the outside of a circle or sphere.

current
A flow of a substance; electrical currents are a flow of electrons or other charged particles.

DC
Short for "direct current," an electric current that flows in one direction continuously.

deceleration
A decrease in velocity that occurs when a force pushes against a moving object in the opposite direction to its direction of motion.

decomposition
To break up into two or more simpler ingredients.

deformation
To be changed in shape by a force, such as being stretched, bent, or squeezed.

density
A quantity of how much matter is held within a known volume of a material.

diffraction
A behavior of waves, in which a wave spreads out in a number of directions after it passes through a small gap, with a width similar to its wavelength.

dipole
A molecule with two poles: one negative and one positive.

displacement
The moving aside of part of a medium by an object placed in that medium. Or the distance between one point and another.

distillation
A process that separates liquid mixtures by boiling away each component in turn, then cooling them back into pure liquids.

DNA
Short for "deoxyribonucleic acid," a complex chained molecule that carries genetic code, the instructions that a cell—and entire body—uses to make copies of itself.

drag
The resistance force formed when an object pushes through a fluid, such as air or water.

dynamic equilibrium
When a reversible reaction takes place at the same rate in both directions so, even though it is continuing in both directions, the overall quantities of the materials involved stay constant.

eclipse
An eclipse occurs when the Earth, Sun, and Moon line up, blocking out the view of one of the objects. In a solar eclipse, the Moon covers up the Sun as seen from Earth. In a lunar eclipse, the Earth sits between the Sun and the Moon.

ecosystem
A collection of living organisms that share a habitat and are reliant on each other for survival.

elasticity
The property of an object that allows it to change shape when forced to but return to its original form when the force is removed.

electrolysis
Dividing compounds into simpler substances using the energy in electricity.

electrolyte
A liquid that conducts electricity.

electromagnet
A magnet that can be turned on by running an electric current through it.

electron
A negatively charged particle that is located around the outside of an atom.

electronics
A field of science and technology that involves using semiconductors to make components for circuits.

element
A natural substance that cannot be divided or simplified into raw ingredients. There are around 90 natural elements on Earth.

endothermy
The ability of an animal to maintain a constant body temperature using energy burned from its food to heat or cool the body.

energy
Energy is what allows things to happen. For example, chemical energy in food allows us to live and move.

enzyme
A protein that is used to control a chemical reaction or other process taking place inside a living body.

evaporate
To turn from a liquid to a gas, such as a puddle drying out.

evolve
A change in the characteristics of a species due to its environment; evolution is driven by a process called natural selection.

exoskeleton
Hard tissue that forms the outer surface of a body, giving shape and structure to it.

exothermic
Describing an animal that does not maintain a constant body temperature but allows it to fluctuate with that of the surroundings.

fat
A solid lipid—a biological material that is used to store energy, insulate nerves, and form membranes. Liquid lipids are called oils.

filtration
The process of passing a substance through a filter to remove solid particles.

fission
Breaking apart; nuclear fission involves radioactive atoms splitting in two, releasing a huge amount of energy.

force
The means that causes a mass to change its momentum.

fossil fuel
A substance that burns easily, releasing heat formed from the remains of ancient plants and other organisms; fossil fuels include coal, natural gas, and oil.

friction
A force that occurs between moving objects, where the surfaces rub against each other, opposing their movement.

fusion
Joining together; nuclear fusion involves two small atoms fusing into a single larger one, releasing huge amounts of energy.

gene
A coded instruction for making a certain body feature that is passed from parent to offspring; the code is stored as a DNA molecule and is translated into proteins, each of which performs a specific job.

generator
A device for converting rotational motion into electric current.

gills
A breathing organ that takes oxygen from the water and releases carbon dioxide. Gills are used by fish and many underwater creatures.

gland
An organ in the body that secretes chemicals in large quantities; endocrine glands release chemicals into the blood stream, exocrine glands secrete onto the surface of the body.

glucose
A simple carbohydrate, or sugar, made by the process of photosynthesis and then used by cells as a source of energy.

gravity
A force that acts between all masses and which tends to pull them together.

habitat
The place where organisms live; every habitat has specific conditions, such as supply of water, range of temperatures, and amount of light.

half-life
The period of time that it takes for a sample of a radioactive element to halve in mass by decaying into other elements.

hormone
A chemical messenger that travels through the bloodstream to control certain life processes; hormones include epinephrine, insulin, and estrogen.

hydrocarbon
A compound composed largely, if not entirely, from hydrogen and carbon.

immiscible
A property where two liquids will not mix with each other because their molecules push away from each other.

indicator
A substance that changes color with pH, the measure of acidity.

induction
The process by which the energy of a moving conductor is converted into an electrical current when it passes through a magnetic field.

inertia
A mass's resistance to changing its state of motion.

insulation
A material with the function of stopping heat moving from a warm object to a colder one; animal insulation, such as hair or blubber, is used to save energy.

interference
The mixing of two or more light waves to produce new, different ones.

invertebrate
An animal with no backbone. Most animals are invertebrates, but are nevertheless not all closely related.

ion
An atom or a molecule that has lost or gained an electron and thus carries a positive or negative charge.

isotope
One of two or more forms of atom all with the same number of protons—and so belonging to the same element—but with varying numbers of neutrons.

keratin
A protein used by vertebrates to cover their bodies; feathers, hair, nails, claws, horns, and reptile scales are all made of keratin.

longitudinal
A wave that is made up of compressions and expansions of a medium.

main sequence star
An average star, like our Sun.

mass
A property of an object that allows it to have weight and be acted on by forces.

matter
Anything that has mass and occupies space.

membrane
A thin layer that surrounds a cell or other body structure; the layer is semipermeable, so only certain substances can cross it.

metabolism
The name used for all processes that support life that take place in a living body; catabolism is all the processes that break things into simpler substances; anabolic processes build simple substances into complex ones.

metal
An element that is likely to react by losing electrons, forming a cation; metals are generally shiny, heavy solids.

micrometer (μm)
A millionth of a meter.

microtubule
A fine fiber of protein that runs through the cytoplasm of a cell and is used to haul larger items around.

mixture
A collection of two or more substances mixed together but which are not chemically connected.

mole
A unit of quantity used to count huge numbers of objects, such as atoms and molecules; for example, one mole of hydrogen atoms is 6.0221415×10^{23} atoms.

molecule
Two or more atoms that are bonded together; the molecule is the smallest unit of a compound; breaking it up into simpler units would destroy the compound.

momentum
The product of the speed of an object and its mass.

nanometer (nm)
A billionth of a meter.

neutron
A neutral particle located in the nuclei of most atoms.

nucleus (plural: nuclei)
The central core of something. An atomic nucleus contains protons and neutrons, while a cell's nucleus contains DNA.

nutrient
A substance that is useful for life as a source of energy or as raw material.

octet
A collection of eight things.

orbit
The path of one mass around another mass under the influence of gravity.

organelle
A structure inside a cell that performs a certain task in the cell's metabolism.

organism
A living thing.

oscillation
A regular vibration around a fixed point.

oxidation
The loss of electrons by an atom, ion, or molecule.

phloem
The vascular tissue that carries sugar fuel around a plant.

pigment
A chemical substance that colors an object.

plasma
A high-energy state of matter where the atoms of a gas have been ripped into their constituent parts.

polarity
Relating to an object, such as a magnet, that has two opposite ends or poles.

polymer
A long chainlike molecule made up of smaller molecules connected together.

precipitation
A solid or liquid that falls from a cloud. Rain, snow, sleet, and hail

are examples of precipitation.

pressure
The amount of force that is applied to a surface per unit of area.

protein
A type of complex chemical found in all living things, used as enzymes and in muscles. A protein is a chain of simple units called amino acids. There are about 20 natural amino acids, and a protein has hundreds of these units connected in a specific order.

protist
A single-celled organism with a complex cell structure, including organelles and a nucleus.

proton
A positively charged particle that is located in the nuclei of all atoms.

pupate
To prepare to change from a larva to an adult form (imago); for example, a caterpillar pupates as a chrysalis before emerging as a butterfly.

radiation
Waves of energy that travel through space. Radiation includes visible light, heat, X-rays, and radio waves; nuclear radiation includes subatomic particles and fragments of atoms.

radicals
Atoms, molecules, or ions with unpaired electrons that cause them to react easily.

radioactive
Relating to atoms that are unstable and break apart, releasing high-energy particles.

rarefaction
A decrease in the pressure and density of molecules along a longitudinal wave.

reactivity
A description of how likely a substance is to become involved in a chemical reaction.

reduction
When a substance gains electrons during a chemical reaction and so its oxidation number is reduced.

reflection
When a wave bounces off a surface.

refraction
When a wave changes direction as it passes from one medium to another.

repulsion
A force that pushes things apart; the opposite of attraction.

respiration
The process occuring in all living cells that releases energy from glucose to power life.

rubisco
Short for "ribulose bisphosphate carboxylase oxidase," an enzyme that is responsible for taking carbon dioxide from the atmosphere and reacting it with water to make glucose as part of photosynthesis.

salt
An ionic compound formed by a reaction between an acid and base (including an alkali).

sedimentary rock
A rock that forms from sediments, which are layers of substances that have settled on the seabed or ground before becoming buried and compressed for millions of years.

solute
A substance that becomes dissolved in another.

solvent
A substance that can have other substances dissolved in it.

speed
The rate of how fast an object is moving.

states of matter
The three main forms of matter that a substance can take are: solid, liquid, or gas. Plasma is a fourth state of matter.

strain
The change of the shape of an object in response to stress.

stress
A force that alters the shape of an object, by stretching, bending, and sometimes breaking it.

subatomic particle
A particle that is smaller than an atom, such as a proton, neutron, and electron.

superconductor
A material that conducts electricity without warming up and so wastes none of the energy it is carrying.

suspension
A mixture in which small solids, blobs of liquid, or gas bubbles are spread throughout a liquid.

temperature
An average measure of the thermal energy or heat of an object.

torque
The turning effect of a force.

torsion
A twist caused by torque.

transformer
A device for altering the voltage of an electrical current.

transverse
A wave that moves by rising and falling perpendicular to the direction of its motion.

vapor
Another word for a gas.

vascular
Concerning vessels, tubes that transport substances around a body.

velocity
A speed of something in a particular direction.

vertebrate
An animal that has a vertebral column, a flexible spine made from a chain of smaller bones called vertebrae; the largest animals are vertebrates, and include fish, amphibians, reptiles, mammals, and birds.

vesicle
A membranous sac that contains a material being processed by a cell; a vesicle may be used to release substances from a cell.

voltage
A measure of the force that pushes electrons around an electric current.

xylem
The vascular tissue that transports water and minerals around a plant.

wavelength
The distance measured between any point on a wave and the equivalent point on the next wave.

weight
The force applied to a mass by gravity.

work
The amount of energy transferred when a force is being applied to a mass over a certain distance.

Index

Acknowledgments

DORLING KINDERSLEY would like to thank: Smiljka Surlja for her design assistance; Fran Baines, Clive Gifford, Clare Hibbert, Wendy Horobin, James Mitchem, Carole Stott, and Victoria Wiggins for their editorial assistance; Nikky Twyman for proofreading; and Jackie Brind for the index.

DORLING KINDERSLEY INDIA would like to thank Sudakshina Basu and Vandna Sonkariya for their design assistance.

The publisher would like to thank the following for their kind permission to reproduce their photographs:

(Key: a-above; b-below/bottom; c-center; f-far; l-left; r-right; t-top)

23 Science Photo Library: Dr. E. Walker (br). 24 Getty Images: Photographer's Choice / Tony Hutchings (br). 34 Corbis: Anup Shah (cr). 37 SuperStock: Stock Connection (bl). 47 Ardea: Alan Weaving (tr). 65 Corbis: Tetra Images (br). 68 Corbis: Owen Franken (cl). 73 Science Photo Library: Mehau Kulyk (br). 79 FLPA: Nigel Cattlin (cl). 81 Corbis: Louie Psihoyos (tr). 85 Science Photo Library: Andrew McClenaghan (bc). 87 Science Photo Library: James King-Holmes (br). 88 Dreamstime.com: Peter Wollinga (bl). 91 Corbis: Richard Chung / Reuters (bc). 99 Corbis: Radius Images (bl). 100 Corbis: Mark Schneider / Visuals Unlimited (bl). 102 Corbis: Bettmann (bc). 107 Corbis: FoodPhotography Eising / the food passionates (c). 108 Science Photo Library: Sheila Terry (cr). 119 Science Photo Library: Charles D. Winters (br). 123 Corbis: Louie Psihoyos / Science Faction (br). 124 Corbis: Thom Lang (br). 126 Alamy Images: Robert Cousins (br). 128 Corbis: Taro Yamada (br). 135 Science Photo Library: Martyn F. Chillmaid (br).

139 Alamy Images: Carol and Mike Werner / PHOTOTAKE Inc. (br). 141 Science Photo Library: Dirk Wiersma (br). 143 SuperStock: imagebroker.net (cra). 145 Dreamstime.com: Cammeraydave (br). 146 Science Photo Library: Cristina Pedrazzini (br). 149 Getty Images: Photolibrary / Wallace Garrison (crb). 161 Corbis: Alex Wild / Visuals Unlimited (bl). 163 Corbis: moodboard (bl). 169 Corbis: Roland Holschneider / DPA (c). 171 Science Photo Library: Middle Temple Library (br). 173 Corbis: Ken Welsh / Design Pics (bl). 175 Corbis: Mike Powell (cr). 180 Corbis: Gene Blevins / LA DailyNews (cra). 183 Corbis: Duomo (br). 187 Alamy Images: Chuck Franklin (bl). 189 Corbis: Barritt, Peter / SuperStock (br). 191 Alamy Images: Ange (br). 192 Corbis: Grantpix / Index Stock (cra). 194 Corbis: Joe McDonald (cra). 199 Science Photo Library: Sinclair Stammers (tr). 201 Corbis: Denis Scott (cra). 205 Science Photo Library: Andy Crump (br). 207 Corbis: Marcus Mok / Asia Images (bl). 209 Science Photo Library: Volker Steger / Peter Arnold Inc. (br). 210 Corbis: Liu Liqun (br). 213 Science Photo Library: David Parker (cra). 215 Alamy Images: Mark Boulton (br). 217 Alamy Images: MShieldsPhotos (br). 219 Science Photo Library: Patrick Landmann (br). 220 Corbis: Chip East / Reuters (br). 222 Corbis: Seth Resnick / Science Faction (bc). 225 Alamy Images: Mark Ferguson (cra). 228 Getty Images: Alan R Moller (cra). 231 Science Photo Library: NASA / JPL (br). 233 Corbis: Heritage Images / Museum of London (br). 237 Corbis: George Steinmetz (br). 238 NASA: ESA and The Hubble Heritage Team (STScI / AURA) (br). 241 Science Photo Library: NASA (cla)

All other images © Dorling Kindersley
For further information see: **www.dkimages.com**